Liposome Technology

Liposome Technology
Third Edition

Volume III
Interactions of Liposomes with the Biological Milieu

Edited by

Gregory Gregoriadis

The School of Pharmacy
University of London
and
Lipoxen PLC
London, U.K.

CRC Press
Taylor & Francis Group
Boca Raton London New York

CRC Press is an imprint of the
Taylor & Francis Group, an **informa** business

CRC Press
Taylor & Francis Group
6000 Broken Sound Parkway NW, Suite 300
Boca Raton, FL 33487-2742

First issued in paperback 2019

© 2010 by Taylor & Francis Group, LLC
CRC Press is an imprint of Taylor & Francis Group, an Informa business

No claim to original U.S. Government works

ISBN-13: 978-0-8493-9725-7 (hbk)
ISBN-13: 978-0-367-39038-9 (pbk)

This book contains information obtained from authentic and highly regarded sources. While all reasonable efforts have been made to publish reliable data and information, neither the author[s] nor the publisher can accept any legal responsibility or liability for any errors or omissions that may be made. The publishers wish to make clear that any views or opinions expressed in this book by individual editors, authors or contributors are personal to them and do not necessarily reflect the views/opinions of the publishers. The information or guidance contained in this book is intended for use by medical, scientific or health-care professionals and is provided strictly as a supplement to the medical or ot her professional's own judge ment, their knowledge of the patient's medical history, relevant manufacturer's instructions and the appropriate best practice guidelines. Because of the rapid advances in medical science, any information or advice on dosages, procedures or diagnoses should be independently verified. The reader is stronglyurged to consult the relevant national drug formulary and the drug companies' and device or material manufacturers' printed instructions, and their websites, before administering or utilizing any of the drugs, devices or materials mentioned in this book. This book does not indicate whether a particular treatment is appropriate or suitable for a particular individual. Ultimately it is the sole responsibility of the medical professional to make his or her own professional judgements, so as to advise and treat patients appropriately. The authors and publishers have also attempted to trace the copyright holders of all material reproduced in this publication and apologize to copyright holders if permission to publish in this form has not been obtained. If any copyright material has not been acknowledged please write and let us know so we may rectify in any future reprint.

Visit the Taylor & Francis Web site at
http://www.taylorandfrancis.com

and the CRC Press Web site at
http://www.crcpress.com

*Dedicated to the memory of my parents,
Christos and Athena*

Preface

The science and technology of liposomes as a delivery system for drugs and vaccines have evolved through a variety of phases that I have been privileged to witness from the very beginning. The initial observation (1) that exposure of phospholipids to excess water gives rise to lamellar structures that are able to sequester solutes led to the adoption of these structures (later to become known as liposomes) as a model for the study of cell membrane biophysics. Solute sequestration into liposomes prompted a few years later the development of the drug delivery concept (2,3) and, in 1970, animals were for the first time injected with active-containing liposomes (3,4). Subsequent work in the author's laboratory and elsewhere worldwide on drug- and vaccine-containing liposomes and their interaction with the biological milieu in vivo culminated in the licensing of a number of injectable liposome-based therapeutics and vaccines. The history of the evolution of liposomes from a structural curiosity in the 1960s to a multifaceted, powerful tool for transforming toxic or ineffective drugs into entities with improved pharmacological profiles today has been summarized elsewhere (5,6).

The great strides made toward the application of liposomes in the treatment and prevention of disease over nearly four decades are largely due to developments in liposome technology; earlier achievements were included in the previous two editions of this book (7,8). The avalanche of new techniques that came with further expansion of liposomology since

v

the second edition in 1992 has necessitated their inclusion into a radically updated third edition. Indeed, so great is the plethora of the new material that very little from the second edition has been retained. As before, contributors were asked to emphasize methodology employed in their own laboratories since reviews on technology with which contributors have no personal experience were likely to be superficial for the purpose of the present book. In some cases, however, overviews were invited when it was deemed useful to reconnoiter distinct areas of technology. A typical chapter incorporates an introductory section directly relevant to the author's subject with concise coverage of related literature. This is followed by a detailed methodology section describing experiences from the author's laboratory and examples of actual applications of the methods presented, and, finally, by a critical discussion of the advantages or disadvantages of the methodology presented vis-a-vis other related methodologies. The 55 chapters contributed have been distributed logically into three volumes. Volume I deals with a variety of methods for the preparation of liposomes and an array of auxiliary techniques required for liposome characterization and development. Volume II describes procedures for the incorporation into liposomes of a number of drugs selected for their relevance to current trends in liposomology. Volume III is devoted to technologies generating liposomes that can function in a "targeted" fashion and to approaches of studying the interaction of liposomes with the biological milieu.

It has again been a pleasure for me to undertake this task of bringing together so much knowledge, experience, and wisdom so generously provided by liposomologist friends and colleagues. It is to be hoped that the book will prove useful to anyone involved in drug delivery, especially those who have entered the field recently and need guidance through the vastness of related literature and the complexity and diversity of aspects of liposome use. I take this opportunity to thank Mrs. Concha Perring for her many hours of help with the manuscripts and Informa Healthcare personnel for their truly professional cooperation.

Gregory Gregoriadis

REFERENCES

1. Bangham AD, Standish MM, Watkins JC. Diffusion of univalent ions across the lamellae of swollen phospholipids. J Mol Biol 1965; 13:238.
2. Gregoriadis G, Leathwood PD, Ryman BE. Enzyme entrapment in liposomes. FEBS Lett 1971; 14:95.
3. Gregoriadis G, Ryman BB. Fate of protein-containing liposomes injected into rats. An approach to the treatment of storage diseases. Eur J Biochem 1972; 24:485.

4. Gregoriadis G. The carrier potential of liposomes in biology and medicine. New Engl J Med 1976; 295:704–765.
5. Gregoriadis G. "Twinkling guide stars to throngs of acolytes desirous of your membrane semi-barriers. Precursors of bion, potential drug carriers...". J Liposome Res 1995; 5:329.
6. Lasic DD, Papahadjopoulos D (Eds), Medical Applications of Liposomes, Elsevier. Amsterdam 1998.
7. Gregoriadis G. Liposome Technology. CRC Press, Boca Raton, Volumes I, II and III, 1984.
8. Gregoriadis G. Liposome Technology 2nd Edition. CRC Press, Boca Raton, Volumes I, II and III, 1992.

Acknowledgments

The individuals listed below in chronological order (1972–2006) worked in my laboratory as postgraduate students, senior scientists, research assistants, post-doctoral fellows, technicians, visiting scholars, and Erasmus or Sandwich students. I take this opportunity to express my gratitude for their contributions to the science and technology of liposomes and other delivery systems, as well as their support and friendship. I am most grateful to my secretary of 14 years, Concha Perring, for her hard work, perseverance, and loyalty.

Rosemary A. Buckland (UK), Diane Neerunjun (UK), Christopher D.V. Black (UK), Anthony W. Segal (UK), Gerry Dapergolas (Greece), Pamela J. Davisson (UK), Susan Scott (UK), George Deliconstantinos (Greece), Peter Bonventre (USA), Isobel Braidman (UK), Daniel Wreschner (Israel), Emanuel Manesis (Greece), Christine Davis (UK), Roger Moore (UK), Chris Kirby (UK), Jackie Clarke (UK), Pamela Large (UK), Judith Senior (UK), Ann Meehan (UK), Mon-Moy Mah (Malaysia), Catherine Lemonias (Greece), Hishani Weereratne (Sri Lanka), Jim Mixson (USA), Askin Tümer (Turkey), Barbara Wolff (Germany), Natalie Garçon (France), Volkmar Weissig (USA), David Davis (UK), Alun Davies (UK), Jay R. Behari (India), Steven Seltzer (USA), Yash Pathak (India), Lloyd Tan (Singapore), Qifu Xiao (China), Christine Panagiotidi (Greece), K.L. Kahl (New Zealand), Zhen Wang (China), Helena da Silva (Portugal), Brenda McCormack (UK), M. Yaşar Ozden (Turkey), Natasa Skalko (Croatia), John Giannios (Greece), Dmitry Genkin (Russia), Maria Georgiou (Cyprus), Sophia Antimisiaris (Greece), Becky J. Ficek (USA), Victor Kyrylenko (Ukraine), Suresh Vyas (India), Martin Brandl (Germany), Dieter Bachmann (Germany), Mayda Gursel (Turkey), Sabina Ganter (Germany), Ishan Gursel (Turkey), Maria Velinova (Bulgaria), Cecilia D'Antuono (Argentina),

Ana Fernandes (Portugal), Cristina Lopez Pascual (Spain), Susana Morais (Portugal), Ann Young (UK), Yannis Loukas (Greece), Vassilia Vraka (Greece), Voula Kallinteri (Greece), Fatima Eraïs (France), Jean Marie Verdier (France), Dimitri Fatouros (Greece), Veronika Müller (Germany), Jean-Christophe Olivier (France), Janny Zhang (China), Roghieh Saffie (Iran), Irene Naldoska (Polland), Sudaxina Murdan (Mauritius), Sussi Juul Hansen (Denmark), Anette Hollensen (Denmark), Yvonne Perrie (UK), Maria Jose Saez Alonso (Spain), Mercedes Valdes (Spain), Laura Nasarre (Spain), Eve Crane (USA), Brahim Zadi (Algeria), Maria E. Lanio (Cuba), Gernot Warnke (Germany), Elizabetta Casali (Italy), Sevtap Velipasaoglu (Turkey), Sara Lauria (Italy), Oulaya Belguenani (France), Isabelle Gyselinck (Belgium), Sigrun Lubke (Germany), Kent Lau (Hong Kong), Alejandro Soto (Cuba), Yanin Bebelagua (Cuba), Steve Yang (Taiwan), Filipe Rocha da Torre Assoreira (Portugal), Paola Genitrini (Italy), Guoping Sun (China), Malini Mital (UK), Michael Schupp (Germany), Karin Gaimann (Germany), Mia Obrenovic (Serbia), Sherry Kittivoravitkul (Thailand), Yoshie Maitani (Japan), Irene Papanicolaou (Greece), Zulaykho Shamansurova (Uzbekistan), Miriam Steur (Germany), Sanjay Jain (India), Ioannis Papaioannou (Greece), Maria Verissimo (Italy), Bruno da Costa (Portugal), Letizia Flores Prieto (Spain), Andrew Bacon (UK).

Contents

Contributors

Michel Adamina Institute for Surgical Research and Hospital Management, University of Basel, Basel, Switzerland

Jill P. Adler-Moore Department of Biological Sciences, California State Polytechnic University, Pomona, California, U.S.A.

Theresa M. Allen Department of Pharmacology, University of Alberta, Edmonton, Alberta, Canada

Carl R. Alving Division of Retrovirology, Department of Vaccine Production and Delivery, U.S. Military HIV Research Program, Walter Reed Army Institute of Research, Silver Spring, Maryland, U.S.A.

Marcel B. Bally Department of Advanced Therapeutics, British Columbia Cancer Agency, Vancouver, British Columbia, Canada

Ande Bao Department of Radiology and Otolaryngology, University of Texas Health Science Center at San Antonio, San Antonio, Texas, U.S.A.

Lajos Baranyi Vaccine and Immunology Research Institute, Washington, D.C., U.S.A.

Eugene A. Bernstein Department of Pharmaceutical Sciences, Northeastern University, Boston, Massachusetts, U.S.A.

Otto C. Boerman   Department of Nuclear Medicine, Radboud University Nijmegen Medical Centre, Nijmegen, The Netherlands

Maria Helena Bueno da Costa   Laboratório de Microesferas e Lipossomas—Centro de Biotecnologia, Instituto Butantan, São Paulo, Brazil

Rolf Bünger   Department of Anatomy, Physiology, and Genetics, Uniformed Services University of the Health Sciences, Bethesda, Maryland, U.S.A.

Myrra G. Carstens   Department of Pharmaceutics, Utrecht Institute of Pharmaceutical Sciences, Utrecht University, Utrecht, The Netherlands

Nejat Düzgüneş   Department of Microbiology, Arthur A. Dugoni School of Dentistry, University of the Pacific, San Francisco, California, U.S.A.

Marcel H. A. M. Fens   Department of Pharmaceutics, Utrecht Institute of Pharmaceutical Sciences, Utrecht University, Utrecht, The Netherlands

Evisa Gjini   Department of Pharmaceutics, Utrecht Institute of Pharmaceutical Sciences, Utrecht University, Utrecht, The Netherlands

Beth A. Goins   Department of Radiology, The University of Texas Health Science Center at San Antonio, San Antonio, Texas, U.S.A.

Heinrich Haas   MediGene AG, Martinsried, Germany

Troy O. Harasym   Celator Pharmaceuticals Corp., Vancouver, British Columbia, Canada

William C. Hartner   Department of Pharmaceutical Sciences, Northeastern University, Boston, Massachusetts, U.S.A.

Andrew S. Janoff   Celator Pharmaceuticals Corp., Vancouver, British Columbia, Canada, and Celator Pharmaceuticals Inc., Princeton, New Jersey, U.S.A.

Gerard M. Jensen   Gilead Sciences, Inc., San Dimas, California, U.S.A.

Sharon A. Johnstone   Celator Pharmaceuticals Corp., Vancouver, British Columbia, Canada

Jan A. A. M. Kamps Medical Biology Section, Department of Pathology and Laboratory Medicine, Endothelial Biomedicine and Vascular Drug Targeting, Groningen University Medical Center, University of Groningen, Groningen, The Netherlands

Peter Laverman Department of Nuclear Medicine, Radboud University Nijmegen Medical Centre, Nijmegen, The Netherlands

Robert J. Lee Division of Pharmaceutics, College of Pharmacy, NCI Comprehensive Cancer Center, and NSF Nanoscale Science and Engineering Center, The Ohio State University, Columbus, Ohio, U.S.A.

Tatayana S. Levchenko Department of Pharmaceutical Sciences, Northeastern University, Boston, Massachusetts, U.S.A.

Montserrat Lopez-Mesas Chemical Engineering Department, EUETIB-Universitat Politecnica de Catalunya, Barcelona, Spain

Gary R. Matyas Division of Retrovirology, Department of Vaccine Production and Delivery, U.S. Military HIV Research Program, Walter Reed Army Institute of Research, Silver Spring, Maryland, U.S.A.

Lawrence D. Mayer Celator Pharmaceuticals Corp., Vancouver, British Columbia, Canada

Luis A. Medina Institute of Physics, Universidad Nacional Autonoma de Mexico, Mexico City, Mexico

Uwe Michaelis MediGene AG, Martinsried, Germany

Seyed M. Moghimi Molecular Targeting and Polymer Toxicology Group, School of Pharmacy, University of Brighton, Brighton, U.K.

Grietje Molema Medical Biology Section, Department of Pathology and Laboratory Medicine, Endothelial Biomedicine and Vascular Drug Targeting, Groningen University Medical Center, University of Groningen, Groningen, The Netherlands

Christien Oussoren Department of Pharmaceutics, Utrecht Institute of Pharmaceutical Sciences, Utrecht University, Utrecht, The Netherlands

Xiaogang Pan   Division of Pharmaceutics, College of Pharmacy, NCI Comprehensive Cancer Center, and NSF Nanoscale Science and Engineering Center, The Ohio State University, Columbus, Ohio, U.S.A.

Kristina K. Peachman   Division of Retrovirology, Department of Vaccine Production and Delivery, U.S. Military HIV Research Program, Walter Reed Army Institute of Research, Silver Spring, Maryland, U.S.A.

Maria C. Pedroso de Lima   Department of Biochemistry, Faculty of Sciences and Technology, and Center for Neuroscience and Cell Biology, University of Coimbra, Coimbra, Portugal

William T. Phillips   Department of Radiology, The University of Texas Health Science Center at San Antonio, San Antonio, Texas, U.S.A.

Richard T. Proffitt   Richpro Associates, Arcadia, California, U.S.A.

Mangala Rao   Division of Retrovirology, Department of Vaccine Production and Delivery, U.S. Military HIV Research Program, Walter Reed Army Institute of Research, Silver Spring, Maryland, U.S.A.

Birgit Romberg   Department of Pharmaceutics, Utrecht Institute of Pharmaceutical Sciences, Utrecht University, Utrecht, The Netherlands

Oleg Rosenberg   Department of Medical Biotechnology, Central Research Institute of Roentgenology and Radiology, Ministry of Health, St. Petersburg, Russia

Raymond M. Schiffelers   Department of Pharmaceutics, Utrecht Institute of Pharmaceutical Sciences, Utrecht University, Utrecht, The Netherlands

Reto Schumacher   Institute for Surgical Research and Hospital Management, University of Basel, Basel, Switzerland

Andrey Seiliev   Department of Medical Biotechnology, Central Research Institute of Roentgenology and Radiology, Ministry of Health, St. Petersburg, Russia

Srikanth Shivakumar   Division of Pharmaceutics, College of Pharmacy, and NCI Comprehensive Cancer Center, The Ohio State University, Columbus, Ohio, U.S.A.

Sérgio Simões   Laboratory of Pharmaceutical Technology, Faculty of Pharmacy, and Center for Neuroscience and Cell Biology, University of Coimbra, Coimbra, Portugal

Giulio C. Spagnoli   Institute for Surgical Research and Hospital Management, University of Basel, Basel, Switzerland

Gert Storm   Department of Pharmaceutics, Utrecht Institute of Pharmaceutical Sciences, Utrecht University, Utrecht, The Netherlands

Janos Szebeni   Division of Retrovirology, Department of Vaccine Production and Delivery, U.S. Military HIV Research Program, Walter Reed Army Institute of Research, Silver Spring, Maryland, U.S.A.

Paul G. Tardi   Celator Pharmaceuticals Corp., Vancouver, British Columbia, Canada

Vladimir P. Torchilin   Department of Pharmaceutical Sciences, Northeastern University, Boston, Massachusetts, U.S.A.

Esther van Kesteren-Hendrikx   Department of Molecular Cell Biology, Vrije University Medical Center, Amsterdam, The Netherlands

Nico van Rooijen   Department of Molecular Cell Biology, Vrije University Medical Center, Amsterdam, The Netherlands

Daya D. Verma   Department of Pharmaceutical Sciences, Northeastern University, Boston, Massachusetts, U.S.A.

Andrey Zhuikov   Department of Medical Biotechnology, Central Research Institute of Roentgenology and Radiology, Ministry of Health, St. Petersburg, Russia

Interactions of Liposomes with Complement Leading to Adverse Reactions

Janos Szebeni and Carl R. Alving

Division of Retrovirology, Department of Vaccine Production and Delivery, U.S. Military HIV Research Program, Walter Reed Army Institute of Research, Silver Spring, Maryland, U.S.A.

Lajos Baranyi

Vaccine and Immunology Research Institute, Washington, D.C., U.S.A.

Rolf Bünger

Department of Anatomy, Physiology, and Genetics, Uniformed Services University of the Health Sciences, Bethesda, Maryland, U.S.A.

INTRODUCTION

The interaction of liposomes with the cellular arm of nonspecific (innate) immunity, i.e., with macrophages of the reticuloendothelial system (RES), has been widely recognized and intensely studied since the birth of liposome science. The humoral innate response to liposomes, manifested by activation of the complement (C) system, has also been early recognized and widely studied, but this effect got much less attention than the interaction of liposomes with phagocytes. Factors hindering progress in this area include the requirement for in vivo measurement of physiological and laboratory end points that are irrelevant from a liposome standpoint (e.g., hemodynamic analysis and C cleavage product assays), the perceived complexity of C reactions, and unpopularity of focusing on adverse effects vis-à-vis clinical benefits.

As a brief reminder, the C system is one of the four homeostatic cascades in blood (coagulation, fibrinolytic, and kinin-kallikrein systems), providing first-line defense against bacteria, viruses, and all other microbes. Its activation, involving highly coordinated limited proteolysis of some 15 plasma proteins, leads to opsonization, lysis, and increased clearance of foreign particles. In addition to this nonspecific defense function, C activation orchestrates the development of specific immunity against the invaders and initiates cellular repair following tissue injury. Figure 1 shows the activation scheme, with three different pathways leading to the central step of C3 activation and subsequent formation of biologically active peptides (anaphylatoxins) and the membrane attack complex (MAC, C5b-9).

Historically, the first analyses of liposomal C activation were described by Kinsky and coworkers, who used liposomes as a model system for studying

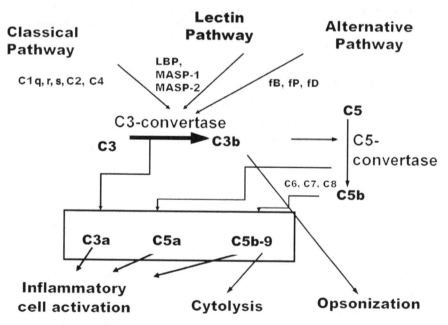

Figure 1 Scheme of complement (C) activation and its biological consequences. Classical pathway activation involves the binding of antibodies to the vesicles with subsequent activation of the C1 complex, C2, C4, and C3, leading to the formations of the C3 and C5 convertases. In the case of alternative pathway activation, formation of the C3 convertase is triggered by covalent attachment of C3b to the membrane. These anaphylatoxins activate mast cells, basophils, platelets, and other inflammatory cells with resultant liberation of inflammatory mediators [histamine, platelet activating factor (PAF), prostaglandins, etc.]. These, in turn, set in motion a complex cascade of respiratory, hemodynamic, and hematological changes, leading to numerous adverse clinical effects.

the mechanism of membrane damage caused by the MAC (1–4). Since then, dozens of studies dealt with various details of C activation by different vesicles, as reviewed earlier (5–8).

As for the mechanism of C activation, liposomes are in the same size range as most microbial pathogens, i.e., small unilamellar vesicles (SUVs) correspond to viruses, whereas large multilamellar liposomes (MLVs) overlap with the sizes of bacteria and yeast cells. In the absence of C control proteins on the surface of liposomes that normally suppress C activation on host cells [e.g., sialic residues and C receptors (CR1, CR2, and CR3), decay accelerating factor, and membrane cofactor protein], activation by and recognition of liposomes represents a physiologic response of the C system. In fact, C activation appears to be the rule rather than the exception, as almost all liposomes can activate C when exposed to plasma for sufficient time. In the authors' experience, for example, incubation of liposomes (final phospholipid concentration: 5–10 mM) with undiluted human or animal serum for 20 to 30 minutes at 37°C results in significant C activation relative to phosphate-buffered saline (PBS) control (i.e., adding PBS instead of liposomes for volume adjustment), regardless of liposome characteristics. There are, of course, great differences in the degree of C activation, with neutral SUVs and negatively charged MLVs with high cholesterol content (i.e., 71%) representing the least and most activating liposome species, respectively (9).

CLINICAL SYMPTOMS AND OCCURRENCE OF COMPLEMENT ACTIVATION-RELATED PSEUDOALLERGY

Complement activation by liposomes becomes a clinical problem only if the vesicles are administered intravenously, whereupon they directly encounter large amounts of C proteins. Even with significant activation, in most people the symptoms remain subclinical, detectable only with sensitive hemodynamic or other physiological monitoring. However, we are not aware of any clinical study devoted to a systematic exploration of the hemodynamic effects of different liposomes. Complement activation may become a serious problem in only a fraction of patients who happen to be hypersensitive to C activation or anaphylatoxin action, a condition whose immunological background has not been clarified. Intravenous (IV) administration of certain liposomes or liposomal drugs in such hypersensitive subjects leads to an acute toxicity referred to as hypersensitivity reaction (HSR) or infusion reaction. The reaction, referred to as C activation–related pseudoallergy (CARPA) (7–12), corresponds to anaphylatoxin toxicity with some of the symptoms common in all allergic diseases, while others are unique to C activation (Table 1).

Table 2 lists some of the liposome- or lipid-based drug products that are marketed or in clinical development for parenteral application. Of these, Doxil (Caelyx) (13–18), AmBisome (19–23), Abelcet (22), Amphocil (22), and DaunoXome (24–32) have been reported to cause unusual HSRs

Table 1   The Clinical Symptoms of In Vivo Complement Activation and Hypersensitivity Reactions to Liposomes in Humans

Common allergic symptoms
   Anaphylactic shock, angioedema, asthma attack, bronchospasm, chest pain, chill,
      choking, confusion, conjunctivitis, coughing, cyanosis, death, dermatitis,
      diaphoresis, dyspnoea, edema, erythema, feeling of imminent death, fever, flush,
      headache, hypertension, hypotension, hypoxemia, low back pain, lumbar pain,
      metabolic acidosis, nausea, pruritus, rash, rhinitis, skin eruptions, sneezing,
      tachypnea, tingling sensations, urticaria, and wheezing
Unique symptoms

Ig-E–mediated allergy	C activation–related pseudoallergy
Reaction arises after repeated exposure to the allergen	Reaction arises at first treatment (no prior exposure to allergen)
Reaction is stronger upon repeated exposures	Reaction is milder or absent upon repeated exposures
Reaction does not ceases without treatment	Spontaneous resolution
Reaction rate is low ($< 2\%$)	High-reaction rate (up to 45%), average 7%, severe 2%

Source: From Ref. 8.

corresponding to CARPA. Actually, the first report of CARPA-like adverse effects to IV infusion of liposomes was published as early as 1986 (33), in one of the pioneer studies on the use of liposomes in cancer chemotherapy. The frequency of similar types of HSRs to liposomal drugs varies between 3% and 45% (6,7).

MECHANISMS OF COMPLEMENT ACTIVATION BY LIPOSOMES

As shown in Figure 1, the central step of C activation, formation of the C3 convertase, can proceed via three pathways with involvement of multiple proteins in each. Within each pathway, there are alternative activation sequences in terms of order and identity of participating proteins, lending substantial redundancy, and variation to the process by which C3 is activated. This multiplicity, taken together with the variety of lipid composition, degree of lamellar structure, size, surface charge and pH and the amount, intravesicular distribution and physicochemical impact of encapsulated material in different liposome preparations, leads to the conclusion that it is very difficult, if at all possible, to generalize any liposome-induced C-activation mechanism. As illustrated in Figure 2, these mechanisms include classical pathway activation triggered by the binding of specific or natural IgG, IgM, C1q, C-reactive protein (CRP), and alternative pathway activation triggered by the binding of C3b, IgG, or C4a3b (6,7,12,34–47).

Table 2 Liposome- or Lipid-Based Products in Market or in Clinical Development

Encapsulated drug	Trade name	Indications	Developer/producer/distributor	Status of development
Doxorubicin (Adriamycin™)	Doxil	Ovarian cancer, Kaposi's sarcoma, metastatic breast cancer	Alza Co./Ortho Biotech, East Bridgewater, New Jersey, U.S.A.	
	Caelyx		Essex Pharma GmBH, Munich, Georgia, U.S.A.	
	Myocet		Elan Corp, Dublin, Ireland	
Daunorubicin	DaunoXome	Advanced HIV-associated Kaposi's sarcoma	Gilead Sciences Inc., Foster City, California, U.S.A.	
Amphotericin B	AmBisome	Systemic fungal infections		Marketed
	Abelcet		Elan Co./Enzon Inc., Piscataway, New Jersey, U.S.A.	
	Amphotec		Intermune Inc., Brisbane, California, U.S.A.	
Verteporfin	Visudyne	Subfoveal neovascularization due to macular degeneration, pathologic myopia, and ocular histoplasmosis	QLT Inc., Vancouver, British Columbia, Canada	
ATRA	ATRA-IV	Non-Hodgkin lymphoma, acute promyelocytic leukaemia, etc.	Antigenics Inc., New York, New York, U.S.A.	Phase I–II
Oxeliplatin Vincristine	Aroplatin Onco TCS	Colorectal and other solid tumors Lymphoma, acute lymphoblastic leukemia, Hodgkin's disease lung cancer, pediatric malignancies	Inex Pharm. Co., Burnaby, British Columbia, Canada	Phase III
Topotecan Vinorelbine	Topotecan TCS Vinorelbine TCS	Various cancers		Preclinical

(Continued)

Table 2 Liposome- or Lipid-Based Products in Market or in Clinical Development (*Continued*)

Encapsulated drug	Trade name	Indications	Developer/producer/distributor	Status of development
Doxorubicin	LED	Advanced cancers, including breast	NeoPharm Lake Forest, Illinois, U.S.A.	Phase I/II
Paclitaxel	LEP-ETU	Advanced cancer, including breast, lung, and ovarian		
Mitoxantrone	LEM-ETU	Advanced cancers, including prostate		
c-raf antisense oligonucleotide	LErafAON	Advanced cancers, including pancreatic		
CPT-11 (irinotecan, Camptosar™)	LE-SN38	Advanced cancer, including colorectal and lung		

Abbreviations: ATRA, all-trans-retinoic acid; TCS, transmembrane carrier system; ETU, easy-to-use; CPT, camptothecin.
Source: From Ref. 8.

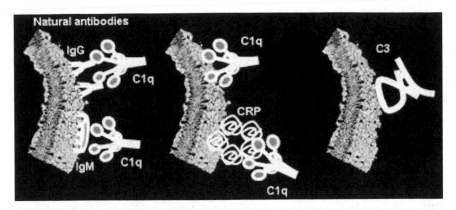

Figure 2 Mechanisms of liposome-induced C activation. The scheme illustrates the various trigger mechanisms of liposome-induced C activation. These include classical pathway activation triggered by the binding to the membrane of IgG, IgM, C1q, or C-reactive protein (CRP). The effects of immunoglobulins and CRP are mediated by C1q. Alternative pathway activation is triggered by direct binding of C3 to the membrane or binding of C3 to the Fab portion of membrane-bound antibodies. Activation of the terminal sequence (C5b, C6, C7, C8, and C9), leading to the formation of membrane attack complex, is identical with all liposomes.

GENERAL APPROACHES OF MEASURING THE INTERACTION OF LIPOSOMES WITH THE COMPLEMENT SYSTEM

Most previous studies on liposomal C activation were in vitro experiments wherein different vesicles were incubated with whole serum, or with purified C components, with or without addition of specific antibodies. The effect of C was quantified by measuring the leakage of an aqueous marker, such as glucose, galactose, radioisotopes (e.g., Rb^+), fluorophores, spin labels, and low-molecular-weight enzymes; by monitoring C consumption in the fluid phase; or by measuring the production of a C-activation marker, such as $C3a_{desarg}$, iC3b, C4d, Bb, and SC5b-9. The present chapter focuses only on the latter two approaches, as liposomal leakage, taken as an index of C activation, was reviewed earlier (6) and has rarely been used recently. In addition to providing specifics of "state-of-the-art" in vitro assays, the goal of this chapter was to present a large animal (porcine) model of liposomal C activation that measures the biological consequences of the reaction.

Assays of Liposome-Induced Changes in Plasma Complement

Plasma CH_{50}/mL

Incubation of liposomes with human or animal serum initiates C activation with consequent depletion of C components which lyze heterologous (sheep) red blood cell (SRBC) sensitized with specific antibodies (hemolysins). The hemolysis assay originally described by Mayer (48) has been adapted to allow measurements in small serum volumes, thereby saving potentially precious test materials and serum. One modification measures CH_{50} in absolute terms (CH_{50}/mL) as defined by Mayer (48), whereas the other allows the estimation of C consumption relative to a baseline value.

Methods: Liposomes are incubated with undiluted sera with constant shaking, typically by adding 50-μL vesicles from a 40 mM phospholipid stock to 200 μL serum in Eppendorf tubes, followed by thorough vortex mixing and incubation at 37°C for 30 to 45 minutes while shaking at 80 rpm. For negative and positive controls, the serum is incubated with PBS and 5 mg/mL zymosan, respectively. After incubation, liposomes are separated from the serum by centrifugation (at >14,000 g for 10 minutes), and the serum is either tested immediately for C levels or are stored at −20°C for later tests.

For measuring absolute CH_{50} in the serum, SRBCs are washed three times and suspended in veronal-buffered saline containing gelatin (VBG) at a cell density of 10^9 cell/mL. The latter can be most easily adjusted by a two-step procedure, as follows. First, one needs to find the exact amount of washed SRBC that gives an optical density (OD) reading of 0.7 at 540 nm, following addition to and thorough mixing with 1 mL 1% Na_2CO_3. In the second step, this volume is added to each mL VBG used in the assay, e.g., one has to estimate how many milliliters of SRBC suspension is needed for the assay and then add to VBG equal times the volume of washed SRBC determined as above. After mixing of SRBC in VBG, hemolysin is added to the cells at a dilution of 1/1000, i.e., 1 μL to each milliliter of SRBC, and the suspension is allowed to stand for 30 minutes at room temperature. The suspension is then aliquoted in 1-mL portions, using Eppendorf tubes, and to each tube increasing amounts of test serum is added in the 2–30-μL range, typically 3, 6, 9, 12, and 16 μL. After thorough vortex mixing, the tubes are incubated in a shaking water bath for 60 minutes at 37°C. Samples are chilled on ice, centrifuged, and the degree of hemolysis is determined photometrically at 540 nm (or 415 nm). CH_{50} is obtained from the intercepts of the regression lines of the log x versus log ($y/100-y$) plots, where x stands for the volume of serum (μL) added to 1 mL SRBC, and y denotes percent hemolysis (48). The regression lines are obtained from at least two to three values in the dynamic range of the assay (20–80% hemolysis). Typically, R^2 values are in the 0.97 to 1.00 range.

Mayer's CH_{50} assay has also been adapted to 96-well plates, allowing further reduction of serum volumes necessary for the test and simultaneous handling of large number of samples. Details of the assay have been described, among others, by Plank et al. (49). In our practice, serum samples are diluted three to fivefold in VBG after incubation with liposomes, as described above, and aliquots with increasing volumes in the 2–20-µL range are added to wells of 96-well enzyme linked immunosorbent assay (ELISA) plates. The volumes are adjusted with PBS to match the highest added volume (e.g., 20 µL). Positive control, i.e., maximum hemolysis is obtained in wells to which the detergent, Triton X-100 is added. Typically, quadruplicate samples from each volume are placed in different columns, using column A for no serum (PBS) baseline and the last column for Triton X-100 positive control. For example, 4×20 mL PBS is placed in column A, $4 \times (3$ mL serum $+ 17$ mL PBS) in column B, $4 \times (6$ mL serum $+ 14$ mL PBS) in C, and so on, and 20 mL 1% triton X-100 in the last column. This is followed by adding 200 µL SRBC from the 10^9 cell/mL VBG suspension prepared as described above. The plate(s) is/are then incubated at 37°C for one hour with shaking, followed by centrifugation at 4000 rpm for 10 minutes, using a plate centrifuge. The supernatants are transferred to another plate and A_{540} is read in a plate reader.

In evaluating the data, one needs to establish for the control serum (i.e., the one that was not incubated with liposomes), which dilution provided readings in the upper part in of the dynamic range of the assay. In other words, one has to select a column in which the OD readings are greater than about 50%, but are lesser than the maximal OD readings in the plate. Depending on the dilution of test serum (i.e., three-, four-, or fivefold), such maxima are obtained in wells that contained the highest amounts of serum, in the positive (Triton X-100 control), or equally in both. For further analysis, values only in these selected columns are used, as they represent hemolysis in the effective dynamic, and therefore, quantifiable range of the test. Relating the mean OD values in different test serum samples to that in the baseline (PBS) serum will provide a relative measure of C consumption, which can be expressed as OD, % of baseline, or % C consumption.

The most important advantages of CH_{50} assays are that they can be used with most mammalian species' sera, and that they utilize commonly available instruments and inexpensive chemicals. As for its limitations, these assays measure the depletion of a C component(s) that is/are rate-limiting to the hemolysis of sensitized sheep red cells; and they are generally less sensitive than most other assays, which detect the liberation of C-activation products. They are labor intensive and require elaborate planning. The data analysis is also complex.

Measurement of Activation Products

ELISAs for anaphylatoxins and other C byproducts: Several enzyme-linked immunoassays have been developed in recent years to measure the levels of anaphylatoxins and other C byproducts in human serum, plasma, or other biological or experimental samples. These assays include C3a-desarg, iC3b, Bb, C4d, and SC5b-9, measuring the cleavage products of C3, C4, factor B, and C5, respectively, using HRP-labeled proprietary monoclonal antibodies. The SC5b-9 terminal complement complex (TCC) assay actually measures a stable, nonlytic form of the TCC that is bound to a naturally occurring regulatory serum protein, the S protein that binds to nascent C5b-9 complexes at the C5b-7 stage of assembly.

Porcine Model of Liposome-Induced C-Activation and HSRs

Young adolescent pigs appear to be uniquely sensitive to liposome-induced C activation allowing their use as a sensitive bioassay to measure the C-activating capability of different liposomes and to study the mechanism of subsequent physiological changes. This sensitivity is manifested in rapid development of massive, easily quantifiable hemodynamic, electrocardiogram (ECG), and laboratory changes, including significant rises in pulmonary arterial pressure (PAP) and heart rate, rises or falls in systemic arterial pressure (SAP) and heart rate, and falls in cardiac output (CO). The ECG changes include arrhythmia with ventricular fibrillation and cardiac arrest, the latter being lethal unless the animal is resuscitated with epinephrine with or without cardiopulmonary resuscitation (CPR) and chest electroshock. This cardiopulmonary distress is also associated with transient declines in blood oxygen saturation, reflecting pulmonary dysfunction (dyspnea), and with transient skin reactions (rash, erythema, or flushing), thus truly mimicking many aspects of the human HSRs to liposomes. These changes were shown to be due to C activation as C-activating substances mimicked, while specific C inhibitors inhibited the reaction.

The reason underlying the high sensitivity of pigs to liposome reactions has not been elucidated to date. Our current hypothesis is that the phenomenon may be due to the presence of macrophages in the lung of pigs, which, like Kupffer cells in the liver, are directly exposed to the blood. These macrophages may have low threshold for anaphylatoxin-induced activation and promptly secrete thromboxane, histamine, and other vasoactive mediators, which mediate the reaction.

Procedures

Pigs are sedated with intramuscular ketamine (Ketalar) and then anesthetized with halothane or isoflurane via nose cone. The subsequent steps are as follows. The trachea is intubated to allow mechanical ventilation with an anesthesia machine, using 1% to 2.5% halothane or isoflurane. A pulmonary artery catheter equipped with thermodilution-based continuous CO detector (TDQ CCO, Abbott Laboratories, Chicago, Illinois, U.S.A.) is advanced via the right internal jugular vein through the right atrium into the pulmonary artery to measure PAP, central venous pressure (CVP), and CO (B). A 6F Millar Mikro-Tip catheter (Millar Instruments, Houston, Texas, U.S.A.) is inserted into the right femoral artery and advanced into the proximal aorta for blood sampling and to measure SAP. A second 6F pig tail Millar Mikro-Tip catheter is inserted through the left femoral artery and placed into the left ventricle to monitor left ventricular end-diastolic pressure (LVEDP). Systemic vascular resistance (SVR) and pulmonary vascular resistance (PVR) are calculated from SAP, PAP, CO, CVP, and LVEDP, using standard formulas (50). Blood pressure values and lead II of the ECG are obtained continually.

Liposome Injections

Liposome stock solutions vary between 5 and 40 mM phospholipid (~50–40 mg/mL lipid) from which 50–200 μL is diluted to 0.5–1 mL with PBS or saline, and injected using 1-mL tuberculin syringes either into the jugular vein, via the introduction sheet, or via the pulmonary catheter, directly into the pulmonary artery. Injections are performed relatively fast (within 10–20 seconds) and are followed by 10 mL PBS or saline injections to wash in any vesicles remaining in the void space of the catheter. Injections are repeated several times over a period of six to eight hours, with 20- to 40-minutes intervals that are necessary for most hemodynamic parameters to return to near baseline levels.

Hemodynamic and ECG Monitoring

Monitoring of hemodynamic parameters (PAP, SAP, LVEDP, and CVP), heart rate, and ECG starts three to five minutes before the injections and continues until all hemodynamic parameters return to baseline, usually within 15 to 30 minutes. Then, baseline monitoring is started for the next injection.

Blood Sampling

Five to 10 mL blood samples are taken from the femoral artery into heparinized tubes before each injection (baseline), and at the top of liposome reactions, usually between 4 and 10 minutes after the injections. Blood is centrifuged immediately at 4°C and the plasma is stored at −20°C until conducting of the various assays.

Typical Results

Figure 3 demonstrates the hemodynamic responses of three different pigs to IV injections of 5 mg (5 μmole phospholipid in 1 mL PBS) large MLVs prepared from dimyristoyl phosphatidylcholine, dimyristoyl phosphatidyl-glycerol, and cholesterol (mole ratio: 50:5:45). These injections caused substantial, although transient hemodynamic changes, including a 50% to 250% increase in PAP (panel A), zero to 80% decline in CO (panel B), a two- to sixfold increase in PVR (panel C), 5% to 10% increase in heart rate (panel D), 20% to 40% fall, rise, or biphasic changes in SAP (panel E), and a zero to 400% rise of SVR (panel F). These responses were observed within

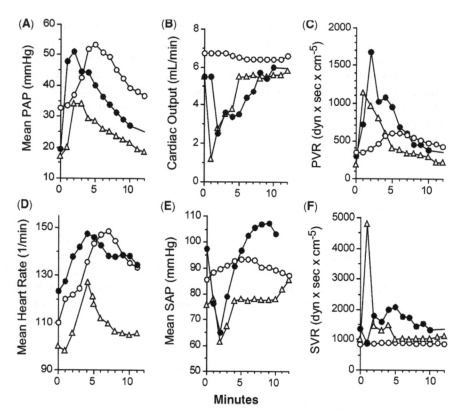

Figure 3 Hemodynamic changes induced by liposomes in pigs. Typical curves from three pigs injected with the liposome boluses. Different symbols designate different pigs. *Abbreviations*: PAP, pulmonary arterial pressure; PVR, pulmonary vascular resistance; SAP, systemic arterial pressure; SVR, systemic vascular resistance. *Source*: From Ref. 10.

the first minute, reached their peak within five to six minutes, and returned to baseline within 10 to 15 minutes.

One remarkable feature of the model is that several liposome injections can be given to the same animal over many hours without tachyphylaxis, i.e., diminution of response. This high reproducibility of the reaction is illustrated by the remarkably low variation in the rise of PAP in response to a same dose of liposomes (5 mg lipid corresponding to the ED_{50}) in 27 pigs, or within one animal, after eight consecutive injections (10). A further unique advantage of using pigs to measure liposome-induced C activation is the capability to assess potential reactogenicity of various liposomes intended for therapeutic use in humans. Essentially, all pigs react to liposomes (in our experience, none out of more than 100 pigs tested to date failed to react), also, because each pig can be used to test multiple preparations, it seems to be economic and convenient, despite the cost and labor-intensive nature of the experiments.

The quantitative nature of this "large animal bioassay" was shown by the linear relationship between liposome dose and submaximal rises of PAP (10), whereas its specificity to C activation became evident from the observations that (i) small unilamellar liposomes, which had negligible C activating effect in vitro, also failed to cause hemodynamic changes in vivo (9), and (ii) nonliposomal C activators (zymosan and xenogeneic immunoglobulins) induced pulmonary pressure changes that were indistinguishable from those caused by MLV (10). These data provide validation of the model in terms of quantifying C-mediated cardiopulmonary reactions with high sensitivity and specificity.

COMPLEMENT ACTIVATION BY DOXIL AND OTHER LIPOSOMAL DRUGS

In Vitro Evidence

Of particular relevance to the role of C activation in liposome-induced HSRs, the authors reported that incubation of Doxil with 10 different normal human sera led to significant rises in C-terminal complex (SC5b-9) levels over PBS control in seven sera, with rises exceeding 100% to 200% (relative to PBS control) in four subjects (11). While providing evidence for activation of the whole C cascade by Doxil in a majority of humans, these data also highlight the significant individual variation of responses. Further experiments showed that in addition to the quantitative variation in SC5b-9 response, Doxil-induced C activation also varied in different individuals in terms of sensitivity to inhibition by 10 mM EGTA/2.5 mM Mg^{2+}, which distinguishes classical versus alternative pathway activation (11). The minimum effective C-activating concentration of Doxil was 0.05–0.10 mg/mL, and there was near linear dose–effect relationship up to about 0.5 mg/mL. The activation curve reached its plateau at doses greater than or equal to 0.6 mg/mL, suggesting saturation of response (11). Doxil also caused variable liberation of Bb, a specific marker of alternative pathway activation, providing further evidence for a role of alternative pathway activation and/or amplification (11).

These and other studies from our laboratories (9,10,51–55) revealed some basic conditions and mechanism of liposomal C activation. Namely, it appears that positive or negative surface charge, size, and polydispersity promote, whereas neutrality and small uniform size reduce, the proneness of liposomes for C activation. The process may involve both the classical and alternative pathways, with the latter acting either as the only activation mechanism, or as a positive feedback mechanism amplifying C activation at the level of C3 convertase. The role of immunoglobulins varies from system to system as an important, although not obligatory trigger. Liposomal C-activation sensitivity depends on the phospholipid composition and cholesterol content of the vesicles. Most importantly, however, there is substantial individual variation in the susceptibility of different sera for C activation by the same liposomes.

In Vivo Evidence

Studies by the authors and their colleagues described that minute amounts (5–10 mg) of MLVs caused significant hemodynamic changes in pigs, including a massive rise in PAP with declines of SAP, CO, and LVEDP (9,10). Similar changes were observed with Doxil and the ^{99m}Tc chelator (HYNIC-PE)–containing pegylated small unilamellar liposomes (55), demonstrating that the phenomenon was not restricted to the use of MLV. In addition to the above hemodynamic changes, we also reported massive ECG changes in pigs treated with HYNIC-PEG liposomes and Doxil (55), attesting to severe transient myocardial ischemia with bradyarrhythmia, ventricular fibrillation, and other ECG abnormalities.

We suggested that the above MLV-induced hemodynamic and cardiac changes in pigs may represent an amplified model for liposome-induced HSRs in man on the following basis: (i) hypotension is one of the major symptoms of acute HSRs in general and of Doxil reactions, in particular; (ii) pulmonary hypertension with consequent decrease of left ventricle filling and coronary perfusion can explain the dyspnea with chest and back pain in man, i.e., typical symptoms of HSRs; (iii) ECG changes observed in the pig exactly correspond to the cardiac electric abnormalities reported in HSRs to liposomes Ambisome (56); and (iv) the vasoactive dose of Doxil (6–840 µg/kg) corresponds to the dose that triggers HSR in humans (17) suggesting that the pigs' sensitivity to Doxil corresponds to that of hypersensitive human subjects.

Clinical Evidence

In addition to the experimental data delineated above, there is ample clinical support for a causal role of C activation in liposome-induced HSRs. The first indirect evidence appeared as early as in 1983, when Coune et al. (57) reported that IV infusion of liposomes containing NSC 251635, a water-insoluble cytostatic agent, led to increased C3d/C3 ratios in the plasma of cancer patients. This study, however, did not address the presence or absence of HSRs. Another indirect proof was communicated by Skubitz et al. (58), who observed transient neutropenia with signs of leukocyte activation in patients who displayed HSRs to Doxil. Neutropenia with leukocyte activation are classical hallmarks of anaphylatoxin action (59–61), yet C activation was not considered in the above study. To the authors' knowledge, the first direct evidence for the causal relationship between C activation and HSRs to liposomes was provided by Brouwers et al. (62)

who reported 16% to 19% decrease of plasma C3, C4, and factor B in the blood of a patient developing HSR to $^{Tc-99m}$-labeled, HYNIC-PE containing pegylated liposomes applied for the scintigraphic detection of infection and inflammation (63). The fact that both C4 and factor B were involved in the consumption of C suggest that C activation proceeded on both the classical and alternative pathways. In a subsequent study by the same group, it was reported that three out of nine patients reacted to pegylated HYNIC liposomes (62). Despite clear benefits in imaging inflammatory bowel disease, the presence of HSRs was considered as unacceptable from a diagnostic agent and the Dutch team temporarily abandoned human trials with pegylated HYNIC liposomes until the hypersensitivity issue could get resolved.

Activation of C as cause of HSRs to Doxil was the subject of a recent clinical study wherein the authors correlated C activation (formation of

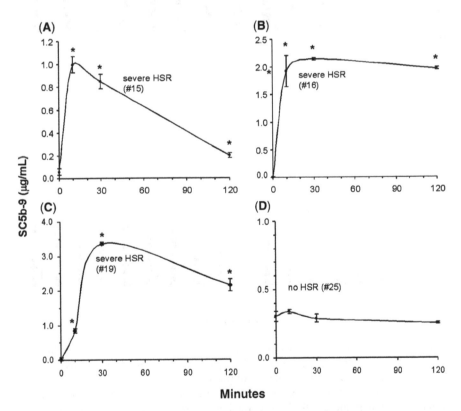

Figure 4 Time course of Doxil-induced changes in plasma SC5b-9 in cancer patients and its individual variation. Data from four subjects displaying different patterns of response (*Panels A–D*). Data are mean ± SD for triplicate determinations. *, significantly different from baseline, $p < 0.05$. *Abbreviation*: HSR, hypersensitivity reaction. *Source*: From Ref. 18.

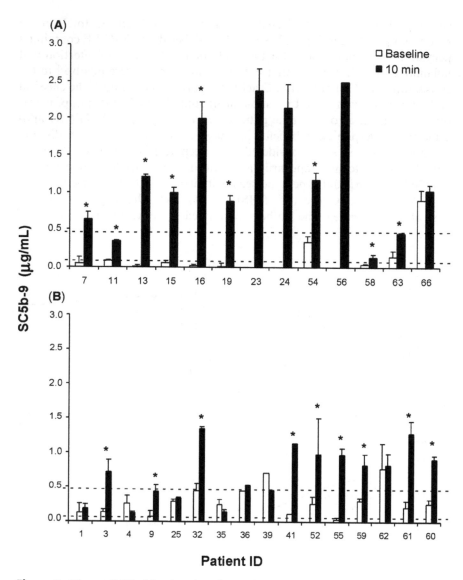

Figure 5 Plasma SC5b-9 levels at baseline and at 10 minutes postinfusion of Doxil in cancer patients displaying (**A**) or not displaying (**B**) hypersensitivity reactions to Doxil. Data are mean $\pm$ SD for triplicate or duplicate determinations. The dashed lines indicate the normal range of SC5b-9, i.e., the normal mean ± 2 SD. $*$, significantly different from baseline ($p < 0.05$). The numbers under the bars are the patient identification. *Source*: From Ref. 18.

SC5b-9) with the frequency and severity of HSRs in cancer patients infused with Doxil for the first time (18). Forty-five percent (13/29) of patients in the study showed grade 2 or 3 HRS, with reactions occurring in men and women in approximately equal proportions. The reactions were not related to the age of patients. Doxil caused C activation in 21 out of 29 patients (72%) as reflected by significant elevations of plasma SC5b-9 levels following infusion of the drug. In addition to these surprising statistics on C activation and HSRs caused by Doxil, the study provided several fundamentally new insights into the mechanism of CARPA.

One new item of information is that the time course of SC5b-9 increase in blood shows substantial individual variation (Fig. 4), including: (i) rapid elevations within 10 minutes with gradual return to near baseline within two hours, (ii) rapid elevation without return within two hours, and (iii) moderately rapid elevation of SC5b-9 until about 30 minutes, followed by partial return to baseline during two hours. The lack of SC5b-9 response is demonstrated in Figure 4D.

Considering the baseline and 10 minutes postinfusion the SC5b-9 values in clinical reactors and nonreactors, the study reported significant

Table 3 The SC5b-9 Assay as Predictor of Hypersensitivity Reactions to Doxil

10 min SC5b-9 (µg/mL)	Sensitivity tp/(tp + fn)	Specificity tn/(fp + tn)	Positive predictive value tp/(tp + fp)	Negative predictive value tn/(fn + tn)
Significant increase[a] (SC5b-9, no limit)	0.92	0.44	0.57	0.88
Significant increase[a] SC5b-9 ≤ 0.98	0.83	0.54	0.45	0.88
0.98 ≤ SC5b-9 ≤ 1.96 (≥2X, ≤4X normal)	0.80	0.70	0.57	0.88
SC5b-9 ≥ 1.96 (≥4X normal)	0.75	1.00	1.00	0.88

Note: Patients were classified into four groups according to the concurrent presence (+) or absence (−) of HSR and C reactivity, as follows: true positive (tp: HSR+, C+), false positive (fp: HSR−, C+), true negative (tn: HSR−, C−), and false negative (fn: HSR+, C−). In addition, laboratory reactors were stratified to three categories on the basis of 10 minutes SC5b-9 values, as specified in column 1. The 0.98 and 1.96 µg/mL cut-off values represent two and four times the upper limit of normal SC5b-9 levels (0.49 µg/mL), respectively, and were chosen arbitrarily. The sensitivity, specificity, and positive and predictive values of the SC5b-9 assay with regard to HSRs were computed as described (18).
[a]Significant increase refers to significant ($p < 0.05$) increase of 10 minutes SC5b-9 relative to baseline.
Abbreviation: HSR, hypersensitivity reaction.
Source: From Ref. 18.

increase of SC5b-9 in 9/10 reactor patients in contrast to 9/16 in the nonreactor group (Fig. 5). Thus, 92% of clinical reactors were also laboratory reactors, while only 56% of clinical nonrectors were laboratory reactors. These data led to the conclusion that C activation and HSR show significant ($p < 0.05$) correlation.

A closer scrutiny of the quantitative relationship between SC5b-9 values at 10 minutes and severity of HSR also revealed that the SC5b-9 assay is highly sensitive in predicting HSRs (Table 3), although the specificity and positive predictive value of the test was relatively low, particularly in patients in whom the rise of SC5b-9 at 10 minutes remained below two times the upper limit of normal SC5b-9 (Table 3, row 2). However, restricting the

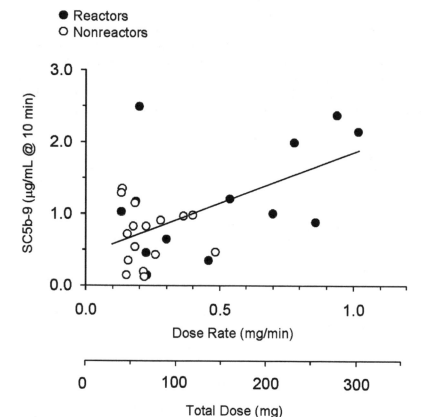

Figure 6 Dependence of C activation and hypersensitivity reactions (HSRs) on Doxil dose rate. The 10-minute SC5b-9 levels were plotted against the initial rate of Doxil administration (total dose/60 min $\times$ 1/5) in cancer patients displaying (*filled circles*) or not displaying HSRs (*empty circles*). The regression line correlates Doxil dose rate with C activation, $R^2 = 0.25$, $n = 29$, and $p < 0.01$. The probability of developing HSRs at different dose rates were quantified by serial computation of odd's ratios in the 0.2 to 0.5 mg/min range. *Source*: From Ref. 18.

criteria for laboratory reactivity to 10-minute SC5b-9 values exceeding two- or fourfold the upper threshold of normal (Table 3, row 3), the specificity and positive predictive value of the C assay remarkably increased with relatively less decrease in sensitivity. Thus, the extent of SC5b-9 elevation was proportional with the specificity and positive predictive value of the C assay with regards to HSRs.

As for the relationships among Doxil dose rate, C activation, and HSRs, Figure 6 shows the 10-minute SC5b-9 values of clinical reactors (filled circles) and nonreactors (empty circles) plotted against the initial rate of Doxil administration. Regression analysis revealed significant correlation between dose rate and SC5b-9 ($p < 0.01$), indicating that C activation at 10 minutes was Doxil dose dependent. Consistent with the correlation between HSRs and Doxil dose and the significant association between C activation and HSRs, the upper right quadrant of the plot contained readings obtained exclusively from clinical reactors. By providing evidence that Doxil caused C activation in a large majority of cancer patients although HSRs was manifested in a smaller portion, the above study confirmed a previous report by Small et al. (64) wherein 19% of patients infused with a RCM displayed HSR, although C activation was detectable in 49%.

CONCLUSIONS

Taken together, the information delineated above strongly suggest that C activation may be an important factor or precondition in eliciting HSRs, but its presence is not sufficient to actually precipitate the reaction. Thus, C activation may be a contributing but not rate-limiting factor in the pathogenesis of "pseudoallergy" to Doxil and other liposomal drugs. A plausible hypothesis explaining this relationship is that reactors differ from nonreactors in at least two criteria: (i) they develop a C reaction to the drug and (ii) their mast cells and basophils have a lower than normal threshold for secretory response to anaphylatoxins. Consistent with this proposal, proneness for HSRs is known to correlate with the presence of other allergies, i.e., with atopic constitution. Also, C5a-induced thromboxane production by leukocytes in vitro was shown to be significantly greater in atopic subjects than in normal controls. In short, CARPA will develop with high probability in those atopic subjects who respond to the drug with C activation and whose mast cells have a low threshold for C5a-induced release reaction.

REFERENCES

1. Haxby JA, Kinsky CB, Kinsky SC. Immune response of a liposomal model membrane. Proc Natl Acad Sci USA 1968; 61:300–307.
2. Kinsky SC, Haxby JA, Zopf DA, Alving CR, Kinsky CB. Complement-dependent damage to liposomes prepared from pure lipids and Forssman hapten. Biochemistry 1969; 8:4149–4158.

3. Alving CR, Kinsky SC, Haxby JA, Kinsky CB. Antibody binding and complement fixation by a liposomal model membrane. Biochemistry 1969; 8: 1582–1587.

4. Haxby JA, Gotze O, Muller-Eberhard HJ, Kinsky SC. Release of trapped marker from liposomes by the action of purified complement components. Proc Natl Acad Sci USA 1969; 64:290–295.

5. Bradley AJ, Devine DV. The complement system in liposome clearance: can complement deposition be inhibited. Adv Drug Deliv Rev 1998; 1(32):19–29.

6. Szebeni J. The interaction of liposomes with the complement system. Crit Rev Ther Drug Carrier Syst 1998; 15:57–88.

7. Szebeni J. Complement activation-related pseudoallergy caused by liposomes, micellar carriers of intravenous drugs and radiocontrast agents. Crit Rev Ther Drug Carr Syst 2001; 18:567–606.

8. Szebeni J. Complement activation-related pseudoallergy: mechanism of anaphylactoid reactions to drug carriers and radiocontrast agents. In: Szebeni J, ed. The Complement System: Novel Roles in Health and Disease. Boston: Kluwer, 2004:399–440.

9. Szebeni J, Baranyi B, Savay S, et al. Liposome-induced pulmonary hypertension: properties and mechanism of a complement-mediated pseudoallergic reaction. Am J Physiol 2000; 279:H1319–H1328.

10. Szebeni J, Fontana JL, Wassef NM, et al. Hemodynamic changes induced by liposomes and liposome-encapsulated hemoglobin in pigs: a model for pseudo-allergic cardiopulmonary reactions to liposomes. Role of complement and inhibition by soluble CR1 and anti-C5a antibody. Circulation 1999; 99: 2302–2309.

11. Szebeni J, Baranyi B, Savay S, et al. The role of complement activation in hypersensitivity to pegylated liposomal doxorubicin (Doxil®). J Liposome Res 2000; 10:347–361.

12. Szebeni J, Baranyi L, Savay S, et al. The interaction of liposomes with the complement system: in vitro and in vivo assays. Meth Enzymol 2003; 373:136–154.

13. Uziely B, Jeffers S, Isacson R, et al. Liposomal doxorubicin: antitumor activity and unique toxicities during two complementary phase I studies. J Clin Oncol 1995; 13:1777–1785.

14. Alberts DS, Garcia DJ. Safety aspects of pegylated liposomal doxorubicin in patients with cancer. Drugs 1997; 54(suppl 4):30–45.

15. Dezube BJ. Safety assessment: DoxilR (Doxorubicin HCl liposome injection) in refractory AIDS-related Kaposi's sarcoma. In: Doxil Clinical Series. Califon, NJ: Gardiner-Caldwell SynerMed, 1996:1–8.

16. Gabizon A, Martin F. Polyethylene glycol-coated (pegylated) liposomal doxorubicin. Rationale for use in solid tumours. Drugs 1997; 54:15–21.

17. Gabizon AA, Muggia FM. Initial clinical evaluation of pegylated liposomal doxorubicin in solid tumors. In: Woodle MC, Austin SG, eds. Long-Circulating Liposomes: Old Drugs, New Therapeutics. Texas: Landes Bioscience, 1998: 155–174.

18. Chanan-Khan A, Szebeni J, Savay S, et al. Complement activation following first exposure to pegylated liposomal doxorubicin (Doxil): possible role in hypersensitivity reactions. Ann Oncol 2003; 14:1430–1437.

19. Levine SJ, Walsh TJ, Martinez A, et al. Cardiopulmonary toxicity after liposomal amphotericin B infusion. Ann Int Med 1991; 114:664–666.
20. Laing RBS, Milne LJR, Leen CLS, Malcolm GP, Steers AJW. Anaphylactic reactions to liposomal amphotericin. Lancet 1994; 344:682.
21. Ringdén O, Andström E, Remberger M, Svahn BM, Tollemar J. Allergic reactions and other rare side effects of liposomal amphotericin. Lancet 1994; 344:1156–1157.
22. de Marie S. Liposomal and lipid-based formulations of amphotericin B. Leukemia 1996; 10(suppl 2):S93–S96.
23. Schneider P, Klein RM, Dietze L, et al. Anaphylactic reaction to liposomal amphotericin (AmBisome). Br J Haematol 1998; 102:1108.
24. Cabriales S, Bresnahan J, Testa D, et al. Extravasation of liposomal daunorubicin in patients with AIDS-associated Kaposi's sarcoma: a report of four cases [comments]. Oncol Nurs Forum 1998; 25:67–70.
25. Eckardt JR, Campbell E, Burris HA, et al. A phase II trial of DaunoXome, liposome-encapsulated daunorubicin, in patients with metastatic adenocarcinoma of the colon. Am J Clin Oncol 1994; 17:498–501.
26. Fossa SD, Aass N, Paro G. A phase II study of DaunoXome in advanced urothelial transitional cell carcinoma. Eur J Cancer 1998; 34:1131 1132.
27. Gill PS, Espina BM, Muggia F, et al. Phase I/II clinical and pharmacokinetic evaluation of liposomal daunorubicin. J Clin Oncol 1995; 13:996–1003.
28. Gill PS, Wernz J, Scadden DT, et al. Randomized phase III trial of liposomal daunorubicin versus doxorubicin, bleomycin, and vincristine in AIDS-related Kaposi's sarcoma. J Clin Oncol 1996; 14:2353–2364.
29. Girard PM, Bouchaud O, Goetschel A, et al. Phase II study of liposomal encapsulated daunorubicin in the treatment of AIDS-associated mucocutaneous Kaposi's sarcoma. AIDS 1996; 10:753–757.
30. Guaglianone P, Chan K, DelaFlor-Weiss E, et al. Phase I and pharmacologic study of liposomal daunorubicin (DaunoXome). Invest New Drugs 1994; 12: 103–110.
31. Money-Kyrle JF, Bates F, Ready J, et al. Liposomal daunorubicin in advanced Kaposi's sarcoma: a phase II study. Clin Oncol (R Coll Radiol) 1993; 5: 367–371.
32. Richardson DS, Kelsey SM, Johnson SA, et al. Early evaluation of liposomal daunorubicin (DaunoXome, Nexstar) in the treatment of relapsed and refractory lymphoma. Invest New Drugs 1997; 15:247–253.
33. Sculier JP, Coune A, Brassinne C, et al. Intravenous infusion of high doses of liposomes containing NSC 251635, a water-insoluble cytostatic agent. A pilot study with pharmacokinetic data. J Clin Oncol 1986; 4:789–797.
34. Masaki T, Okada N, Yasuda R, Okada H. Assay of complement activity in human serum using large unilamellar liposomes. J Immunol Meth 1989; 123:19–24.
35. Mold C, Gewurz H. Activation of human complement by liposomes: serum factor requirement for alternative pathway activation. J Immunol 1980; 125: 696–700.
36. Okada N, Yasuda T, Tsumita T, Okada H. Membrane sialoglycolipids regulate the activation of alternative complement pathway by liposomes containing

trinitrophenylaminocaproyldipalmitoylphosphatidylethaolamine. Immunology 1983; 48:129–140.

37. Ozato K, Ziegler HK, Henney CS. Liposomes as model membrane systems for immune attack. II. The interaction of complement and K cell populations with immobilized liposomes. J Immunol 1978; 121:1383–1388.

38. Richards RL, Habbersett RC, Scher I, et al. Influence of vesicle size on complement-dependent immune damage to liposomes. Biochim Biophys Acta 1986; 855:223–230.

39. Roerdink F, Wassef NM, Richardson EC, Alving CR. Effects of negatively charged lipids on phagocytosis of liposomes opsonized by complement. Biochim Biophys Acta 1983; 734:33–39.

40. Alving CR, Urban KA, Richards RL. Influence of temperature on complement-dependent immune damage to liposomes. Biochim Biophys Acta 1980; 600:117.

41. Harashima H, Sakata K, Funato K, Kiwada H. Enhanced hepatic uptake of liposomes through complement activation depending on the size of liposomes. Pharm Res 1994; 11:402–406.

42. Liu D, Liu F, Song YK. Recognition and clearance of liposomes containing phosphatidylserine are mediated by serum opsonin. Biochim Biophys Acta 1995; 1235:140–146.

43. Bonte F, Juliano RL. Interactions of liposomes with serum proteins. Chem Phys Lipids 1986; 40:359–372.

44. Chonn A, Cullis PR, Devine DV. The role of surface charge in the activation of the classical and alternative pathways of complement by liposomes. J Immunol 1991; 146:4234–4241.

45. Cunningham CM, Kingzette M, Richards RL, et al. Activation of human complement by liposomes: a model for membrane activation of the alternative pathway. J Immunol 1979; 122:1237–1242.

46. Funato K, Yoda R, Kiwada H. Contribution of complement system on destabilization of liposomes composed of hydrogenated egg phosphatidylcholine in rat fresh plasma. Biochim Biophys Acta 1992; 1103:198–204.

47. Harashima H, Sakata K, Funato K, Kiwada H. Enhanced hepatic uptake of liposomes through complement activation depending on the size of liposomes. Pharm Res 1994; 11:402.

48. Mayer MM. Complement and complement fixation. In: Kabat EA, Mayer MM, eds. Kabat Mayer's Experimental Immunochemistry. 2nd ed. Springfield, IL: Charles C. Thomas, 1961:133–240.

49. Plank C, Tang MX, Wolfe AR, Szoka FCJ. Branched cationic peptides for gene delivery: role of type and number of cationic residues in formation and in vitro activity of DNA polyplexes. Hum Gene Ther 1999; 10:319–332.

50. Fontana JL, Welborn L, Mongan PD, et al. Oxygen consumption and cardiovascular function in children during profound intraoperative normovolemic hemodilution. Anesth Analg 1995; 80:219–225.

51. Szebeni J, Wassef NM, Spielberg H, Rudolph AS, Alving CR. Complement activation in rats by liposomes and liposome-encapsulated hemoglobin: evidence for anti-lipid antibodies and alternative pathway activation. Biochem Biophys Res Comm 1994; 205:255–263.

52. Szebeni J, Wassef NM, Rudolph AS, Alving CR. Complement activation in human serum by liposome-encapsulated hemoglobin: the role of natural anti-phospholipid antibodies. Biochim Biophys Acta 1996; 1285:127–130.
53. Szebeni J, Wassef NM, Hartman KR, Rudolph AS, Alving CR. Complement activation in vitro by the red blood cell substitute, liposome-encapsulated hemoglobin: mechanism of activation and inhibition by soluble complement receptor type 1. Transfusion 1997; 37:150–159.
54. Szebeni J, Spielberg H, Cliff RO, et al. Complement activation and thromboxane A2 secretion in rats following administration of liposome-encapsulated hemoglobin: inhibition by soluble complement receptor type 1. Art Cells Blood Subs and Immob Biotechnol 1997; 25:379–392.
55. Szebeni J, Baranyi L, Savay S, et al. Role of complement activation in hypersensitivity reactions to Doxil and HYNIC-PEG liposomes: experimental and clinical studies. J Liposome Res 2002; 12:165–172.
56. Aguado JM, Hidalgo M, Moya I, et al. Ventricular arrhythmias with conventional and liposomal amphotericin. Lancet 1993; 342(8881):1239.
57. Coune A, Sculier JP, Frühling J, et al. IV administration of a water-insoluble antimitotic compound entrapped in liposomes. Preliminary report on infusion of large volumes of liposomes to man. Cancer Treat Rep 1983; 67:1031–1033.
58. Skubitz KM, Skubitz AP. Mechanism of transient dyspnea induced by pegylated-liposomal doxorubicin (Doxil). Anticancer Drugs 1998; 9:45–50.
59. Cheung AK, Parker CJ, Hohnholt M. Soluble complement receptor type 1 inhibits complement activation induced by hemodialysis membranes in vitro. Kidney Int 1994; 46:1680–1687.
60. Skroeder NR, Jacobson SH, Lins LE, Kjellstrand CM. Acute symptoms during and between hemodialysis: the relative role of speed, duration, and biocompatibility of dialysis. Artif Organs 1994; 18:880–887.
61. Skroeder NR, Kjellstrand P, Holmquist B, Kjellstrand CM, Jacobson SH. Individual differences in biocompatibility responses to hemodialysis. Int J Artif Organs 1994; 17:521–530.
62. Brouwers AH, De Jong DJ, Dams ET, et al. Tc-99m-PEG-Liposomes for the evaluation of colitis in Crohn's disease. J Drug Target 2000; 8:225–233.
63. Dams ET, Oyen WJ, Boerman OC, et al. [99]mTc-PEG liposomes for the scintigraphic detection of infection and inflammation: clinical evaluation. J Nucl Med 2000; 41:622.
64. Small P, Satin R, Palayew MJ, Hyams B. Prophylactic antihistamines in the management of radiographic contrast reactions. Clin Allergy 1982; 12:289–294.

2

Fixed Drug Ratio Liposome Formulations of Combination Cancer Therapeutics

Troy O. Harasym, Paul G. Tardi, Sharon A. Johnstone, and Lawrence D. Mayer
Celator Pharmaceuticals Corp., Vancouver, British Columbia, Canada

Marcel B. Bally
Department of Advanced Therapeutics, British Columbia Cancer Agency, Vancouver, British Columbia, Canada

Andrew S. Janoff
Celator Pharmaceuticals Corp., Vancouver, British Columbia, Canada, and Celator Pharmaceuticals Inc., Princeton, New Jersey, U.S.A.

INTRODUCTION

When chemotherapy agents are used to treat cancer, they are most effective when administered in combination. The development of drug combinations evolved from the pioneering work by medical oncologists in the 1950s and 1960s. For example, clinical studies led by Frei and Freireich demonstrated that dramatic improvements in the treatment of childhood leukemia could be achieved through the use of increasing numbers of drugs (1,2). Specifically, response rates in the range of 40% and no cures with methotrexate alone increased to >95% complete response and 35% cure rates with the inclusion of 6-mercaptopurine, prednisone, and vincristine into the treatment regimen. Eventually, cure rates increased to 75% to 80% with the inclusion of asparaginase, daunorubicin, and cytarabine. The principle

25

underlying this approach was to administer combinations of chemotherapeutic drugs with nonoverlapping toxicities in full doses as early as possible in the disease (3).

In practical terms, executing the development of combination chemotherapy regimens relies on escalating the drugs to their maximum tolerated dose (MTD). This principle has remained largely unchanged from the trials conducted in the 1960s to today. Clinical evaluation of drug combinations typically establishes the recommended dose of one agent and then adds subsequent drugs to the combination, increasing the dose until the aggregate effects of toxicity are considered to be the maximum tolerated (4). The efficacy of such combinations in patients is then determined in postmarketing trials under the assumption that maximum therapeutic activity will be achieved with maximum dose intensity for all drugs in the combination. As we will describe below, this assumption may be incorrect due to the ways in which combinations of chemotherapy drugs interact when exposed to tumor cells.

Clinicians and research scientists have been actively searching for synergistic anticancer drug combinations since the concept of combination chemotherapy was adopted into widespread use (5). Combining different antitumor agents with distinct and independent therapeutic effects can improve patient responses. However, these benefits can be significantly enhanced if the agents interact synergistically where the responses are greater than predicted based on the contribution of individual agents. In contrast, combined drugs can interact antagonistically so that the combination is less active than predicted for additive activity of the individual agents. In reality, it is very difficult to determine whether drug combinations are acting in a synergistic, additive, or antagonistic fashion in cancer patients. Ultimately, one can only determine whether a new combination provides a statistically significant increase in an efficacy end point such as response rate, time to progression, or survival. As a result, one cannot resolve whether such combinations are truly optimized for potential synergistic interactions in the clinic.

Because of the difficulties associated with evaluating drug synergy in a clinical setting, researchers have utilized in vitro tumor cell lines to determine how drug combinations interact. Although this approach has the disadvantage of working with immortalized tumor cells in nonphysiological conditions, it provides a well-controlled environment where the cytotoxic effects of anticancer drug combinations can be carefully studied and analyzed by any one of a number of methods available to quantify whether drug combinations interact in a synergistic, additive, or antagonistic fashion. Also, the availability of a wide range of human tumor cell lines allows for common trends to be established, thereby enhancing the reliability of synergy predictions based on preclinical behavior. The following section provides a summary of several methods currently utilized for the quantitative evaluation of drug–drug interactions.

DETERMINATION OF DRUG–DRUG INTERACTION EFFECTS: SYNERGY VS. ANTAGONISM

Before discussing the results of drug interaction analyses, it is important that proper definitions of synergy and antagonism be defined. In order to evaluate synergy, the quantification of additivity is first required. Simply put, additivity is defined as the combined effect of two drugs predicted from the sum of the quantitative effects of the individual components. Synergy is therefore defined as a more than expected additive effect, and antagonism as a less than expected additive effect when the drugs are evaluated in combination. Although these definitions are basic, there are many complexities of synergy evaluations as described by Chou (6,7).

A variety of mathematical methods have been proposed to evaluate drug combination effects in the context of synergy and antagonism, ranging in complexity from general techniques requiring simple manual calculations to sophisticated algorithms aided by computers (8–11). Table 1 highlights some of the more commonly used methods. The underlying principles of

Table 1 Various Drug Interaction Methods Used for Evaluating Synergy

Evaluation models
Approaches for continuous response data
   Isobologram (1870)
   Loewe additivity (1926)
   Bliss independence (1939)
   Fractional product method of Webb (1963)
   Method of Valeriote and Lin (1975)
   Method of Drewinko et al. (1976)
   Interaction index calculation of Berenbaum (1977)
   Method of Steel and Peckman (1979)
   Median-effect method of Chou and Talalay (1984)
   Method of Berenbaum (1985)
   Bliss independence response surface approach
   Method of Pritchard and Shipman (1990)
Nonparametric response surface approaches
   Bivariate spline fitting (Sühnel, 1990)
Parametric response surface approaches
   Models of Greco et al. (1990)
   Models of Weinstein et al. (1990)
Approaches for discrete success/failure data
   Approach of Gessner (1974)
Parametric response surface approaches
   Method of Greco and Lawrence (1988)
   Multivariate linear logistic model

Source: From Ref. 8.

the numerous models do vary but many of the models are modifications on existing models. Consequently, prior to selecting a method for data analysis, a thorough understanding of a particular model's origin as well as its strengths and weakness is required for proper data interpretation.

By far the most prevalent model used for drug combination analysis is the median-effect method of Chou and Talalay (12). The advantages of this method include the following: (i) the fundamental equations used were derived from basic mass action enzyme kinetic models; (ii) the experimental design efficiently utilizes experimental data points compared to other methods; and (iii) the analysis method is available as a software package allowing for easy data entry and modeling. After a comprehensive evaluation, the median-effect model was the primary model that we have chosen to apply to drug-combination analysis. However, to reduce any bias that could be incurred from the use of a single data analysis method and associated assumptions we also analyze selected data using isobologram and response surface methods. For a comprehensive review of the median effect, isobologram, response surface, and other various methods, the reader is referred to Greco et al. (8).

METHODOLOGICAL APPROACHES TO DETERMINING SYNERGY/ANTAGONISM

When studying in vitro drug combinations for antitumor activity, two parameters inevitably must be established, namely drug concentration (dose) and drug:drug ratio. As described above, the different methodologies that one uses to evaluate synergy/antagonism relationships for drug combinations may lead to different drug doses and ratios evaluated, but nonetheless, distinct ratios and doses will result. Considering the application of carriers to deliver drug combinations, we recognized that drug:drug ratios will be fixed in the carrier, and consequently drug-ratio effects on synergy must be evaluated in vitro. This leads to the selection of the median-effect analysis method developed by Chou where different fixed-ratio combinations can be compared as a function of drug concentration employing the commercially available software CalcuSyn (13). This mimics the application of carrier-based drug combinations in vivo where the amount of the two drugs will be fixed in the delivery vehicle that maintains the ratio after administration and the dose injected can then be escalated at that fixed ratio.

The median-effect model introduced by Chou and Talalay (12) is the most widely utilized method for synergy determinations by investigators. A key element in its wide use can be attributed to its commercial availability as a software package, CalcuSyn (BioSoft). The original model was based upon the derivation of hundreds of enzyme kinetic models from mass-action law principles using methods of mathematical induction and deduction.

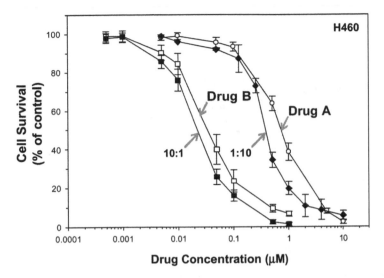

Figure 1 Cell viability curves for two drugs (*Drug A and Drug B*) exposed to tumor cells and assessed for cell viability (relative to control cells) using the MTT assay. The drugs were combined at 1:10 and 10:1 molar ratios and simultaneously exposed to the tumor cells for 72 hours.

The result was the generation of the median-effect equation, $f_a/f_u = (D/D_m)^m$, correlating the *dose* and the *effect* (tumor cell growth inhibition) in the simplest possible form, where D is the drug dose, D_m is the median effect dose, f_a is the fraction of cells affected, and f_u is the fraction of cells unaffected ($f_a + f_u = 1$), and m signifies the sigmoidicity or shape of the dose–effect curve. The median-effect plot is a plot of $x = \log (D)$ versus $y = \log (f_a/f_u)$ and was introduced by Chou in 1976 (14,15). Further refinements in 1981 introduced the concept of the combination index (CI) and the latest refinements in 1988 introduced the dose-reduction index (16,17).

The initial analysis of drug combinations is performed by acquiring viability curves (cell viability as a function of drug concentration) for the individual drugs and drug combinations in tumor cell lines using the MTT assay (Fig. 1). The sigmoidal dose–effect curves are subsequently transformed to linear data using the logarithmic form of the median-effect equation, $\log (f_a/f_u) = m \log (D) - m \log (D_m)$ (Fig. 2). For each linearized plot two parameters are obtained, the x-intercept (D_m) and the slope (m). These parameters are subsequently used with two alternative forms of the median-effect equations: the first allows the effect f_a to be determined at any dose (D), $f_a = 1/[1 + (D_m/D)]^m$; the second allows the dose D for any effect f_a to be determined, $D = D_m[f_a/(1 - f_a)]^{1/m}$. The latter calculations allow drug doses (D) and effect (f_a) values to be tabulated. Subsequently, the tabulated data are used to determine the CI values at each f_a using

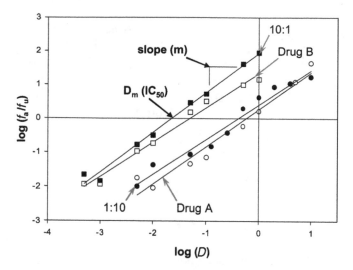

Figure 2 Median-effect plot of the cell viability data presented in Figure 1. Cell viability curves are linearized using the median-effect equation.

$CI = (D)_1/(D_x)_1 + (D)_2/(D_x)_2$ (Fig. 3) where $(D)_1$ and $(D)_2$ are the doses of Drugs 1 and 2 in combination for a given effect and $(D_x)_1$ and $(D_x)_2$ are the doses of Drugs 1 and 2 alone for a given effect. Based on this equation, additive drug interactions provide a CI value of 1.0 whereas synergistic

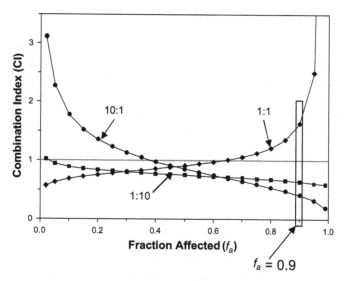

Figure 3 A plot of combination index (CI) as a function of the fraction of affected cells for Drug A:Drug B molar ratios of 1:10, 1:1, and 10:1 as calculated by the median-effect relationship utilizing the commercial software CalcuSyn. The box highlights the relationship between CI and drug:drug ratio at high tumor cell kill ($f_a = 0.9$).

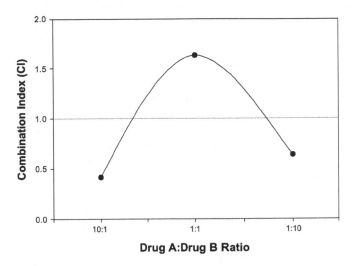

Figure 4 Combination index values for Drug A:Drug B ratios of 1:10, 1:1, and 10:1 obtained at an f_a value of 0.9 as a function of drug ratio.

interactions result in $CI < 1$ and antagonistic interactions lead to CI values >1. When CI is plotted versus f_a (fraction of tumor cell growth inhibition), we often observe not only drug ratio–dependent synergy, but also changes in CI as a function of drug concentration that is exponentially related to f_a (Fig. 3). Because we are working with anticancer agents, we focus our attention on f_a values $\geq 50\%$ owing to the fact that tumor growth inhibition below this value is not clinically meaningful. As shown in Figure 4, we can highlight the large ratio dependence reflected at high f_a values, e.g., $f_a = 0.9$, and replot as CI versus drug:drug ratio. In this example, it is evident that the 10:1 and 1:10 ratios are synergistic, i.e., CI values <1.0 whereas the 1:1 ratio is antagonistic ($CI > 1$).

IN VITRO INFORMATICS: THE FOUNDATION OF FIXED-RATIO FORMULATIONS

Examples of synergy dependence on drug:drug ratio and drug concentration for drug combinations have been documented previously (18–21). However, these phenomena are underappreciated and the implications of such relationships on therapeutic applications have not been considered because, until now, no one has attempted to control drug:drug ratios after injection in vivo. Early in our drug screening activities, we observed that many combinations of approved anticancer drugs used in chemotherapy regimens today display drug-ratio dependency in addition to concentration dependency. While some ratios are synergistic, others can be additive or even antagonistic. Furthermore, the synergy observed for a specific drug:drug ratio can vary as a function of concentration.

Table 2 presents some examples of drug combinations that display ratio-dependent synergy/antagonism behavior. For this purpose, we are showing CI values at high drug concentrations (ED75, $f_a = 0.75$) that are most relevant for in vivo efficacy. Two features become very apparent when the data are presented in this fashion. First, there is a significant degree of drug:drug ratio dependency for a number of drug combinations and for each drug combination in a number of cell lines. Of even greater interest is that patterns of similar ratio dependency emerge when multiple cell lines are examined and this pattern is different for different drug combinations. Together, these observations indicate that drug ratio–dependent antitumor activity is a phenomenon of broad relevance to chemotherapy combinations. What is not evident in Table 2 is the fact that, in addition to combinations exhibiting drug ratio dependency, the CI for individual fixed drug ratios can vary significantly as a function of concentration (f_a). This is shown for one example in Figure 5 where CI values observed for floxuridine:carboplatin exposed to HCT-116 human colorectal cancer cells increase dramatically to antagonism at $f_a > 0.6$ for floxuridine:carboplatin molar ratios of 10:1, 5:1 and 1:1, whereas ratios of 1:5 and 1:10 are strongly synergistic.

The dependence of antitumor activity on drug:drug ratio and drug concentration for anticancer drug combinations has profound implications on the in vivo application of drug combination therapy. When anticancer drug combinations in conventional aqueous-based formulations are administered to patients (or laboratory animals in the case of preclinical studies), the individual agents distribute and are metabolized independently and as a result there is little that can be done to control the drug ratio that is exposed to tumor cells. For example, although a Drug A:Drug B molar ratio of 5:1 may be optimal for in vitro synergy, administration of that ratio to patients will most certainly not result in a 5:1 ratio being exposed to the tumor over any significant amount of time, if at all. Furthermore, drug concentrations vary dramatically over short periods of time after administration of conventional drug cocktails. Consequently, not only is it difficult, if not impossible, to control drug ratios after administration, but the concentration (and consequently f_a) cannot be controlled. This no doubt results in tumors being exposed to drug ratios and concentrations that are therapeutically inferior when drug combinations are coadministered in conventional aqueous-based formulations.

Given the above considerations, it is possible that many drug combinations developed based on tolerability may in fact be using drug ratios that are inferior from a synergy/antagonism standpoint. However, because it is extremely difficult to control the drug:drug ratio for chemotherapy cocktails, the utility of in vitro synergy information is of very limited value

Table 2 Drug Mixture Combination Index Profiles Displaying Drug-Ratio Dependency

Drug combinations (A+B)	Cell lines screened	Tumor type	~Dm (µM)		CI @ ED75 for ratios of Drugs A and B				
			A	B	10:1	5:1	1:1	1:5	1:10
Floxuridine + carboplatin	HCT-116	Colon/carcinoma	3.2	36	2.1	2.6	2.5	0.1	0.1
	Capan-1	Pancreas/ carcinoma	0.12	29	1.7	1.5	1.2	0.7	0.6
	IGROV-1	Ovary/carcinoma	0.3	10	1.3	1.3	1.4	0.6	0.2
	N87	Gastric	0.41	88	1.3	0.9	0.7	0.9	0.8
	MCF-7	Breast	0.004	50	1.2	1.0	1.0	1.1	0.8
	H322	Lung/carcinoma	0.6	175	0.6	1.7	0.8	0.7	1.1
	HT-29	Colon/carcinoma	1.5	63	1.1	1.4	0.6	0.2	0.1
	L1210	Leukemia	0.0013	8.5	0.8	0.6	1.1	0.8	1.2
Floxuridine + oxaliplatin	HCT-116	Colon	7	2	1.0	1.0	0.6	2.4	1.5
	HT-29	Colon	800	4	0.7	1.2	1.2	1.5	1.1
	LS174T	Colon	0.5	0.7	1.5	0.8	0.4	0.5	1.0
Paclitaxel + cisplatin	IGROV-1	Ovary/carcinoma	0.03	0.65	1.3	1.6	2.0	0.4	0.3

(Continued)

Table 2 Drug Mixture Combination Index Profiles Displaying Drug-Ratio Dependency (*Continued*)

Drug combinations (A + B)	Cell lines screened	Tumor type	~Dm (µM)		CI @ ED75 for ratios of Drugs A and B				
			A	B	10:1	5:1	1:1	1:5	1:10
Etoposide + carboplatin	SF-268	CNS/glioma	4	22	1.4	–	0.7	–	0.9
Irinotecan + carboplatin	LS174T	Colon	2.1	19	1.1	1.6	1.0	1.2	0.7
	A549	Lung/carcinoma	22	37	0.7	0.9	0.5	1.0	1.3
	BXPC-3	Ovarian	11.3	12.4	1.0	1.0	0.8	0.8	1.2
	SW620	Colon/carcinoma	14	85	1.1	0.9	0.4	0.3	0.4
Cisplatin + topotecan	H460	Lung/carcinoma	0.66	0.05	0.8	–	0.7	–	8.0
Taxol + doxorubicin	IGROV-1	Ovary/carcinoma	0.03	1	1.3	–	2.0	–	0.3

Note: CI > 1.1; CI = 0.9–1.1; CI < 0.9
Abbreviations: CI, combination index; CNS, central nervous system.

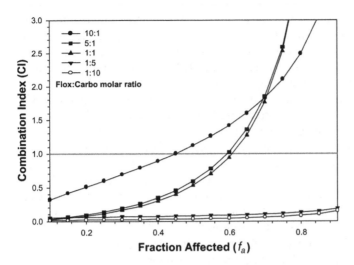

Figure 5 Combination index values as a function of f_a observed for floxuridine: carboplatin exposed to HCT-116 human colorectal cancer cells at Flox:Carbo molar ratios of 1:10 (o), 1:5 (▼), 1:1 (▲), 5:1 (■), 10:1 (•).

for in vivo applications. Two important tenets arise from these concepts. First, the ability to maintain drug ratios after administration in vivo could allow one to fix the most therapeutically active ratio and avoid antagonistic ratios. This would expose tumor cells to the optimized drug ratio, theoretically resulting in significantly enhanced antitumor activity. Second, maximizing the concentration of the optimal drug ratio at the tumor site will amplify synergistic effects in vivo by pushing the response to higher f_a values. This is important because reports have documented that the degree of synergy often increases with increasing f_a (22,23). Taken together, this information indicates that translating in vitro informatics for drug combinations to in vivo treatments in order to capture the benefits of synergistic interactions relies on fixing the appropriate drug ratio in vivo and delivering high concentrations of this drug ratio directly to tumor cells.

FINDING, FIXING, AND DELIVERING SYNERGISTIC DRUG COMBINATIONS

Once the implications of drug synergy being dependent on drug ratios and concentrations were recognized, we speculated that drug delivery vehicles may provide a means to capture in vivo the synergy effects observed in vitro. It is well documented that small (100–200-nm diameter) drug-delivery vehicles such as liposomes preferentially accumulate at sites of tumor growth due to the enhanced permeability and retention effects associated with tumors (24–26). Specifically, tumor vasculature is poorly formed and has gaps or fenestrea through which delivery vehicles can pass and extravasate into the tumor interstitium (27). This principle has been exploited to deliver a variety of single anticancer drugs and several liposome-based cancer drug products are currently marketed in the United States and Europe (28–32). Taking this concept one step further, we have integrated drug-delivery technology with in vitro synergy information in order to control drug ratios in vivo. In addition, the tumor uptake properties of delivery vehicles provide much higher tumor drug levels compared to conventional anticancer drugs, thereby pushing the drug combination toward higher f_a ranges. We predicted that together these effects could dramatically improve the efficacy of anticancer drug combinations utilizing liposome-based drug combinations, a strategy we refer to as CombiPlex^{TM}.

The CombiPlex approach for developing drug combinations is to: (i) *find* drug combinations that exhibit ratio-dependent synergistic antitumor activity using in vitro screening; (ii) *fix* the drug ratio that optimizes activity as well as avoids antagonism in drug delivery vehicles; and (iii) *deliver* the drug combination in vivo so that the synergistic drug ratio is maintained and exposed to tumors. By encapsulating drug combinations in delivery vehicles, the two agents no longer are metabolized and eliminated independently but rather distribute as a unit, dictated by the characteristics of the drug carrier.

Drug combinations are screened in vitro against a range of tumor types utilizing standard cytotoxicity assays. The cell lines selected will typically include representation from major tumor groups (e.g., colon, breast, and lung) and may also be enriched with tumor types that represent a clinical indication for which a combination may already be approved. Drug combinations are evaluated for synergy in vitro at molar drug ratios ranging from 10:1 to 1:10. The MTT assay or alternate cell proliferation assays are used for quantitative evaluation of cell viability over a three-day continuous drug exposure time period. Synergy evaluations of drug combinations are performed in various murine and human tumor cell lines. Initial prescreening studies of individual drugs are performed in each cell line to identify respective IC_{50} values. For viability curves, all individual drugs and drug combinations are assayed at eight concentrations. Data are then compiled for CalcuSyn analysis for quantifying synergy.

The in vitro information on drug ratios is utilized to formulate the drug combinations in delivery vehicles. Specifically, as patterns of ratio-dependent synergy emerge from results on multiple tumor cell lines, drug ratio ranges are identified that most frequently optimize synergistic interactions and avoid antagonism. Within this range, the final drug ratio is selected such that it maximizes total drug dosing intensity. Formulation research then focuses on encapsulating and retaining the drugs at this ratio inside liposomes. It should be noted that although our current focus is on the use of liposomes, polymer nanoparticles and other delivery systems of similar size (50–200 nm) should theoretically be compatible with this approach provided that they are designed to control the pharmacokinetics of the two drugs in a coordinated fashion. Drugs in a combination are first formulated individually in liposomes in order to elucidate conditions and formulation attributes that govern encapsulation and drug retention properties. Alterations in lipid composition and entrapped buffers are made and their effects on drug retention both in vitro and in vivo are assessed. This information is then combined and formulations are generated in which the two drugs are maintained after systemic administration (ideally for 24 hours) at the drug:drug ratio identified in vitro to be optimal. Once formulations have been developed that can deliver the synergistic drug ratio in vivo, expanded preclinical studies are undertaken to establish the therapeutic benefits of the CombiPlex formulation. Dose range finding studies are first completed in mice to establish the MTD for the CombiPlex formulation, drug cocktail, individual free drugs and individual carrier formulated drugs. This information is utilized to establish dosing regimens in efficacy studies. Subsequently, therapeutic activity of the CombiPlex formulation is compared to dose matched and MTD treatments with the free drug cocktail in selected human xenograft and murine tumor models.

DESIGNING LIPOSOMAL SYSTEMS TO MAINTAIN FIXED DRUG RATIOS IN VIVO

Anticancer drug combinations typically utilized for treating cancer are often comprised of agents with very different chemical compositions and physical properties. This is evident in the combinations presented in Table 2 where drugs vary in molecular weight from 246 to nearly 1000 and exhibit water solubilities ranging from <1 mg/mL to >100 mg/mL. Similarly, oil–water partition coefficients vary by orders of magnitude as do membrane permeability coefficients. Consequently, the concept of formulating drug combinations with such disparate physicochemical properties into a single, delivery-vehicle–based pharmaceutical product that can release both drugs at the same rate after administration in vivo represents a significant technical challenge. Fortunately, over the past 20 years a large body of liposome encapsulation technology has been generated for a wide range of individual drugs. Coordinating the pharmacokinetics of liposome formulated fixed drug-ratio combinations therefore relies upon innovative integration of multiple carrier features in a manner that facilitates the independent control of drug-release kinetics for the individual agents.

The most straightforward approach to coordinating the pharmacokinetics of liposomal drug combinations is to encapsulate each drug independently into different carriers that provide the required drug retention properties and subsequently mix the liposomes at the desired drug:drug ratio in a single dosing solution. This is illustrated in Figure 6 which demonstrates that by encapsulating two drugs in separate liposomes, mixed formulations can be readily constructed to provide either synergistic or antagonistic drug ratios based on drug-ratio dependency observed in vitro (10:1 and 1:1 in the example presented). In this manner, encapsulation and drug retention properties can be optimized for the two drugs separately by manipulations in lipid composition and/or internal buffer compositions so that the plasma drug elimination kinetics are matched for both agents. To accomplish this, liposomes must be selected that exhibit similar circulation times and tissue distribution properties so that drug elimination kinetics can be correlated to bioavailable drug exposure (e.g., drug released from liposomes). This can be achieved by utilizing liposomes with diameters on the order of 100 nm composed of lipids that are not readily recognized and removed from the circulation by the reticuloendothelial system (33).

Suitable liposome formulations for this drug delivery application include conventional compositions of uncharged phospholipids such as phosphatidylcholine in combination with near equimolar amounts of cholesterol (Chol). These liposomes can also contain relatively inert negatively charged phospholipids such as phosphatidylglycerol to stabilize the membrane (34). The drug retention properties of the formulation are dramatically influenced

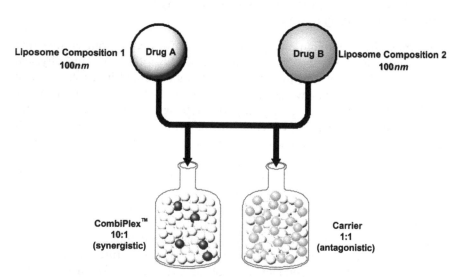

Figure 6 Illustration of the use of individually encapsulated drugs to create fixed-ratio drug liposome formulations that maintain the drug ratio after intravenous administration.

by the acyl chain length (for saturated chains) of the phospholipid (34,35). As the acyl chain length is increased from C14 to C18, the transition temperature of the phospholipid increases from 23°C to 55°C and this corresponds to a significant increase in drug retention. Recently, reports have shown that Chol content can also be manipulated to regulate drug retention properties (36,37). For example, when the drug idarubicin was encapsulated in Chol free, DSPC/DSPE-PEG (95:5, mol:mol) liposomes, drug retention was dramatically improved when compared to conventional DSPC/Chol (55:45, mol:mol) liposomes. In this case, the use of low Chol content in combination with high transition temperature saturated lipids generates a gel-state formulation that can provide a useful tool for enhancing the retention properties of numerous agents. This feature is particularly useful in matching the release of drug combinations from liposomes because different drugs can display opposite trends in their Chol content dependence for membrane permeability (37).

Drug release rates can also be controlled by the method in which the drug is loaded into the liposome. When liposomes are formed in a solution containing the active agent, only a small percentage (typically less than 10%) of the drug becomes trapped within the liposome, a method referred to as passive encapsulation (38). Although easy to perform, the drug-to-lipid molar ratios are typically low and the drugs are often poorly retained when injected in vivo (38). An alternate approach has been to actively accumulate drugs inside liposomes exhibiting an appropriate transmembrane gradient following the addition of free drug to a preparation of preformed liposomes. The sequestration of drug into the liposome can be driven by pH, ion, or metal gradients (38–41). The most common method of drug loading involves pH gradients that are typically established with citrate or ammonium sulphate (39,40). When a membrane permeable anticancer drug containing a protonatable amine is added to the liposome solution, it readily passes though the liposome membrane and becomes protonated and trapped inside the liposome. This method typically provides trapping efficiencies >90% and in vivo drug retention that is superior to the same drug that is passively encapsulated (39). We have recently developed a new method for the active loading of anticancer drugs based on complexation of drugs with metal ion salts (42). This method provides another means of controlling the release of suitable anticancer drugs from liposomes independent of pH gradient and lipid permeability effects.

Ideally, one would prefer to coencapsulate drug combinations inside a single liposome, thereby alleviating potential uncertainties about coordinated biodistribution characteristics for different liposome compositions. This presents additional challenges for matching drug release kinetics of chemotherapy combinations because one is now limited to a single lipid composition that will provide a single permeability barrier to the two drugs inside the liposomes. As a result, multiple features must be engineered into the formulation that is able to differentially affect the release of different chemical classes of drugs from liposomes. It is here where integration of active encapsulation methods with manipulations of lipid composition such as low Chol liposomes can provide comparable release rates for two different drugs and thereby maintain the encapsulated drug:drug ratio after systemic administration. However, one must ensure that the intraliposomal conditions utilized are compatible with both agents in order to avoid drug degradation or adverse drug–drug chemical interactions. An example of this approach is presented in Figure 7 for carboplatin and daunorubicin encapsulated inside liposomes. When these drugs were entrapped inside DSPC/Chol (55:45, mol:mol) liposomes, carboplatin was not released over 24 hours after IV administration to mice whereas >90% of the originally encapsulated daunorubicin was released from circulating liposomes over this time period (Fig. 7A). In contrast, when low Chol liposomes were utilized (DSPC/SM/DSPG, 75:5:20, mol:mol), daunorubicin release decreased while carboplatin release increased, resulting in coordinated plasma pharmacokinetics of the two drugs (Fig. 7B).

EXAMPLES OF THE BENEFITS PROVIDED BY COMBIPLEX™ TECHNOLOGY

Preclinical evaluation of the CombiPlex approach was determined using animal models of cancer where treatment with a CombiPlex drug combination was compared with the identical drug combination given as a free drug cocktail. Initially, we utilized drug combinations for which there was previous scientific evidence supporting synergy in vitro. For example, literature reports indicated that the combination of the approved anticancer agents cisplatin and topotecan were synergistic when combined to treat cancer cells in tissue culture (22). In vitro drug synergy analysis revealed that this combination had significant synergy (CI < 1) when tested against human lung cancer cells (H460 cell line) at a 10:1 molar ratio, whereas the same drug combination was antagonistic at a 1:10 ratio (Fig. 8).

A CombiPlex formulation comprising these two drugs at a 10:1 molar ratio was developed by encapsulating the two drugs individually in liposomes that provided comparable plasma drug pharmacokinetics for the two agents and mixing the two liposome preparations at a 10:1 cisplatin:topotecan concentration ratio. Cisplatin was passively entrapped inside saline containing 100 nm DMPC/Chol (55:45, mol:mol) liposomes whereas topotecan was actively loaded into 100 nm DSPC/Chol (55:45, mol:mol) liposomes utilizing a transmembrane pH gradient (inside acidic). When this putatively synergistic fixed ratio formulation was administered *iv* to mice, the plasma drug:drug ratio was maintained at approximately a 10:1 molar ratio over 24 hours postinjection (Fig. 9). This coordinated drug pharmacokinetic behavior was a result of both similar liposome elimination from the plasma and similar drug release from circulating liposomes.

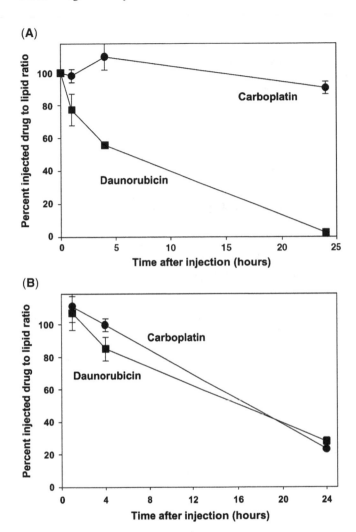

Figure 7 Release of carboplatin and daunorubicin from liposomes in the plasma after IV administration to mice. Drug release was determined by monitoring liposomal lipid in the plasma utilizing ^{3}H-cholesterylhexadecylether as a lipid label and drugs by atomic absorption and fluorescence spectroscopy, respectively. Decreases in the plasma drug-to-lipid ratio relative to the original formulation are proportional to the extent of drug release. In Panel A, the drugs were encapsulated inside DSPC/Chol (55:45, mol:mol) 100-nm liposomes utilizing a citrate-based pH gradient to load daunorubicin. In Panel B, the drugs were encapsulated inside DSPC/SM/DSPG (75:5:20, mol:mol) 100-nm liposomes again utilizing a pH gradient for daunorubicin loading.

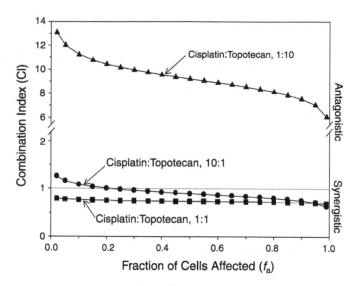

Figure 8 Combination index values as a function of f_a for cisplatin:topotecan combined at molar ratios of 10:1, 1:1, and 1:10 and simultaneously exposed to H460 human non–small cell lung cancer cells for 72 hours.

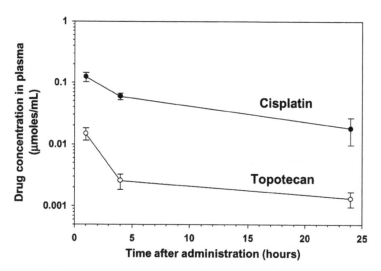

Figure 9 Coordinated plasma drug elimination of a CombiPlex™ formulation of cisplatin:topotecan in Scid/Rag-2 mice (three mice per time point). Cisplatin encapsulated inside 100-nm diameter DMPC/Chol (55:45, mole ratio) liposomes and topotecan encapsulated inside DSPC/Chol (55:45, mole ratio) liposomes were mixed at a 10:1 cisplatin:topotecan molar ratio and administered IV to mice at 1.6 mg/kg of cisplatin and 0.25 mg/kg of topotecan. Cisplatin in plasma was monitored by atomic absorption spectrometry for total platinum and topotecan was monitored utilizing tritiated topotecan.

The CombiPlex formulation containing cisplatin and topotecan at a 10:1 ratio was then tested for in vivo antitumor activity in the human H460 non–small cell lung cancer solid tumor xenograft model. In this model, tumor cells were inoculated subcutaneously (SC) and treatments were administered IV on days 13, 17 and 21 post tumor cell inoculation. Cisplatin:topotecan administered at 1.6:0.25 mg/kg as the CombiPlex formulation resulted in significant tumor growth inhibition where tumors did not grow beyond 0.05 cm^3 for up to 40 days compared to control tumors that grew to over 0.5 cm^3 within 30 days after tumor cell inoculation (Fig. 10). In contrast, the same doses of cisplatin and topotecan administered as a free drug cocktail in saline provided negligible antitumor activity (Fig. 10).

Given that liposome encapsulation of cisplatin and topotecan lead to significantly elevated plasma drug concentrations and extended drug circulation times compared to free drugs, it was possible that the efficacy improvements obtained with the CombiPlex formulation reflected primarily alterations in biodistribution properties due to the liposome delivery vehicles rather than drug ratio effects. This was examined by comparing the antitumor activity of the CombiPlex formulation with that provided by the individual liposomal drugs. As shown in Figure 11A, a cisplatin dose

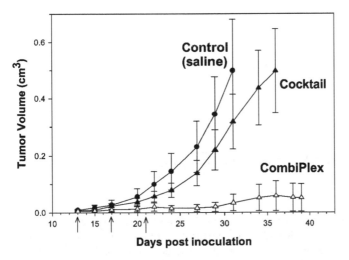

Figure 10 Antitumor activity of the 10:1 fixed molar ratio cisplatin:topotecan CombiPlex^{TM} formulation versus free drug cocktail in the H460 human non–small cell lung tumor model grown SC in Scid/Rag-2 mice (six mice per group). Free drug cocktail was prepared by mixing the drugs in saline just prior to injection. Treatments were administered IV on days 13, 17, and 21 post tumor cell inoculation. Cisplatin and topotecan doses in both treatment groups were 1.6 and 0.25 mg/kg, respectively.

(A)

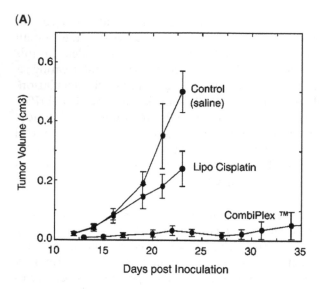

(B)

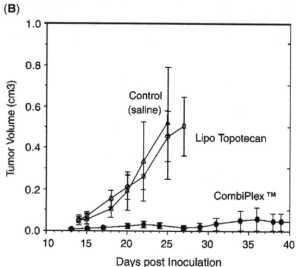

Figure 11   Antitumor activity of liposomal cisplatin and liposomal topotecan administered individually to Scid/Rag-2 mice bearing H460 human non–small cell lung
tumors. Drug doses were matched to those administered as the CombiPlex^{TM} formulation; 1.6 mg/kg liposomal cisplatin **(A)** and 0.25 mg/kg liposomal topotecan **(B)**.
Treatments were administered on days 13, 17, and 21 post tumor inoculation.

of 1.6 mg/kg encapsulated inside DMPC/Chol liposomes provided only
modest antitumor activity compared to the CombiPlex formulation. This
was particularly striking considering that liposomal topotecan administered
at 0.25 mg/kg provided no therapeutic effect (Fig. 11B). Clearly, the efficacy

of the CombiPlex formulation was far greater than predicted for by the additive antitumor activity of the individual components. This confirmed that the enhanced efficacy seen with the CombiPlex formulation reflected synergy and was not simply a liposome delivery effect.

ESTABLISHING A NEW PARADIGM FOR DEVELOPING DRUG COMBINATIONS

To date only minor attention has been given to understanding the role that drug ratios play in governing drug–drug interactions. This is not surprising given that the clinical development of drug combinations pushes the doses of the individual agents to maximum tolerability and hence the final drug ratio arises empirically. Also, until now there was no mechanism to effectively control the drug ratio after administration in vivo and consequently information obtained on drug ratio effects in vitro would have limited utility in prospectively designing dosing regimens in clinical trials.

The novelty of the CombiPlex approach to developing drug combinations lies in the concept that one can fix the optimal drug ratio and maintain this ratio in vivo. This opens the possibility to harvest in vitro drug combination informatics and apply it directly to clinically viable drug products in a manner previously not achievable. As such, this strategy represents a new paradigm for developing anticancer drug combinations. In this paradigm, optimally effective drug ratios can be identified and developed prospectively during preclinical development and subsequently tested in clinical trials as a single formulation. This has the advantage of greatly reducing the complexity associated with optimizing drug combinations in large, later stage clinical trials as is done today. The combination can be utilized upon initiation of Phase I trials by virtue of the fact that the agents behave pharmacokinetically as a single unit. This alleviates problems associated with nonlinear pharmacokinetics and dose-dependent adverse drug–drug interactions that can arise from combining drugs in conventional (non–carrier-based) formulations.

CombiPlex formulations have three features working together to increase the likelihood of success in the clinic. First, fixing the most effective drug ratio ensures that tumors will not be exposed to deleterious effects of antagonistic drug ratios. Second, the use of small particulate delivery vehicles increases the level of drug exposure to tumors, thereby amplifying synergistic effects. Third, because the drug combination is utilized throughout clinical development, it is reasonable to expect that signs of significant activity will be observed earlier in clinical evaluation than expected for single agents. Taken together, the CombiPlex approach opens up the possibility to shift drug combination development in a manner that will enhance our ability to tap the extensive informatics obtained in vitro and link it more directly to clinical product development. Ultimately this should yield high-value oncology products that provide significant benefit to cancer patients.

REFERENCES

1. Frei E III, Freireich EJ. Leukemia. Sci Am 1964; 210:88–96.
2. Freireich EJ, Frei E III. Recent advances in acute leukemia. Prog Hematol 1964; 27:187–202.
3. Frei E III. Clinical studies of combination chemotherapy for cancer. In: Chou TC, Rideout DC, eds. Synergism and Antagonism in Chemotherapy. San Diego, CA: Academic Press, 1991:103–108.
4. DeVita VT Jr. Principles of cancer management: Chemotherapy. In: DeVita Jr., Hellman S, Rosenberg SA, eds. Cancer: Principles and Practice of Oncology. Vol. 1. Philadelphia, NY: Lipinicott-Raven, 1997:333–347.
5. Rideout DC, Chou TC. Synergism, antagonism, and potentiation in chemotherapy: an overview. In: Chou TC, Rideout DC, eds. Synergism and Antagonism in Chemotherapy. San Diego, CA: Academic Press, Inc., 1991:3–53.
6. Chou TC. The median-effect principle and the combination index for quantitation of synergism and antagonism. In: Chou TC, Rideout DC, eds. Synergism and Antagonism in Chemotherapy. San Diego, CA: Academic Press, Inc., 1991:61–102.
7. Chang TT, Chou TC. Rational approach to the clinical protocol design for drug combinations: a review. Acta Paediatr Taiwan 2000; 41:294–302.
8. Greco WR, Bravo G, Parsons J. The search for synergy: a critical review from a response surface perspective. Pharmacol Rev 1995; 47:332–385.
9. Berenbaum MC. Isobolographic, algebraic, and search methods in the analysis of multiagent synergy. J Am Coll Toxicol 1988; 7:927–938.
10. Prichard MN, Shipman C Jr. A three dimensional model to analyze drug-drug interactions (review). Antiviral Res 1990; 14:181–206.
11. Pöch G, Reiffenstein RJ, Unkelbach HD. Application of the isobologram technique for the analysis of combined effects with respect to additivity as well as independence. Can J Physiol Pharmacol 1990; 68:682–688.
12. Chou TC, Talalay P. Quantitative analysis of dose-effect relationships: the combined effects of multiple drugs or enzyme inhibitors. Adv Enz Regul 1984; 22:27–55.
13. Chou TC. CalcuSyn: Windows software for dose effect analysis. In: Manual and Software. Cambridge, U.K.: Biosoft, 1996.
14. Chou TC. Derivation and properties of Michaelis-Menton type and Hill type equations for reference ligands. J Theoret Biol 1976; 39:253–276.
15. Chou TC, Talalay P. A simple generalized equation for the analysis of multiple inhibitions of Michaelis-Menton kinetic systems. J Biol Chem 1977; 252: 6438–6442.
16. Chou TC, Talalay P. Generalized equations for the analysis of inhibitors of Michaelis-Menton and higher order kinetic systems with two or more mutually exclusive and nonexclusive inhibitors. Europ J Biochem 1981; 115:207–216.
17. Chou J, Chou TC. Computerized simulation of dose reduction index (DRI) in synergistic drug combinations. Pharmacologist 1988; 30:231.
18. Pavillard V, Kherfellah D, Richard S, Robert J, Montaudon D. Effects of the combination of camptothecin and doxorubicin or etoposide on rat glioma cells and camptothecin-resistant variants. Br J Cancer 2001; 85:1077–1083.

19. Parsels LA, Parsels JD, Tai DCH, Coughlin DJ, Maybaum J. 5-fluoro-2-deoxyuridine-induced cdc25A accumulation correlates with premature mitotic entry and clonogenic death in human colon cancer cells. Cancer Res 2004; 64:6588–6594.
20. Goldwasser F, Shimizu T, Jackman J, et al. Correlations between S and G2 arrest and the cytotoxicity of camptothecin in human colon carcinoma cells. Cancer Res 1996; 56:4430–4437.
21. Lee MS, Zimmermann GR, Lehár J, et al. Systematic discovery of novel anti-cancer combination therapies. Proc Amer Ass Can Res 2005; 46:5010.
22. Kaufmann SH, Peereboom D, Buckwalter CA, et al. Cytotoxic effects of topotecan combined with various anticancer agents in human cancer cell lines. NCI 1996; 88:734–741.
23. Maurer BJ, Melton L, Billups C, Myles CC, Reynolds CP. Synergistic cytotoxicity in solid tumor cell lines between *N*-(4-hydroxyphenyl)retinamide and modulators of ceramide metabolism. JNCI 2000; 92:1897–1909.
24. Allen TM, Cullis PR. Drug delivery systems: entering the mainstream. Science 2004; 303:1818–1822.
25. Maeda H, Wu J, Sawa T, Matsumura Y, Hori K. Tumor vascular permeability and the EPR effect in macromolecular therapeutics: a review. J Control Release 2000; 65:271–284.
26. Hashizume H, Baluk P, Morikawa S, et al. Openings between defective endothelial cells explain tumor vessel leakiness. Am J Pathol 2000; 156:1363–1380.
27. Jain RK. Transport of molecules across tumor vasculature. Cancer Metastasis Rev 1987; 6:559–593.
28. Krown SE, Northfelt DW, Osoba D, Stewart JS. Use of liposomal anthracyclins in Kaposi's sarcoma. Semin Oncol 2004; 31(6 suppl 13):36–52.
29. Glantz MJ, Jaeckle KA, Chamberlain MC, et al. A randomized controlled trial comparing intrathecal sustained-release cytarabine (DepoCyt) to intrathecal methotrexate in patients with neoplastic meningitis from solid tumors. Clin Cancer Res 1999; 5:3394–3402.
30. Bressler NM. Verteporfin therapy of subfoveal choroidal neovascularization in age-related macular degeneration: two-year results of a randomized clinical trial including lesions with occult with no classic choroidal neovascularization-verteporfin in photodynamic therapy report 2. Am J Ophthalmol 2002; 133:168–169.
31. Adler-Moore J. AmBiosome targeting to fungal infections. Bone Marrow Transplant 1994; 14(suppl 5):3–7.
32. Batist G, Ramakrishnan G, Rao CS, et al. Reduced cardiotoxicity and preserved antitumor efficacy of liposome-encapsulated doxorubicin and cyclophosphamide compared with conventional doxorubicin and cyclophosphamide in a randomized, multicenter trial of metastatic breast cancer. J Clin Oncol 2001; 19:1444–1454.
33. Mayer LD, Krishna R, Bally MB. Liposomes for cancer therapy applications. In: Dumitriu S, ed. Polymeric Biomaterials. New York, NY: Marcel Dekker, 2001:823–841.
34. Tari A, Huang L. Structure and function relationship of phosphatidylglycerol in the stabilization of the phosphatidylethanolamine bilayer. Biochem 1989; 28(19):7708–7712.

35. Boman NL, Mayer LD, Cullis PR. Optimization of the retention properties of vincristine in liposomal systems. Biochim Biophys Acta 1993; 1152(2):253–258.
36. Dos Santos N, Waterhouse D, Masin D, et al. Substantial increases in idarubicin plasma concentration by liposome encapsulation mediates improved antitumor activity. J Control Release 2005; 20:89–105.
37. Dos Santos N, Mayer LD, Abraham SA, et al. Improved retention of idarubicin after intravenous injection obtained for cholesterol-free liposomes. Biochim Biophys Acta 2002; 1561:188–201.
38. Mayer LD, Cullis PR, Bally MB. The use of transmembrane pH gradient-driven drug encapsulation in the pharmacodynamic evaluation of liposomal doxorubicin. J Liposome Res 1994; 4:529–553.
39. Cullis PR, Hope MJ, Bally MB, et al. Influence of pH gradients on the trans-bilayer transport of drugs, lipids, peptides, and metal ions into large unilamellar vesicles. Biochim Biophys Acta 1997; 1331:187–211.
40. Haran G, Cohen R, Bar LK, Barenholz Y. Transmembrane ammonium sulfate gradients in liposomes produce efficient and stable entrapment of amphipathic weak bases. Biochim Biophys Acta 1993; 1151:201–215.
41. Fenske DB, Wong KF, Maurer E, et al. Ionophore-mediated uptake of ciprofloxacin and vincristine into large unilamellar vesicles exhibiting transmembrane ion gradients. Biochim Biophys Acta 1998; 1414:188–204.
42. Ramsay E, Alnajim J, Anantha M, et al. A novel approach to prepare a liposomal irinotecan formulation that exhibits significant therapeutic activity in vivo. Proc Amer Assoc Cancer Res 2004; 45:639.

3

Pharmacokinetics and Biopharmaceutics of Lipid-Based Drug Formulations

Theresa M. Allen

Department of Pharmacology, University of Alberta, Edmonton, Alberta, Canada

INTRODUCTION

A number of liposomal and lipid-based products have reached the market, with a full pipeline of similar products in clinical trials or preclinical development (1). In spite of this, there is not a good understanding of how the physical properties of the various drug formulations affect the pharmacokinetics (PK) and biopharmaceutics (BP) of their associated drugs relative to free (conventional) drugs.

The terminology used in this field has evolved over time, which has caused some confusion, e.g., in the interpretation of intellectual property claims, and there is still considerable variability across the field of lipid-based formulations in the use of certain terms such as liposome-entrapped drug. In order to discuss the topic of PK and BP fully it is necessary to define the terminology that is used in this paper (see Section on "Glossary"). It is hoped that these definitions will be widely adapted.

BACKGROUND

The physical properties of lipids and drugs govern how they interact with each other, and the variety of choices possible for combinations of drugs and lipids results in an enormous versatility in the composition and physical

Table 1 Versatility of Lipid-Based Drug Formulations

Lipid properties amenable to manipulation	Drug properties that influence lipid-based formulations	Combined lipid and drug properties influence the following parameters
Liposome size and degree of lamellarity	Molecular weight (size)	Encapsulation efficiency of drugs
Headgroup composition(s) and charge	Stability	Rate of drug clearance
Fatty acyl chain length(s)	Charge	Rate of drug release
Inclusion of other lipids such as cholesterol	Hydrophobicity or hydrophilicity, e.g., as measured by the octanol to water partition coefficient	Tissue localization of the formulation
		Formulation stability

properties of liposomal and lipid-based formulations. It has been known for many years that the properties of either the liposomes or the drugs could be manipulated to control the stability of the drugs and the rate at which associated contents partitioned out of the liposomes or other lipid-based formulations. This versatility gives us considerable control over the PK and biodistribution (BD) of lipid-based drug formulations (Table 1).

PK OF SUSTAINED RELEASE LIPID-BASED FORMULATIONS VS. FREE DRUGS ADMINISTERED INTRAVENOUSLY

Sustained release formulations are those formulations where drug release occurs over a period of at least a few hours. Commonly, lipid-based sustained release formulations are liposomes with the drug sequestered in the aqueous interior. Sustained-release formulations change the PK/BD of the entrapped drugs to a considerable extent, leading to increased therapeutic indices for the entrapped drugs through their ability to increase the efficacy of the drug and/or reduce their side effects [reviewed in Ref. (2)]. The therapeutic activity of the formulation will depend on the rate of drug release and the concentration of drugs achieved at the site of drug action, i.e., the rate and extent of bioavailability of the drug (see Section on "Bioavailability of Liposomal Drugs"). The biological properties of entrapped drugs compared to released drugs are given in Table 2.

The interpretation of PK data for sustained release formulations is not always straightforward; therefore, a detailed analysis of the PK curves for free drug compared to liposomal drug is given below.

One Compartment Model for Free Drugs

In a one compartment model for free drugs, after intravenous (IV) administration, free (conventional) drug equilibrates rapidly between the plasma and

Table 2 Properties of Liposome-Entrapped Drug vs. Released Drug

Liposome-entrapped drug	Released drug
Biologically inactive (until it is released)	Biologically active
Protected from degradation and metabolism	Degraded and/or metabolized by the usual routes for the free drug
Has the same PK/BD as the liposomes themselves, if the release rate is slow	Has the same PK/BD as free drug given by the same route, and at the same site of administration

Abbreviations: PK, pharmacokinetics; BD, biodistribution.

tissue compartments and the PK analysis treats these as a single compartment (Fig. 1A). The decrease in plasma levels are log-linear and are due to elimination of the drug from the body (B) at a constant rate K. Changes in tissue drug levels are proportional to changes in plasma drug levels (C) (3).

Log-Linear Clearance of Liposomal Drugs

For particulate carriers such as liposomes, the situation is more complicated (Fig. 2). PK plots have been generated for plasma clearance of small

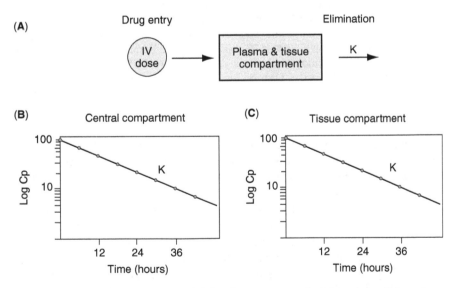

Figure 1 One compartment model for intravenous administration of free drug. (A) The drug rapidly equilibrates between plasma (central compartment) and tissues (tissue compartment) and is eliminated from the plasma (B) and the tissues (C) with log-linear pharmacokinetics. The rate of elimination is the elimination rate constant "K," which equals the slope of the line.

(A)

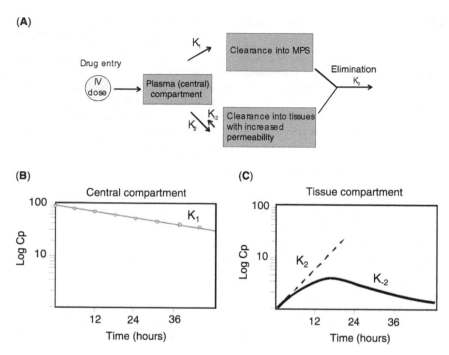

Figure 2 Log-linear pharmacokinetics (PK) for the central compartment for sterically stabilized liposomal sustained release drugs is not due, as in Figure 1, to rapid equilibration of the liposomes and their drugs between the central and the tissues compartment. Instead, the log-linear PK is due to the constant, slow rate of clearance of the liposomes into the mononuclear phagocyte systems (MPS) (**A**, **B**), at a rate that does not saturate MPS uptake. Distribution to the tissue compartment (**C**), e.g., solid tumors, is slow and occurs to a lesser extent than clearance into the MPS. *Abbreviation*: K, elimination rate constant.

(<150 nm in diameter) sterically stabilized liposomal drugs that look very similar to the log-linear plots generated for one compartment models for free drugs (4–7), but the underlying processes are very different. Liposomes such as those stabilized with poly(ethylene glycol) (PEG)-lipid derivatives display log-linear, or close to log-linear, PK in the plasma (central) compartment when the rate of drug release is slow relative to the rate of clearance. The liposomes are confined mainly to the central compartment, with little distribution to tissue compartments other than those with increased vascular permeability, e.g., solid tumors undergoing angiogenesis or sites of inflammation. The PK plots approaches log-linear for the central compartment because the rate of clearance into the mononuclear phagocyte systems (MPS), K_1, occurs at a slow rate that does not saturate the MPS over a wide concentration range. Uptake into the MPS of sterically stabilized liposomes occurs to a greater degree than tissue uptake and so MPS clearance dominates tissue distribution (K_2). Also, unlike free drugs, the rate at which particular carriers leave tissues such as solid tumors (K_{-2}) is hypothesized to be much slower than the rate at which they enter (K_2) due to the high osmotic pressure in tumors. A substantial portion of liposomes and their associated drugs are taken up into cells such as macrophages, where they enter normal metabolic pathways, and very little elimination of liposomes takes place via kidney filtration. Therefore, the rate of elimination of lipids and/or encapsulated drugs from the body (K_3) tends to be much slower than for free drugs.

To summarize, for sterically stabilized liposomes, tissue distribution K_2 is low relative to the capacity of the MPS to take up liposomes. Drug retained in liposomes remains to a very large extent in the central compartment and is not available for redistribution to tissues, which is a dominating feature of Figure 1. Released drug will be redistributed to tissues at the same rate and to the same extent as free drug. The degree to which redistribution of released drugs will impact the plasma PK for sterically stabilized drugs depends on the rate of drug release. For drugs like Doxil®, which have very slow rates of drug release, the effects on PK of the redistribution of released drug will be minimal. The log-linear decrease in plasma drug levels in the central compartment is due to dose-independent clearance of liposomes into the MPS, and a portion of the injected dose is slowly distributed to tissues with increased vascular permeability.

MULTICOMPARTMENT MODEL FOR FREE DRUGS

For most free drugs, the PK in the central compartment is best analyzed by a two compartment (or higher) model (Fig. 3). The drug is distributed into various tissue compartments from the central compartment at various rates (the distribution phase) depending on the perfusion of the tissue and the affinity of the drug for the particular tissue (3). After equilibration of the drug occurs throughout the body, first-order elimination occurs (K_2).

Michaelis Menten PK for Liposomal Drugs

Classical liposomes, lacking steric stabilization, are cleared from plasma with saturable, Michaelis-Menten-type PK (Fig. 4). The clearance, K_1, for sterically stabilized liposomes is substantially faster than K_1 for classical liposomes (Fig. 2), which results in saturation of the uptake of liposomes into the MPS as the dose increases (5). In other words, classical liposomes exhibit dose-dependent PK. Although the plasma clearance curves for liposomes with dose-dependent PK and slow drug release have a similar appearance to multi-compartment clearance for free drugs, the underlying process is very different.

As the dose increases, the PK plots approach log-linear PK (Fig. 4C). The PK plots change from biexponential (or multiexponential) to log-linear

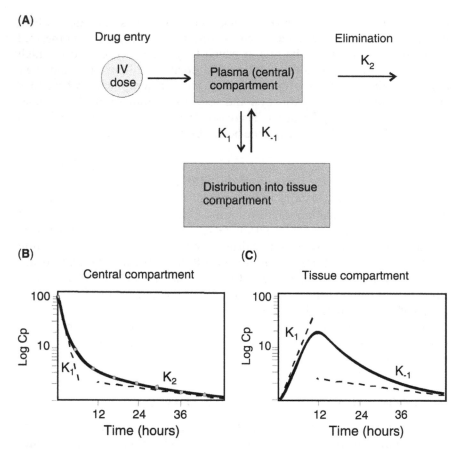

Figure 3 Two compartment model for free drugs. Drug is redistributed from the central compartment into the tissue compartment at a rate K_1. (**A–C**) With time, equilibrium between the central (**B**) and tissue compartment (**C**) is reached. Drug is eliminated from plasma at a constant rate K_2 (via metabolism and kidney filtration), which drives the drug back out of the tissues (K_{-1}).

with increasing lipid dose; therefore, it is important to know the administered lipid dose for the interpretation of the results (5).

In summary, for dose-dependent liposomal sustained release formulations, $K_{1,1}$ is high relative to the capacity of the MPS to clear liposomes from circulation. PK changes from biexponential to log-linear as the dose increases. As the MPS becomes saturated, clearance slows and becomes first order. Because a higher portion of the dose is cleared into the MPS relative to sterically stabilized liposomes, a smaller portion of the lipid-drug package is available for distribution to tissues with enhanced vascular permeability. Finally, a biexponential clearance, as in Figure 4B, could be due to either

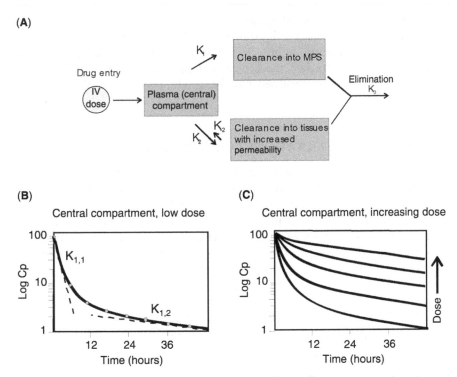

Figure 4 Classical liposomes at low doses (**A, B**) are initially cleared rapidly into the mononuclear phagocyte systems (MPS) ($K_{1,1}$) and, as the MPS saturates, the clearance into the MPS begins to slow down ($K_{1,2}$). Please note that $K_{1,2}$ does not correspond to the elimination phase K_2 for free drugs in Figure 3. As the dose increases (**C**), the contribution of the rapid component of clearance, $K_{1,1}$, toward the pharmacokinetics plot decreases relative to the contribution of the slow component of clearance, $K_{1,2}$. This results from increasing MPS saturation with increasing dose.

rapid clearance of a slow-release formulation, or fast drug release from either sterically stabilized or classical formulations, with the released drug cleared via a two compartment model (Fig. 3B). In most cases, the PK plots, like those represented in Figure 4B, are likely the sum of both liposome clearance and released drug clearance.

SUSTAINED-RELEASE FORMULATIONS: ALTERNATE ROUTES OF ADMINISTRATION

Although most PK data is available for IV administration of lipid-based carriers, some data has been generated for other routes, in particular the intraperitoneal and subcutaneous routes (8,9). From these data, it is apparent that the rate of absorption of the liposomes from the route of administration into the central compartment is dependent primarily on liposome size, with charge also playing a role. For liposomes with diameters above approximately 120 nm, very little absorption into the central compartment takes place. Once the liposomes have reached the central compartment, their PK/BD is the same as if the amount absorbed had been administered via the IV route of administration.

RAPID-RELEASE FORMULATIONS OF LIPID-BASED DRUGS

Liposomes and other lipid-based carriers such as lipid micelles, emulsions, complexes, and aggregates are effective at solubilizing hydrophobic drugs (10–12). Associating hydrophobic drugs with lipid-based carriers has, in large part, not resulted in substantial changes in the PK/BD of hydrophobic drugs relative to the same drug solubilized in vehicles such as Cremaphor® (13). In effect, the lipids are not functioning as a *carrier* for hydrophobic drugs but as a *vehicle* for the solubilization of the drugs. Lipid-based carriers of hydrophobic drugs have not, to date, resulted in significant changes in the therapeutic index for the drugs (13–15). Nevertheless, these types of formulations are clinically useful because the lipid-based carriers can result in improvements of the toxicity profile relative to cremaphor or other vehicles (16).

EFFECT OF FORMULATION AND DRUG RELEASE ON PK PROFILES

The PK profile for liposomal formulations of the same drug can vary over a wide range depending on the liposome properties and the rate of drug release. This is illustrated in Table 3 for the drug doxorubicin. Even without

Table 3 Comparison of the Pharmacokinetic Profiles for Different Formulations of Doxorubicin at a Dose of 25 mg/m^2

	Free drug (6)	Doxil® (6)	Myocet® (17)	Other[a]
$T_{1/2\alpha}$ (hr)	0.07	3.2	0.3	0.1
$T_{1/2\beta}$ (hr)	8.7	45.2	6.7	10
AUC, μg hr/mL	1.0	609	19.7	2.0
CL, mL/min	755	1.3	388	500
V_D, L	254	4.1	18.8	240

Note: $T_{1/2\alpha}$, initial phase of clearance. $T_{1/2\beta}$, second phase of clearance. In the case of free drug this is the terminal half-life.
[a]Hypothetical formulation where doxorubicin is not retained in the liposomes by a remote loading, drug sequestration process, so the drug rapidly equilibrates out of the liposomes.
Abbreviations: AUC, area under the time versus concentration curve; CL, clearance; V_D, volume of distribution.

knowing much about the compositions and sizes of the various formulations, some conclusions can be immediately drawn from the PK parameters. The free drug has a high volume of distribution, so it rapidly redistributes from the central compartment to the tissues. This results in a low $T_{1/2\alpha}$ and a high rate of clearance and a low area under the time versus concentration curve (AUC). As the free drug is eliminated from the plasma, drug in the tissues diffuses down its concentration gradient to re-enter plasma, resulting in a longer elimination half-life, $T_{1/2\beta}$.

For Doxil, the V_D is the same as the plasma volume in humans, 4 L. This demonstrates that the liposomes are confined primarily to the circulation and that the rate of drug release is low (4,6). A faster rate of drug release would result in an increase in the V_D, as is seen for Myocet® (17). For very fast rates of drug release, the V_D would be similar to that of the free drug, as seen for the "other" formulation. For most liposomal formulations a small component of rapid clearance is often seen. For Doxil, the $T_{1/2\alpha}$ is 3.2 hours. However, the large AUC for Doxil, which is almost three orders of magnitude larger than that of the free drug, indicates that the majority of the clearance takes place via the second half-life, $T_{1/2\beta}$, and the overall plasma PK has a log-linear appearance (7).

The lower $T_{1/2\alpha}$ and $T_{1/2\beta}$ of Myocet relative to Doxil are due to differences in liposome composition and size. Myocet is not a sterically stabilized liposome, whereas Doxil is, so the clearance of myocet is more rapid. The larger size of Myocet (160 nm in diameter vs. 100 nm) also contributes to the more rapid clearance of the carrier. The PK of Myocet is a good example of PK that are a sum of the PK of the carrier plus the PK of the released drug. The PK profile for the "other" formulation is a good example of the minor changes in PK that are seen when the lipid is behaving as a vehicle and drug release is rapid. The closer the ratio of the V_D of the free drug to that of the liposomal drug is to 1.0, the more rapid the rate of drug release.

EFFECT OF DRUG-RELEASE RATE ON THERAPEUTIC EFFECT

Until now, there have been very few studies on the implications of drug-release rates for the therapeutic effects of lipid-based formulations. The drug-delivery community is now beginning to try to design experiments that will shed light on this important concept. It is useful to consider the two extremes. If the drug is released rapidly from the formulation (see Section on "Rapid-Release Formulations of Lipid-Based Drugs"), then the therapeutic effect will be approximately equivalent to that of the free drug. However, if the formulation can reduce the toxicity of the vehicle, allowing increased dosing, then the therapeutic effect can increase due to the higher drug dose. At the other extreme, if the drug is released from the formulations very slowly, or not at all, then the drug concentration at the site of action may not reach the minimum effective concentration (MEC) (Fig. 5) and little or no therapeutic effect will be seen.

In order to begin addressing some of the issues surrounding the appropriate rate of drug release, we have looked at accumulation of liposomal doxorubicin in orthotopically implanted mouse mammary tumours for liposomes composed of phosphatidylcholine and cholesterol having three different drug-release rates, achieved by altering the fatty acyl chain composition of the phospholipid component (18). We examined accumulation of both liposomal lipid and drug in tumor tissues. The accumulation of doxorubicin in the tumors was inversely proportional to the drug-release rate. Liposomes with the slowest release rate had the highest tumor accumulation and vice versa. Almost no drug accumulated in tumors for liposomes with the fastest drug-release rate. Notably, liposomal lipid accumulated to the same extent in tumor tissue, independent of drug-release rate, i.e., drug-depleted liposomes accumulated in tumors. This demonstrates one of the pitfalls of relying on lipid labels to predict tissue accumulation of drugs.

In the same experiments, we also measured the therapeutic effect of the various formulations against the tumors (18). Liposomes having the highest

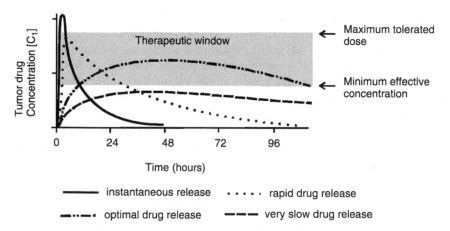

Figure 5 Tumor drug concentration for formulations with different rates of drug release. The therapeutic window is the range of drug concentrations between the minimum effective concentration (MEC) and the maximum tolerated dose. For drugs released instantaneously or rapidly, the tissue exposure to the drug [i.e., tissue area under the time vs. concentration curve (AUC) for tissue] will be similar to that achieved for free drugs. For drugs released very slowly, the MEC necessary for activity may not be reached. Optimal drug release rates are those in which released (bioavailable) drug concentrations in the tumor, or other tissue of interest, exceed the MEC and are sustained there for extended periods of time. This applies in particular to anticancer drugs that have a cell-cycle-specific mechanism of action, such as the vinca alkaloids or the camptothecins.

tumor accumulation had the best therapeutic effect, liposomes with intermediate release rates had increased toxicity, and liposomes with the fastest
release rates were not toxic but had reduced therapeutic effects (18). In other
words, for tissues such as tumors, where accumulation of the carrier into the
tissue is slow, the drug has to stay in the liposomes long enough for substantial levels of liposomal drug accumulation to occur, but the drug needs to
be released at a rate which will allow the MEC to be exceeded in the tumor
for sustained periods of time. Alternatively, after tissue accumulation of the
liposomal drug has occurred, release of a bolus of drug at or near the site of
drug action could be triggered via ultrasound, temperature, pH-sensitive
release, etc. (19–21). Triggered release may be most effective for anticancer
drugs that work by a cell-cycle independent mechanism of action such as the
anthracyclines.

BIOAVAILABILITY OF LIPOSOMAL DRUGS

Drugs entrapped in liposomal sustained release systems are not bioavailable,
are protected from degradation and metabolism, and have no biological
activity until the drug is released from the carrier. An understanding of the
rate and extent of bioavailability of liposomal drugs at their site of action
is important for the optimal design of liposomal carriers. However, it is
the total drug levels in tissues of interest such as tumors that are normally
reported, and total drug consists of both entrapped (nonbioavailable) and
released (bioavailable) drug. For example, in prostate cancer xenografts,
Vaage et al. showed that the tumor area under the time vs. concentration
curve (AUC) for total liposomal DXR was more than 1.5 orders of magnitude higher than the tumor AUC for the free drug (22). However, this number does not tell us anything about the amount of active (bioavailable) drug
in the tumor for the liposomal preparation. Because the therapeutic activity
is related to the time versus concentration curves for bioavailable drug, it is
important to develop methodologies to measure bioavailable drug. This can
be technologically difficult, but several investigators have started to develop
methods for measuring both total and bioavailable drugs. Harashima et al.
have attempted to model the PK/BD of liposomal drugs in vitro and in vivo
(23). They assumed that cytotoxic effects were due to DXR released from the
liposomes. Their work represents an important starting point for PK/BD
modeling of liposomal drugs. Eliaz et al. have modeled the kinetics of
doxorubicin cell kill in vitro for DXR and DXR entrapped in hyaluronan-
targeted liposomes (24). They assumed that the rate of cell killing is
dependent on the concentration of drug within the cell. They concluded that
the uptake of the targeted encapsulated drug was greater than the uptake of
the free drug and that for given amounts of intracellular DXR, the encapsulated drug was more efficient in killing cells than the free drug.

The Allen lab has attempted over the years to correlate the rate of intracellular release of DXR to cytotoxicity for targeted and triggered release systems (25,26). We have demonstrated that more rapid rates of intracellular drug release in vitro correlate with greater cytotoxicity. Recently, we have developed a method for estimating the rate and extent of bioavailable drug in vivo in mammary tumors in mice. We hypothesized that the tumor cell nuclei serves as an irreversible sink for drugs such as DXR that interchelate strongly with nuclear DNA. Only released drug can reach the nuclei; therefore, DXR extracted by acidified isopropanol from isolated tumor nuclei provides a reasonable estimate of bioavailable drug in tumors in vivo (27). For free DXR, this technique shows a high degree of bioavailability, as expected. Almost 100% of the total drug that is found in tumors is found to be associated with the tumor nuclei within an hour or two after the administration of the drug. We have also applied this assay to slow-release and rapid-release liposomal formulations of DXR. The total amount of DXR in tumors is substantially higher than that of free DXR for slow-release formulations. The levels of drug in tumor nuclei do not peak until approximately three days after administration, which is consistent with slow drug release, but the tumor AUC for bioavailable (nuclear) drug is significantly higher than that see for free DXR. Liposomes with more rapid drug release have tumor PK profiles that were intermediate between the free drug and the slow-release formulation (27). We are now extending these assays to targeted formulations.

Other investigators have used the tumor window flap model to estimate the location of released DXR based on fluorescence dequenching of the drug upon release from liposomes (28). Collection of tumor exudates and analysis of free drug or metabolite levels by high performance liquid chromotography (HPLC) is another way to determine levels of bioavailable drug (6).

EFFECT OF LIPOSOME SIZE, DOSING SCHEDULE, AND DOSE INTENSITY

Quite a lot of work has been done in the past on the effect of liposome size on the plasma PK of liposomes [reviewed in Ref. (2)]. We have recently extended these studies to examine the effect of size on uptake of liposomal DXR into various tissues, including tumor tissue (29). Liposomes and other macromolecular structures accumulate in solid tumors via the enhanced permeability and retention effect (30). The vasculature of solid tumors is reported to have increased permeability to macromolecules due to the existence of pores of up to 600 nm in diameter in the tumor vascular endothelium (31). In murine mammary cancer, we found that liposomes with diameters between 80 and 155 nm had similar accumulation in tumors, while liposomes of approximately 250 nm in diameter had reduced tumor accumulation relative to smaller liposomes. The reduced tumor accumulation of larger liposomes resulted in lesser control of tumor growth when the mice were treated with sterically stabilized liposomal DXR (29).

We have also examined the effect of dose, dose intensity, and dosing schedule in a murine mammary carcinoma model for multiple dosing regimens (32). Weekly IV administration of sterically stabilized liposomal doxorubicin (DXR) at a dose of 9 mg/kg given once a week for a total of 4 doses, resulted in accumulation of doxorubicin in cutaneous tissues of mice and development of lesions resembling palmar-plantar erythrodysesthesia (PPE). Lengthening the dose interval to every two weeks for a total of 4 doses, reduced the accumulation of DXR and lowered the incidence of PPE-like lesions. A dose interval of every four weeks again with 4 doses, resulted in complete clearance of doxorubicin from tissues between subsequent doses and a negligible incidence of PPE-like lesions. Therapeutic effects for doses of 9 mg DXR/kg of sterically stabilized liposomal DXR, given at every week (two doses) or every two weeks (two doses) had similar therapeutic activities, whereas prolonging the dose interval to every four weeks 9 (two doses) reduced therapeutic activity (32). PK, BD, and therapeutic activity were also studied in tumor-bearing mice for three dose schedules having the same dose intensity (4.5 mg/kg every three days × four doses, 9 mg/kg every week × two doses, or 18 mg/kg every two weeks × one dose). For these schedules, larger doses administered less often tended to be superior therapeutically to smaller doses given more often (32).

MAXIMUM TOLERATED DOSE

The effect of association of drugs with liposomes or other lipid-based carriers on the maximum tolerated dose of the drug (MTD) is not easily predictable. For rapid drug release systems, the MTD will be close to that of the free drug. However, for hydrophobic drugs, if vehicle toxicity reduces the MTD for the free drug, then using a less toxic vehicle, such as a lipid-based carrier, may increase the MTD. For sustained-release systems the MTD may stay relatively the same, e.g., liposomal vincristine (33), may increase, e.g., liposomal doxorubicin (34), or may decrease, e.g., liposomal cytosine arabinoside (35). For free drugs that are rapidly degraded in vivo, such as cytosine arabinoside or the camptothecins, entrapping the drugs in sustained-release liposomes protects the drug from degradation and lowers the MTD.

CONCLUSIONS

In conclusion, liposome size, composition, charge, drug-release rates, bioavailability, dose, dose intensity, and dose schedule can all affect the PK/BD of liposomal drugs, their tissue accumulation, side effects, and therapeutic activities. Many drug properties also affect the PK/BD of liposome-associated drugs. The more we understand about how these various properties interact with each other and impinge on the therapeutic effect of the formulations, the more we can begin to use rational rather than empirical design for new lipid-based therapeutics.

GLOSSARY

Liposomes: Phospholipid bilayer spheres consisting of one or more concentric bilayer membranes surrounding an aqueous interior compartment(s).

Lipid-based formulations: Formulations that include phospholipids or other lipids that form particulates of various types. These include liposomes, lipid colloidal dispersions, and other lipid structures such as micelles, inverted micelles, and discoid structures.

Associated contents as in "liposome-associated drug": This is a term with broad meaning, which does not draw any conclusions about the location of drugs or other compounds relative to the lipid components, liposomal bilayer, or enclosed aqueous volumes of the delivery systems. Drugs or other chemicals may be inserted between the hydrophobic fatty acyl chains in the lipid bilayer, located between the fatty acyl chains of micelles, may be in the liposome aqueous interior or associated with the interfacial region between lipid headgroups and tails.

Entrapped or encapsulated drug: This term should be reserved for drugs or other compounds that are sequestered within the aqueous interior compartment of a liposome, unless qualified by the modifier "membrane" (see below).

Membrane-entrapped or membrane-associated drug: The drug is either partly or fully buried between the fatty acyl chains of a liposome. Membrane-associated drug can be used when drug is bound to the bilayer surface, e.g., through charge-charge interactions.

Excipient: An inactive or inert substance that is added to a drug formulation, usually to provide stability or bulk.

Vehicle: A substance, usually inactive therapeutically, used in a medicinal preparation as the agent for carrying or solubilizing the active ingredient.

Sink: A large reservoir such as plasma lipoproteins where the thermodynamics is such that a drug will quickly move from its lipid-based formulation to the sink.

REFERENCES

1. Allen TM, Cullis PR. Drug delivery systems: entering the mainstream. Science 2004; 303:1818–1822.
2. Allen TM, Stuart D. Liposome pharmacokinetics: classical, sterically stabilized, cationic liposomes and immunoliposomes. In: Janoff AS, ed. Liposomes: Rational Design. New York: Marcel Dekker, Inc., 1998:63–97.

3. Shargel L, Yu ABC. Applied biopharmaceutics and pharmacokinetics. 4th ed. McGraw Hill, 1999:768.
4. Gabizon A, Shiota R, Papahadjopoulos D. Pharmacokinetics and tissue distribution of doxorubicin encapsulated in stable liposomes with long circulation times. J Natl Cancer Inst 1989; 81:1484–1488.
5. Allen TM, Hansen CB. Pharmacokinetics of stealth versus conventional liposomes: effect of dose. Biochim Biophys Acta 1991; 1068:133–141.
6. Gabizon A, Catane R, Uziely B, et al. Prolonged circulation time and enhanced accumulation in malignant exudates of doxorubicin encapsulated in polyethyleneglycol coated liposomes. Cancer Res 1994; 54:987–992.
7. Gabizon A, Shmeeda H, Barenholz Y. Pharmacokinetics of pegylated liposomal doxorubicin: review of animal and human studies. Clin Pharmacokinet 2003; 42:419–436.
8. Allen TM, Hansen CB, Guo LSS. Subcutaneous administration of liposomes: a comparison with the intravenous and intraperitoneal routes of injection. Biochim Biophys Acta 1993; 1150:9–16.
9. Oussoren C, Zuidema J, Crommelin DJA, et al. Lymphatic uptake and biodistribution of liposomes after subcutaneous injection I. Influence of the anatomical site of injection. J Liposome Res 1997; 7:85–99.
10. Perez-Soler R, Zou Y. Liposomes as carriers of lipophilic antitumor agents. In: Lasic D, Papahadjopoulos D, eds. Medical Applications of Liposomes. Amsterdam: Elsevier, 1998:283–296.
11. Torchilin VP, Weissig V. Polmeric micelles for the deliver of poorly soluble drugs. In: Park K, Mrsny RJ, eds. Controlled Drug Delivery. Washington, D.C.: American Chemical Society, 2000:297–313.
12. Zhang JX, Hansen CB, Allen TM, et al. Lipid-derivatized poly(ethylene glycol) micellar formulations of benzoporphyrin derivatives. J Control Release 2003; 86:323–338.
13. Cabanes AKEB, Gokhale PC, et al. Comparative in vivo studies with paclitaxel and liposome-encapsulated paclitaxel. Int J Oncol 1998; 12:1035–1040.
14. Sharma A, Straubinger RM, Ojima I, et al. Antitumor efficacy of taxane liposomes on a human ovarian tumor xenograft in nude athymic mice. J Pharm Sci 1995; 84:1400–1404.
15. Sharma A, Sharma US, Straubinger RM. Paclitaxel-liposomes for intracavitary therapy of intraperitoneal P388 leukemia. Cancer Lett 1996; 107:265–272.
16. Cabanes A, Reig F, Garcia-Anton JM, et al. Evaluation of free and liposome-encapsulated gentamycin for intramuscular sustained release in rabbits. Res Vet Sci 1998; 64:213–217.
17. Uziely B, Jeffers S, Isacson R, et al. Liposomal doxorubicin: Antitumor activity and unique toxicities during two complementary phase I studies. J Clin Oncol 1995; 13:1777–1785.
18. Charrois GJR, Allen TM. Drug release rate influences the pharmacokinetics, biodistribution, therapeutic activity, and toxicity of pegylated liposomal doxorubicin formulations in murine breast cancer. Biochim Biophys Acta 2004; 1663:167–177.
19. Needham D, Anyarambhatla G, Kong G, et al. A new temperature-sensitive liposome for use with mild hyperthermia: characterization and testing in a human tumour xenograft model. Cancer Res 2000; 60:1197–1201.

20. Choi JS, MacKay JA, Szoka FCJ. Low-pH-sensitive PEG-stabilized plasmid-lipid nanoparticles: preparation and characterization. Bioconjug Chem 2003; 14:420–429.
21. Zhang JX, Zalipsky S, Mullah N, et al. Pharmaco attributes of dioleoylphosphatidylethanolamine/cholesterylhemisuccinate liposomes containing different types of cleavable lipopolymers. Pharmacol Res 2004; 49:185–198.
22. Vaage J, Barbera-Guillem E, Abra R, et al. Tissue distribution and therapeutic effect of intravenous free or encapsulated liposomal doxorubicin on human prostate carcinoma xenografts. Cancer 1994; 73:1478–1484.
23. Harashima H, Tsuchihashi M, Iida S, et al. Pharmacokinetic/pharmacodynamic modeling of antitumor agents encapsulated into liposomes. Adv Drug Deliv Rev 1999; 40:39–61.
24. Eliaz RE, Nir S, Marty C, et al. Determination and modeling of kinetics of cancer cell killing by doxorubicin and doxorubicin encapsulated in targeted liposomes. Cancer Res 2004; 64:711–718.
25. Lopes de Menezes DE, Pilarski LM, Allen TM. In vitro and in vivo targeting of immunoliposomal doxorubicin to human B-cell lymphoma. Cancer Res 1998; 58:3320–3330.
26. Kirchmeier MJ, Ishida T, Chevrette J, et al. Correlations between the rate of intracellular release of endocytosed liposomal doxorubicin and cytotoxicity as determined by a new assay. J Liposome Res 2001; 11:15–29.
27. Laginha K, Verwoert S, Charrois GJR, Allen TM. Determination of bioavailable drug levels from pegylated liposomal doxorubicin in murine orthotopic breast tumor. Clin Cancer Res 2005; 11:6944–6949.
28. Zu NZ, Da D, Rudoll TL, et al. Increased microvascular permeability contributes to preferential accumulation of Stealth liposomes in tumor tissue. Cancer Res 1993; 53:3765–3770.
29. Charrois GJR, Allen TM. Rate of biodistribution of STEALTH® liposomes to tumor and skin: influence of liposome diameter and implications for toxicity and therapeutic activity. Biochim Biophys Acta 2003; 1609:102–108.
30. Maeda H, Wu J, Sawa T, et al. Tumor vascular permeability and the EPR effect in macromolecular therapeutics: a review. J Control Release 2000; 65: 271–284.
31. Yuan F, Dellian M, Fukumura D, et al. Vascular permeability in a human tumor xenograft: Molecular size dependence and cutoff size. Cancer Res 1995; 55:3752–3756.
32. Charrois GJR, Allen TM. Multiple injections of pegylated liposomal doxorubicin: Pharmacokinetics and therapeutic activity. J Pharm Exp Ther 2003; 306:1058–1067.
33. Gelmon KA, Tolcher A, Diab AR, et al. Phase I study of liposomal vincristine. J Clin Oncol 1999; 17:697–705.
34. Northfelt DW, Martin FJ, Kaplan LD, et al. Pharmacokinetics, tumour localization and safety of Doxil (liposomal doxorubicin) in AIDS patients with Kaposi's sarcoma (Meeting abstract). Proc Am Soc Clin Oncol 1993; 12:A8.
35. Allen TM, Mehra T, Hansen CB, et al. Stealth liposomes: an improved sustained release system for 1-b-D-arabinofuranosylcytosine. Cancer Res 1992; 52:2431–2439.

4

Optimization Strategies in Lymph Node Targeting of Interstitially Injected Immunoglobulin G–Bearing Liposomes

Seyed M. Moghimi

Molecular Targeting and Polymer Toxicology Group, School of Pharmacy, University of Brighton, Brighton, U.K.

INTRODUCTION

Subcutaneously injected liposomes are drained rapidly into the initial lymphatic system through patent junctions in the lymphatic capillaries (Fig. 1) and are conveyed to the regional lymph nodes via the afferent lymph (1). In the lymph node, macrophages of medullary sinuses and paracortex are mainly responsible for liposome capture (1–4). Littoral cells and polymorphonuclear granulocytes may also participate in liposome extraction from the lymph (2). The above-mentioned means of liposome transportation from interstitium and sequestration by lymph node scavengers can be taken into experimental and clinical advantage. The potential medical applications, therefore, include lymphoscintigraphic tracing, lymph node mapping, treatment of macrophage infections, antigen delivery to macrophages and dendritic cells, and treatment or prevention of tumour metastases (1,5,6).

Among the key factors controlling liposome drainage through the ground substance of interstitium into the lymphatic system as well as capture by lymph node scavenger cells are vesicle size and surface characteristics and these have been studied extensively (1–3,7–10). A common reported observation, however, is that only a small fraction of the rapidly drained

Figure 1 Electron micrograph of a rat lymphatic capillary (L) located in the subcutaneous footpad region. In lymphatic capillaries, numerous endothelial cells (*arrowhead*) overlap extensively at their margins. Following interstitial injection, many of the overlapped endothelial cells are separated and thus passageways, referred to as patent junctions, are provided between the interstitium and the lymphatic lumen. The micrograph also shows the presence of some inflammatory cells in the region (*arrow*).

liposomes is retained by the regional lymph nodes; this is a reflection of the small weight of the lymph nodes (e.g., the weight of a healthy popliteal lymph node may vary between 4 and 7 mg in a 150-g rat) (3,7–9). For instance, although up to 50% of the injected liposome dose is drained into the lymphatic system within 8 to 10 hours, liposome retention in the regional nodes rarely exceeds 2% of this fraction (3,7–9). Noncaptured liposomes subsequently gain access into the systemic circulation via thoracic duct and are cleared by scavengers of both liver and spleen (3,7–9,11).

To date, two approaches have been applied to further enhance lymph node retention of interstitially injected liposomes. The first approach was based on surface modification of liposomes with macrophage ligands such as antibodies and mannose (1,7,8,10). Such modifications have dramatically improved the drainage rate of the liposomes from the injection site into the lymphatic system, but have resulted in modest increases (two- to three-fold increase over the unmodified liposome injection) in their lymph node retention. The second approach was described recently by Phillips et al. (12), taking advantage of biotin-coated liposomes. These liposomes were injected subcutaneously, followed by an adjacent subcutaneous injection of multivalent avidin. As the biotin-coated liposomes and avidin migrate through the ground substance of the interstitium and subsequently the lymphatic vessels, the avidin encounters the liposomes and causes vesicle aggregation. Aggregated vesicles are prone to rapid filtration in regional lymph nodes. Indeed, this approach yielded liposome retention in the regional nodes by several folds when compared to the unmodified liposome formulation (12). As a result of avidin-mediated vesicle aggregation, a significant fraction of biotin-coated liposomes may remain at the injection site; aggregated vesicles are susceptible to macropinocytosis by interstitial macrophages.

In light of these observations, a new approach for enhancing lymph node retention of interstitially injected liposomes is described here. This approach is based on liposome surface engineering with both poly(ethylene glycol) (PEG) and immunoglobulin G (IgG), resulting not only in rapid lymphatic uptake of interstitially injected liposomes but also in improved vesicle retention among the associated regional lymph nodes (13).

RATIONALE

Earlier, the concept of steric stabilization was applied in this laboratory to successfully control lymphatic uptake and lymph node localization of interstitially injected model polystyrene nanoparticles of 40 to 60 nm in size (14,15). For steric stabilization, ethylene oxide/propylene oxide–based copolymers (poloxamers and poloxamines) were used. The transport efficiency into lymphatic capillaries and the extent of particle uptake by lymph node macrophages was dependent on the density and the molecular architecture of the surface-exposed ethylene oxide chains (15). Thus, by careful surface manipulation, 40% of the injected dose of nanospheres was delivered to macrophages in the regional lymph nodes within six hours of administration (15). Because surface modification with ethylene oxide–based copolymers modulated the lymphatic fate and distribution of interstitially injected model polystyrene particles, an attempt was directed to translate these finding to liposome engineering with PEG. Recently, biophysical studies have established that the bilayer concentration of methoxyPEG-lipid (mPEG-lipid) controls conformation of the surface exposed mPEG chains (16). For example, with incorporation of 5 to 7 mol% $mPEG_{2000}$-phospholipid into liposomal bilayer, the exposed mPEG chains predominantly assume a "mushroom-brush transition" conformation, whereas at concentrations up to 4 mol% mPEG projections are in a "nonoverlapped mushroom" regimen. It is also known that particles with surface exposed $mPEG_{2000-5000}$ in "mushroom-brush" or "brush" conformation are resistant to clearance by nonstimulated macrophages, whereas particles with surface mPEG molecules in a "nonoverlapped mushroom" display are susceptible to ingestion by phagocytic cells (17,18). Therefore, the primary aim of this work was to examine the effect of bilayer concentration of a candidate $mPEG_{2000}$-lipid, and therefore its conformation, on lymphatic uptake and target-binding of interstitially injected immuno-PEG-liposomes, where nonspecific rat IgG is coupled to the terminus of PEG using a functionalized PEG_{2000}-phospholipid.

METHODOLOGY

Liposomes were composed of egg phosphatidylcholine (egg PC) and cholesterol (chol) with or without various amounts of $mPEG_{2000}$-distearoyl-phosphatidylethanolamine ($mPEG_{2000}$-DSPE). In addition, liposomes contained different amounts of either the negatively charged lipid dicetyl-phosphate (DCP) or phosphatidylserine (PS) or phosphatidylglycerol (PG). Some preparations contained N-(4'-(4''-maleimidophenyl)butyroyl)-phosphatidylethanolamine (MPB-PE), or MPB-PEG_{2000}-DSPE as linker lipids for IgG attachment. Liposomes were prepared by hydrating the dried lipid film with a buffer (25 mM Hepes, 25 mM Mes, 135 mM NaCl), pH 6.7, containing $[^{125}I]$-poly(vinylpyrrolidone)$[(^{125}I)$-PVP], and extruded through polycarbonate nuclopore filters of 100 nm in pore diameter using a high-pressure extruder. Nonspecific rat IgG was thiolated, using N-succinimi-dyl-3-(2-pyridyldithio)proprionate (SPDP) followed by reduction with dithiothreitol (DTT) prior to coupling to liposomes (antibody to phospho-lipid molar ratio of 1:1000) via the linker lipids (13,19). Any nonreacted maleimide can be blocked with DTT or cysteine; free DTT/cysteine and unbound antibody was finally removed by passing the liposome suspension over a Sepharose CL-4B column. Alternatively, IgG can be thiolated using N-succinimidyl-S-acetylthioacetate. The advantage with this approach over SPDP/DTT thiolation is that the endogenous disulfide bridges in IgG are not broken down. The composition and characteristics of key liposome preparations are summarized in Table 1.

For lymphatic distribution studies, groups of three male Wistar rats, weighing 150 ± 10 g, were injected subcutaneously into the dorsal surface of the left footpad with $[^{125}I]$-PVP–encapsulated liposomes (2.8 μmol phospholipid). The $[^{125}I]$-PVP is an established and reliable label for determining the fate and distribution of liposomes following both intravenous and subcutaneous injections (8). In free form or when coinjected with empty liposomes, not more than 0.1% of the administered dose of $[^{125}I]$-PVP was recovered in the regional lymph nodes (the whole popliteal or primary and iliac or secondary nodes). Rats were sacrificed at various time intervals and associated radioactivity was measured in the footpad (whole foot), regional lymph nodes, and the blood using a gamma counter.

LYMPHATIC DISTRIBUTION OF IgG-PEG$_{2000}$-LIPOSOMES

Liposome-Macrophage Interaction

The target-binding ability of immuno- and immuno-PEG-liposomes was first tested in vitro, using peritoneal macrophages. The results in Figure 2

Table 1 Liposome Classification, Composition, and Characteristics

Liposome type	Composition (mole ratio)	mPEG-lipid (mol%)	MPB-PEG-lipid (mol%)	IgG/phospholipid (µg/µmol)	Size after IgG grafting (nm)
A	PC:chol:DCP:MPB-PE (6.925:6.925:1:0.15)	None	None	76±18	120±34
B	PC:chol:DCP:mPEG$_{2000}$-DSPE:MBP-PEG$_{2000}$-DSPE (6.5:6.5:1:0.85:0.15)	5.66	1.0	69±18	129±32
C	PC:chol:DCP:mPEG$_{2000}$-DSPE:MBP-PEG$_{2000}$-DSPE (6.7125:6.7125:1:0.425:0.15)	2.83	1.0	64±15	127±30
D	PC:chol:DCP:mPEG$_{2000}$-DSPE:MBP-PEG$_{2000}$-DSPE (6.925:6.925:1:0.0:0.15)	None	1.0	71±14	117±23

Abbreviations: PEG, poly(ethylene glycol); mPEG-lipid, methoxyPEG-lipid; MPB-PEG, *N*-(4'-(4''-maleimidophenyl)butyroyl)-PEG; IgG, immunoglobulin G; DCP, dicetylphosphate; MPB-PE, *N*-(4'-(4''-maleimidophenyl)butyroyl)-phosphatidylethanolamine; PC, phosphatidylcholine; DSPE, distearoylphosphatidylethanolamine; chol, cholesterol.

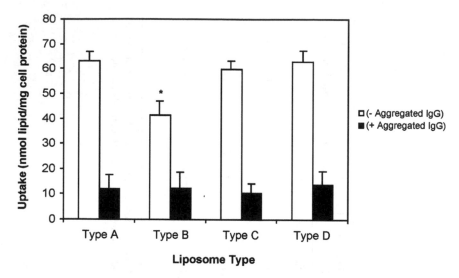

Figure 2 Interaction of IgG-grafted vesicles and IgG-poly(ethylene glycol)-liposomes with rat peritoneal macrophages in the absence and presence of aggregated IgG. See Table 1 for liposome classification and type. Two-sided t-test: $^*p < 0.05$ compared to immunoliposome (type A). *Abbreviation*: IgG, immunoglobulin G.

confirm macrophage recognition of all types of IgG-bearing liposomes. The binding of IgG-bearing vesicles to the macrophage is predominantly via Fc receptors because in the presence of aggregated IgG, an inhibitor of Fc receptor-mediated interactions (20), liposome uptake is dramatically suppressed. The extent of liposome-macrophage interaction is also controlled by bilayer concentration of mPEG-lipid. Macrophages interact more favourably with type A liposomes than that of type B vesicles. Possible steric interference to antibody-Fc receptor interaction arising from free $mPEG_{2000}$ chains, which are predominantly in a "mushroom-brush transition" conformation (16), could explain this observation. As the bilayer concentration of mPEG-lipid is reduced to 2.83 mol%, thus shifting mPEG conformation to a "nonoverlapped mushroom" regimen, immuno-PEG-liposomes interact with macrophages to the same extent as type A vesicles.

The Effect of mPEG-Lipid Concentration

The results in Figure 3 shows the drainage of IgG-grafted negatively charged (DCP incorporated) small unilamellar vesicles (type A liposomes) from the interstitial injection site into the initial lymphatic over a period of 48 hours,. Over 50% of the initial liposome dose leaves the dermal site within the initial four hours of injection. Afterwards, the rate of drainage is slower; this may indicate the movement of larger (or aggregated) vesicles into the lymphatic channels. Between 24 and 48 hours, the rate of liposome

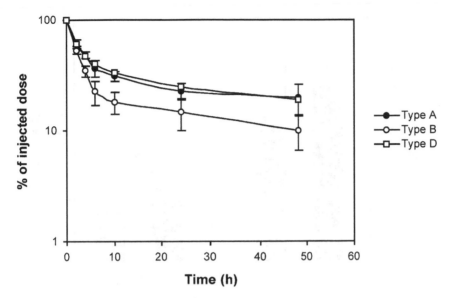

Figure 3 The drainage kinetics of IgG-grafted vesicles (type A) and IgG-poly (ethylene glycol) liposome (types B and D) from the interstitial space of the footpad into the lymphatic system. For clarity, the drainage kinetic of type C liposome is not shown, but it was similar to that of type B liposomes. For liposome composition, see Table 1. *Abbreviation*: IgG, immunoglobulin G.

drainage into the lymphatic system is rather static and could be indicative of phagocytosed vesicles at interstitial spaces (1). Liposome localization to both 1° and 2° nodes was highest at six hours postinjection, expressed as percentage of injected dose/g of node tissue (Fig. 4). This translates to approximately 2% of the injected dose per popliteal node and 1% of dose per iliac node, respectively. Thereafter, liposome levels in regional nodes did not change significantly.

The drainage pattern of immuno-PEG-liposomes (type D vesicles), bearing a similar surface IgG density to type A vesicles, is also similar to type A liposomes (Fig. 3). However, their localization to the popliteal node was significantly better than type A vesicles (Fig. 4), thus indicating that PEG may function as an spacer arm for better exposure of IgG to Fc receptors under lymph flow conditions. Incorporation of mPEG-lipid into the bilayer of immuno-PEG-liposomes (types B and C) dramatically accelerates their drainage from the injection site. This can be attributed to the steric effect of projected mPEG chains in suppressing vesicle aggregation in interstitial spaces. Also, their localization to both 1° and 2° nodes are considerably higher than that of IgG-grafted vesicles and immuno-PEG-liposomes without mPEG-lipid in their bilayer (Fig. 4). In the absence of free $mPEG_{2000}$-phospholipid, while maintaining the concentration of MBP-PEG-phospholipid at 1 mol%, IgG-PEG projections assume a "nonoverlapped mushroom"

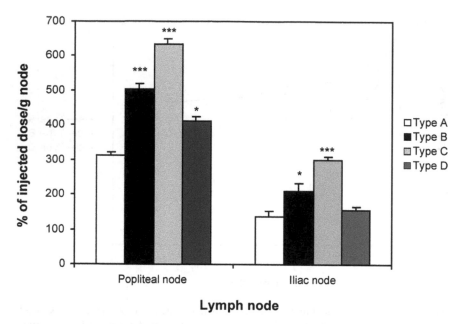

Figure 4 Liposome uptake by primary (popliteal) and secondary (iliac) lymph nodes at six hours postsubcutaneous injection. Classification and composition of liposomes are defined in Table 1. Two-sided *t*-test: $^{*}p < 0.05$, $^{***}p < 0.001$ compared to type A liposome in the respective lymph node.

conformation. Therefore, such vesicles (type D) may be more prone to aggregation and as a result drain at a slower rate into the lymphatic system than type B and C liposomes; subsequently their retention in the regional lymph nodes is comparatively lower. Because of close proximity of IgG-PEG chains to the liposome surface, IgG molecules may even interact with or protrude to some extent through the lipid bilayer during vesicle drainage into the lymphatic system. This could further reduce their target-binding capability and explain lower lymph node retention levels of drained type D vesicles. In the presence of intermediate levels of mPEG (2.8 mol%), immuno-PEG-liposomes (type C) cannot only drain rapidly into the lymphatic system and at a similar rate to that of type B immuno-PEG-liposomes, but their lymph node retention exceeds that of all liposome types (Fig. 4). A nearly "overlapped mushroom" conformation of mPEG and IgG-PEG chains on the liposome surface may explain these observations (16). Thus, the assumed surface PEG conformation efficiently suppresses interaction between vesicles and interstitial elements, but avoids interference with the binding of conjugated IgG to Fc receptors. In support of the latter, following incubation with peritoneal macrophages the uptake of type C liposomes was found to be of similar levels to that of native immunoliposomes (type A) (Figure 2).

The Effect of Anionic Lipid Type and Concentration

Increasing the bilayer concentration of DCP to 20 mol% in IgG-bearing and type C immuno-PEG-liposomes had marginal effect on lymph node targeting (Fig. 5). Also, replacement of DCP with PG had no stimulating effect of lymph node retention of liposomes. However, the extent of liposome uptake by both popliteal and iliac nodes was significantly increased when PS was incorporated into the bilayer of both IgG-bearing and immuno-PEG-liposomes containing 2.83 mol% mPEG-DSPE. The effect was more significant when PS concentration reached 20 mol%. The improved uptake is presumably due to liposome recognition by both Fc and PS receptors. Indeed, PS–exposure is known to stimulate liposome recognition by professional phagocytes and also serves as a signal for triggering macrophage recognition of apoptotic cells (21). Lymph node retention of interstitially injected PS-incorporated nonimmune liposomes is three- to four-fold higher than other negatively charged vesicles (e.g., DCP- and PG-containing liposomes, data not shown). Partial shielding of the PS headgroup by mPEG and IgG-PEG chains as well as the bulky MPB linker moiety may explain why PS incorporation fails to enhance lymph node retention of IgG-bearing vesicles and IgG-PEG-liposomes by several folds. Nevertheless, by this approach improved liposome retention in regional lymph nodes can be achieved without inducing liposome aggregation during the process of lymphatic absorption.

The Effect of Liposome Aggregation

Similar to the biotin-avidin approach described by Phillips et al. (12), an effort was made to induce immuno-liposome aggregation during lymphatic absorption, thus enhancing vesicle retention in the regional lymph nodes. IgG-grafted or type C immuno-PEG-liposomes were injected subcutaneously, followed by an adjacent subcutaneous injection of pentavalent IgM against rat IgG. In control experiments, an irrelevant IgM was used. Indeed, this approach dramatically enhanced liposome retention in the regional nodes (Fig. 6). Interestingly, the results were superior with type C liposomes, presumably as a result of their faster drainage when compared to IgG-grafted vesicles (Fig. 3) and/or better aggregate formation.

CONCLUSIONS

Simultaneous inclusion of PS and an appropriate quantity of $mPEG_{2000}$-lipids into the liposomal bilayer can maximize the rate of $IgG\text{-}PEG_{2000}$-liposome drainage from the interstitial site of injection into the lymphatic system and subsequent vesicle uptake by the regional lymph nodes. Surface mPEG concentration affects mPEG and IgG-PEG conformation; these biophysical characteristics, concomitantly, control the extent of liposome

(A)

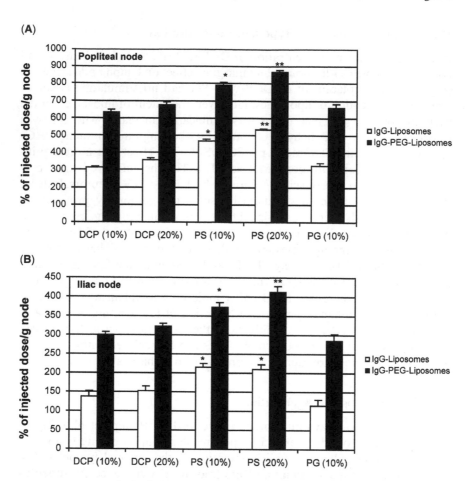

(B)

Anionic lipid type and content (mol%)

Figure 5 The effect of anionic lipid type and bilayer concentration on lymph node localization of interstitially injected IgG-grafted vesicles and IgG-PEG-liposomes. Lymph node localization of liposomes was assessed at six hours postinjection. Immuno-liposomes and immuno-PEG-liposomes contained 1 mol% of N-(4'-(4''-maleimidophenyl)butyroyl)-phosphatidylethanolamine and N-(4'-(4''-maleimidophenyl)butyroyl)-PEG$_{2000}$-distearoylphosphatidylethanolamine, respectively, in their bilayer prior to antibody conjugation. Immuno-PEG-liposomes contained 2.83 mol% mPEG$_{2000}$-lipid in their bilayer. Antibody/phospholipid ($\mu g/\mu mol$) were 76 ± 18, 71 ± 22, 68 ± 15, 74 ± 12, and 56 ± 17 for IgG-liposomes containing DCP (10%), DCP (20%), PS (10%), PS (20%), and PG (10%), respectively. The corresponding values for immuno-PEG-liposomes containing DCP (10%), DCP (20%), PS (10%), PS (20%), and PG (10%) were 64 ± 15, 68 ± 19, 71 ± 15, 73 ± 13, and 60 ± 14, respectively. All liposome preparations were in the range of 120 ± 35 nm in diameter. Two-sided t-test: $^*p < 0.05$, $^{**}p < 0.01$ compared to the respective preparation containing DCP (10 mol%). *Abbreviations*: DCP, dicetylphosphate; PS, phosphatidylserine; PG, phosphatidylglycerol; PEG, poly(ethylene glycol); IgG, immunoglobulin G.

(A)

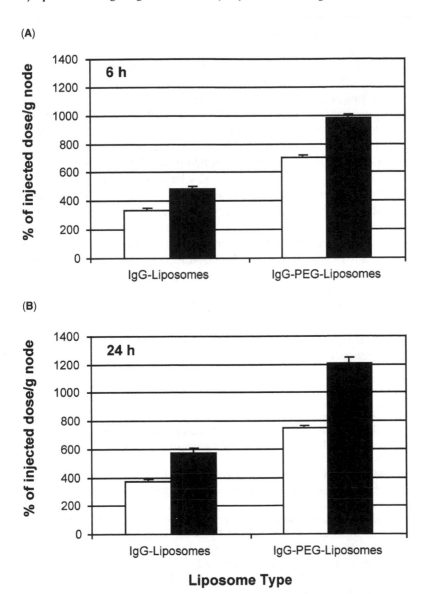

(B)

Liposome Type

Figure 6   The effect of subcutaneous injection of antirat IgG on popliteal lymph node localization of IgG-bearing vesicles and type C liposomes. Open columns represent injection of an irrelevant IgM (control), whereas black columns represent injection of a sheep-derived IgM, which interacts with rat IgG. IgM reactivity with rat IgG was conformed by immunodiffusion. *Abbreviations*: IgG, immunoglobulin G; PEG, poly(ethylene glycol).

interaction with both the ground substance of the interstitium and Fc receptor-bearing cells in the lymphatic system. This work may have important applications in lymphatic biology and medicine, as in for lymphatic imaging and for efficient antigen delivery to the lymph nodes, particularly through harnessing those receptors that affects immunogenicity or adjuvanticity using appropriate ligands. Although, the described antibody coupling procedures result in a random orientation of IgG on the liposomes, thus exposing the Fc region and facilitating Fc receptor recognition, IgG orientation on the liposome surface may be controlled through biochemical and/or chemical modification. For instance, IgG may be converted to $F(ab)_2$ fragments by pepsin digestion, which following reduction with DTT yields Fab fragments containing free sulfhydryl groups allowing reaction with liposomes containing MPB-PEG-lipid. Alternatively, oxidation of the carbohydrate moiety of the Fc region prior to coupling to the distal ends of an activated PEG-lipid (e.g., hydrazide-PEG_{2000}-phospholipid) could diminish Fc region exposure (19). Thus, such engineered vesicles may serve as useful tools for targeting nonmacrophage elements of the regional lymph nodes following subcutaneous injection. However, it should be emphasized that maleimido benzoyl residue of MPB may act as an immunogenic determinant in it own right, and may interfere with the stimulation of immune responses by engineered liposomes.

REFERENCES

1. Moghimi SM, Rajabi-Siahboomi AR. Advanced colloid-based systems for efficient delivery of drugs and diagnostic agents to the lymphatic tissues. Prog Biophys Mol Biol 1996; 65:221.
2. Velinova M, et al. Morphological observations on the fate of liposomes in the regional lymph nodes after footpad injection in rats. Biochim Biophys Acta 1996; 1299:207.
3. Oussoren C, et al. Lymphatic uptake and biodistribution of liposomes after subcutaneous injection. II. Influence of liposomal size, lipid composition and lipid dose. Biochim Biophys Acta 1997; 1328:261.
4. Oussoren C, et al. Lymphatic uptake and biodistribution of liposomes after subcutaneous injection. IV. Fate of liposomes in regional lymph nodes. Biochim Biophys Acta 1998; 1370:259.
5. Moghimi SM, Bonnemain B. Subcutaneous and intravenous delivery of diagnostic agents to the lymphatic system: applications in lymphoscintigraphy and indirect lymphography. Adv Drug Deliv Rev 1999; 37:295.
6. Moghimi SM, Hunter AC, Murray JC. Nanomedicine: current status and future prospects. FASEB J 2005; 18:311.
7. Wu MS, et al. Modified in vivo behaviour of liposomes containing synthetic glycolipids. Biochim Biophys Acta 1981; 674:19.
8. Mangat S, Patel HM. Lymph node localization of non-specific antibody-coated liposomes. Life Sci 1985; 36:1917.

9. Oussoren C, Storm G. Lymphatic uptake and biodistribution of liposomes after subcutaneous injection: III. Influence of surface modification with poly(ethylene glycol). Pharm Res 1997; 14:1479.
10. Dufresne I, et al. Targeting lymph nodes with liposomes bearing anti-HLA-Dr Fab fragments. Biochim Biophys Acta 1999; 1421:284.
11. Hirnle P. Liposomes for drug targeting in the lymphatic system. Hybridoma 1997; 16:127.
12. Phillips WT, Klipper R, Goins B. Novel method of greatly enhanced delivery of liposomes to lymph nodes. J Pharmacol Exp Ther 2000; 295:309.
13. Moghimi SM. The effect of methoxyPEG chain length and molecular architecture on lymph node targeting of immuno-PEG-liposomes. Biomaterials 2006; 27:136.
14. Moghimi SM, et al. Surface engineered nanospheres with enhanced drainage into lymphatics and uptake by macrophages of the regional lymph nodes. FEBS Lett 1997; 244:25.
15. Moghimi SM. Modulation of lymphatic distribution of subcutaneously injected poloxamer 407-coated nanospheres: the effect of ethylene oxide chain configuration. FEBS Lett 2003; 540:241.
16. Tirosh O, et al. Hydration of polyethylene glycol-grafted liposomes. Biophys J 1998; 74:1371.
17. Gbadamosi JK, Hunter AC, Moghimi SM. PEGylation of microspheres generates a heterogeneous population of particles with differential surface characteristics and biological performance. FEBS Lett 2002; 532:338.
18. Moghimi SM, Szebeni J. Stealth liposomes and long circulating nanoparticles: critical issues in pharmacokinetics, opsonization and protein-binding properties. Prog Lipid Res 2003; 42:463.
19. Hansen CB, et al. Attachment of antibodies to sterically stabilized liposomes: evaluation, comparison and optimization of coupling procedures. Biochim Biophys Acta 1995; 1239:133.
20. Derksen JTP, et al. Interaction of immunoglobulin-coupled liposomes with rat liver macrophages in vitro. Exp Cell Res 1987; 168:105.
21. Moghimi SM, Hunter AC. Recognition by macrophages and liver cells of opsonized phospholipid vesicles and phospholipid headgroups. Pharm Res 2001; 18:1.

5

Observations on the Disappearance of the Stealth Property of PEGylated Liposomes: Effects of Lipid Dose and Dosing Frequency

Myrra G. Carstens, Birgit Romberg, Christien Oussoren, and Gert Storm

Department of Pharmaceutics, Utrecht Institute of Pharmaceutical Sciences, Utrecht University, Utrecht, The Netherlands

Peter Laverman and Otto C. Boerman

Department of Nuclear Medicine, Radboud University Nijmegen Medical Centre, Nijmegen, The Netherlands

INTRODUCTION

The key issue in targeted drug delivery is selectively enhancing drug accumulation at target sites. The usefulness of long-circulating particles within the nanoscale size range ("stealth" particles) to this favorable property has been extensively documented (as, for example, in the relevant chapters in this book series). From a clinical application perspective, formulations of polyethylene glycol (PEG)-coated liposomes represent the prime example of such particles. These stealth liposomes circulate in the blood for prolonged periods of time, with reported half-lives up to 24 hours in mice and rats and about two days in humans. This enables significant localization in tumors and other pathological tissues with increased vascular permeability (1–4). Several formulations based on these long-circulating PEG-liposomes are commercially available and/or in clinical testing (2,4). Despite this maturity, there are several clinically important issues not yet addressed sufficiently. In the clinical setting, the lipid dose needed depends on the intended application. Also, multiple dosing is often required. Remarkably, until recently, the effects of lipid dose and repeated administration on the pharmacokinetics of (empty) stealth liposomes have been largely neglected. Though in several studies the pharmacokinetics of liposomal cytostatics upon repeated administration were investigated, these studies focused on the fate of the drug rather than the carrier (2). This chapter summarizes the evidence obtained recently in our group and other laboratories showing that the long circulation property of PEG-liposomes is certainly not guaranteed under all circumstances.

EFFECT OF LIPID DOSE

PEGylated liposomes are generally considered to be long-circulating, irrespective of the lipid dose. Several studies have indeed shown that the circulation time of PEG-liposomes is independent of the administered lipid dose within a broad dose range (4–400 µmol/kg) (3,5). Utkhede and Tilcock were the first to describe dose-dependent blood clearance of PEG-liposomes at lipid doses below 1µmol/kg in rabbits (6). These low doses are used for the application of liposomes in scintigraphic imaging (7–10). Lipid doses of 0.50 and 0.16 µmol/kg resulted in a 5- to 10-fold reduction of the circulatory half-life as compared to a dose of 2.12 µmol/kg. This was accompanied by a slight increase in liver uptake (6). Laverman et al. confirmed this phenomenon by demonstrating the loss of the stealth property of PEG-liposomes at lipid doses lower than 0.03 µmol/kg in rats and lower than 0.02 µmol/kg in rabbits (11). In accordance with the results of Utkhede and Tilcock, enhanced hepatic uptake was observed. Furthermore, the uptake of liposomes by the spleen was shown to be increased (11). However, the dose levels at which significant differences in kinetics were detected were not the same in both studies, which may relate to the difference in liposome composition (11). Importantly, the disappearance of the stealth property of PEG-liposomes has also been observed in humans (11). Figure 1 shows the whole-body scintigrams of two patients who received intravenous (IV) Tc-99m–labeled PEG-liposomes. This was part of an ongoing clinical trial to investigate the application of these liposomes for diagnostic imaging of infections. One patient received a lipid dose of 0.5 µmol/kg, the other patient 0.1 µmol/kg. The scintigram of the latter patient clearly shows increased elimination from the bloodstream at the lower dose level. This is reflected by a decreased radioactivity in the heart region (cardiac blood pool) and an increased radioactivity in the liver and the spleen as compared to the other patient. To determine the hepatosplenic uptake [as percentage of the injected dose, (%ID)] at four hours postinjection in these patients, regions of interest were drawn over the liver and spleen. At the high lipid dose, the liver uptake was

(A)   **(B)**

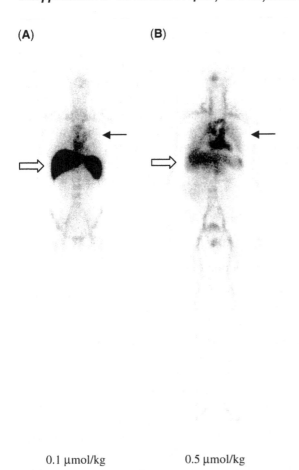

0.1 μmol/kg   0.5 μmol/kg

Figure 1 Anterior whole-body scintigram of two patients after treatment with different doses of Tc-99m-PEG-liposomes, four hours postinjection (p.i.). →: indicates the heart region, ⇒: indicates the hepatosplenic region. *Source*: From Ref. 11.

14%ID and the splenic uptake was 3%ID. At the low lipid dose, these values were increased to 23%ID and 17%ID, respectively.

At present, the mechanism behind the enhanced blood clearance of the PEG-liposomes from the circulation at lower lipid doses is not completely understood. Depletion of the liver and spleen macrophages in rats by IV. injection of liposomal clodronate resulted in the usual long circulation times, even at low lipid dose. This observation suggests a role for hepatosplenic macrophages in the enhanced blood clearance effect (11). Interestingly, predosing rats with a dose of PEG-liposomes (0.1 μmol/kg) one hour before the administration of Tc-99m-PEG-liposomes at a low lipid dose also resulted in long circulation times (11). It was hypothesized that a limited amount of some type of opsonic protein is present in the bloodstream causing the rapid clearance of PEG-liposomes. Above a critical threshold dose, this pool of opsonins is saturated, leading to the appearance of the prolonged circulation characteristic of PEG-liposomes (11). A similar mechanism was proposed to explain the dose-dependent blood clearance of conventional (non-PEGylated) liposomes that show increased circulation times at high lipid doses up to 1 mmol/kg (12). The speculated presence of a limited blood pool of yet unidentified opsonins able to bind to circulating PEG-liposomes is in good agreement with the often overlooked observation of the rapid clearance of a fraction of the injected dose of stealth liposomes within the first hour after administration (5,11,13,14).

EFFECT OF REPEATED ADMINISTRATION

The Accelerated Blood Clearance Phenomenon

The majority of possible diseases to be treated by liposomal drugs are life-threatening or chronic diseases. Therefore, adequate treatment will often require repeated injections of liposomal drugs. Tables 1 and 2 present an overview of the type of liposomes studied and the data reported on the effect of repeated administration on the pharmacokinetics of PEG-liposomes. Goins et al. were the first to report on the circulation kinetics of empty PEG-liposomes labeled with Tc-99m upon repeated injection in rabbits (15). At the dosing interval studied (six weeks), no difference in circulation time was noted (15). Oussoren and Storm studied the influence of repeated injections of [^{3}H]-labeled PEG-liposomes at much shorter dosing intervals (one and two days) in rats. No changes in the pharmacokinetics were seen (16). The first observation on pharmacokinetic irregularities upon repeated administration of PEG-liposomes was reported by Dams et al. (17). It was observed that the circulation kinetics and biodistribution of Tc-99m–labeled PEG-liposomes in rats can be greatly affected by repeated administration. A second dose of PEG-liposomes, given five days up to four weeks after the first injection, yielded a dramatically decreased circulation time and elevated hepatosplenic uptake (17). The effect was most pronounced at a dosing interval of one week (Fig. 2). At subsequent weekly injections, the intensity of the accelerated blood clearance (ABC) effect attenuated. After the fourth dose, the pharmacokinetics of the PEG-liposomes had almost returned to normal (Fig. 3). Increasing the dosing interval from one week up to four weeks still resulted in shorter circulation times (17). In addition to rats, the ABC phenomenon was also observed in mice and rabbits (14) and in a rhesus monkey (17), thus suggesting that this phenomenon is relevant for a broad range of species. Further characterization of the ABC phenomenon in rats showed that not only the dosing interval but also the lipid dose of the second injection plays a role. No changes in pharmacokinetics were

(*Text continued on page 87.*)

Table 1 Overview of the Liposome Types Used in the Various Publications

Liposome type	Liposome composition (refs.)	Reported size (nm)
PEG-liposomes	DSPC/CH/PEG5000-DSPE/α-tocopherol, 50/43/5/2 (15)	140
	DPPC/CH/PEG2000-DSPE, 1.85/1/0.15 (16)	120
	PHEPC/CH/PEG2000-DSPE/HYNIC-DSPE, 1.85/1/0.15/0.07 (13,17)	85–90 400 (large)
	HEPC/CH/PEG2000-DSPE, 1.85/1/0.15 (18–21), 1.70/1/0.3 (20), 1.55/1/0.45 (20)	90–110 50 (small)
	HEPC/CH/PEG5000-DSPE, 1.85/1/0.15 (20)	110
	DSPC/CH/PEG2000-DSPE, 50/45/5 (22)	110–140
	DSPC/CH/DODAP/PEG-Cer20, 25/45/20/10 (22)	200
	DSPC/CH/DODAP/PEG-Cer14, 25/45/20/10 (22)	200
Exchangeable PEG-liposomes Doxil®	Doxorubicine in PEGliposomes (13)	100
Non-PEG-liposomes	PHEPC/CH, 1.85/1 (17)	100
	DPPC/CH/HYNIC-DSPE, 1.85.1/0/07 (13)	600 (large) 115–120 60 (small)
	HEPC/CH, 2/1 (20,21)	110–140
Anionic liposomes	DSPC/CH, 55/45 (22)	105
	DPPC/CH/DPPG, 54/40/6 (20)	310–680 (large) 110 55 (small)
	HEPC/CH/DCP, 5/4/1 (21)	110
Cationic liposomes	DPPC/CH/SA, 52/40/8 (20)	440–900 (large) 105
	HEPC/CH/SA, 5/4/1 (21)	60 (small)

Abbreviations: DSPC, distearoyl phosphatidylcholine; CH, cholesterol; PEG5000-DSPE, poly(ethyleneglycol) with a molecular weight of 5000 conjugated to distearoylphosphatidylethanolamine; PhL, phospholipid; DPPC, dipalmitoyl phosphatidylcholine; PHEPC, partially hydrogenated egg phosphatidylcholine; HYNIC, *N*-Hydroxysuccinimidyl hydrazine nicotinate hydrochloride; HEPC, hydrogenated egg phosphatidylcholine; DPPG, dipalmitoyl phosphatidylglycerol; SA, stearylamine; DCP, dicetylphosphate; DODAP, 1,2-dioleoyl-3-*N*,*N*-dimethylammoniumpropane; PEG-Cer20, PEG-conjugated arachidoylsphingosine; PEG-Cer14, PEG-conjugated myristoylsphingosine.

Table 2 Overview of Results in the Various Publications on the Pharmacokinetics of PEG-Liposomes upon Repeated Administration

First injection: liposome type, lipid dose	Second injection: liposome type, lipid dose	Animal species	Dosing interval	ABC-effect (refs.)
PEG-liposomes, 17 µmol PL/kg	PEG-liposomes, 17 µmol PL/kg	Rabbit	6 wks	No (15)
PEG-liposomes, 45 µmol TL/kg	PEG-liposomes, 45 µmol TL/kg	Rat	24 or 48 hrs	No (16)
PEG-liposomes, 5 µmol PL/kg	PEG-liposomes, 5 µmol PL/kg	Rat	1–4 days	No (17)
PEG-liposomes, 5 µmol PL/kg	PEG-liposomes, 5 µmol PL/kg	Rat	5 days–4 wks	Yes (13,17)
PEG-liposomes, 5 µmol PL/kg	PEG-liposomes, 5 µmol PL/kg	Mouse	1 wk	No (17)
PEG-liposomes, 5 µmol PL/kg	PEG-liposomes, 5 µmol PL/kg	Monkey	1 wk	Yes (17)
Large PEG-liposomes, 5 µmol PL/kg	PEG-liposomes, 5 µmol PL/kg	Rat	1 wk	Yes (17)
NonPEG-liposomes, 5 µmol PL/kg	PEG-liposomes, 5 µmol PL/kg	Rat	1 wk	Yes (13,17)
PEG-liposomes, 5 µmol PL/kg	NonPEG-liposomes, 5 µmol PL/kg	Rat	1 wk	Yes (13,17)
(Large) nonPEG-liposomes, 5 µmol PL/kg	NonPEG-liposomes, 5 µmol PL/kg	Rat	1 wk	Yes (13)
PEG-liposomes, 5 µmol PL/kg	PEG-liposomes, 15–50 µmol PL/kg	Rat	1 wk	No (13,17)
Doxil®, 5 µmol PL/kg	Doxil®, 5 µmol PL/kg	Rat	1 wk	No (13)
PEG-liposomes, 5 µmol PL/kg	Doxil®, 5 µmol PL/kg	Rat	1 wk	Yes (13)
PEG-liposomes, 25 µmol PL/kg	PEG-liposomes, 25 µmol PL/kg	Mouse	10 days	Yes[a] (18,20)
PEG-liposomes, 25 µmol PL/kg	PEG-liposomes, 25 µmol PL/kg	Mouse	5, 7, 14 days	No (18,20)

First injection	Second injection	Animal	Time interval	ABC-effect
PEG-liposomes, 5 μmol PL/kg	PEG-liposomes, 5 μmol PL/kg	Rat	3–21 days	Yes (19)
PEG-liposomes, 5 μmol PL/kg	PEG-liposomes, 5 μmol PL/kg	Rat	4–5 wks	No (19)
PEG-liposomes, 0.0001–0.01 μmol PL/kg	PEG-liposomes, 25 μmol PL/kg	Mouse	10 days	No (20)
PEG-liposomes, 0.05–25 μmol PL/kg	PEG-liposomes, 25 μmol PL/kg	Mouse	10 days	Yes (20)
NonPEG-liposomes, (large) cationic liposomes, 25 μmol PL/kg	PEG-liposomes, 25 μmol PL/kg	Mouse	10 days	No (20)
(Small) PEG-liposomes, (small) nonPEG-liposomes, (small) cationic liposomes, (small) anionic liposomes, 0.1 μmol PL/kg	PEG-liposomes, 5 μmol PL/kg	Rat	5 days	Yes[b] (21)
PEG-liposomes, 0.1 μmol PL/kg	(Small) nonPEG-liposomes, 5 μmol PL/kg	Rat	5 days	No (21)
PEG-liposomes, 70 μmol TL/kg	PEG-liposomes, 70 μmol TL/kg	Mouse	1 wk	No (22)
PEG-liposomes with ODN, 70 μmol TL/kg	PEG-liposomes with ODN, 70 μmol TL/kg	Mouse	1 wk	Yes (22)
NonPEG-liposomes, exchangeable PEG-liposomes, with ODN, 70 μmol TL/kg	NonPEG-liposomes, exchangeable PEG-liposomes, with ODN, 70 μmol TL/kg	Mouse	1 wk	No (22)

[a]The induction of the ABC-effect decreased with increasing PEG-content and increasing molecular weight of the PEG.
[b]The induction of the ABC-effect decreased in the following order: small PEG-liposomes ≈ small anionic liposomes > PEG-liposomes > small cationic liposomes > small non-PEG-liposomes ≈ non-PEG-liposomes ≈ anionic liposomes ≈ cationic liposomes.
Abbreviations: PL, phospholipid; TL, total lipid; ODN, antisense oligodeoxynucleotide.

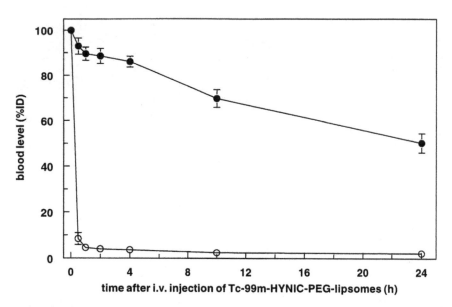

Figure 2 Blood levels of Tc-99m-PEG-liposomes in rats after the first (*closed dots*) and second (*open dots*) injection. *Abbreviation*: PEG, polyethylene glycol. *Source*: From Ref. 17.

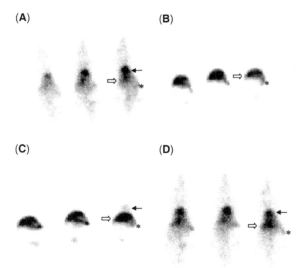

Figure 3 Scintigraphic images of rats after weekly injections with Tc-99m-PEG-liposomes on day 0 (**A**), day 7 (**B**), day 14 (**C**), and day 21 (**D**), four hours p.i. →: indicates the heart region, ⇒: indicates the liver region and ∗: indicates the spleen region. *Abbreviation*: PEG, polyethylene glycol. *Source*: From Ref. 17.

observed at a higher lipid dose of 50 μmol/kg, as compared to the earlier used dose of 5 μmol/kg (13). In the ABC effect, two phases can be distinguished: the induction phase, following the first injection with liposomes accompanied with the usual long circulation times, and the effectuation phase, following the second or subsequent injections in which the PEGylated liposomes show enhanced blood clearance (13,14). It appeared that neither the lipid dose of the first injection (0.05–5 μmol/kg) (13), the liposome size (85 or 400 nm), nor the presence of the radiolabel and/or chelator were critical factors in the induction of the ABC effect after the first injection (17). In addition, it was demonstrated that the presence of the PEG-coating (with a molecular weight of PEG of 2000) is not critical either. Small (100 nm) non-PEGylated liposomes, composed of partially hydrogenated egg-phosphatidylcholine could induce changes in the pharmacokinetics and bio distribution of subsequently injected PEG-liposomes as well (17). Small (100 nm) nonPEGylated dipalmitoylphospatidylcholine (DPPC)-liposomes, which are known to possess the long-circulating property, and large (600 nm), short-circulating DPPC-liposomes also induce the ABC effect (13). The presence of the PEG-coating was not crucial during the effectuation phase of the ABC effect either: Small DPPC-liposomes were cleared rapidly after a second injection in rats when they were injected one week earlier with either PEG-liposomes or small DPPC-liposomes (13).

Recent studies by Ishida et al. partly confirmed and further specified our observations (18–21). The ABC effect observed by this group in rats was less pronounced, but small changes were already seen at a dosing interval of three days (19). Surprisingly, in mice, a significantly increased clearance of a second dose of [^{3}H]-labeled PEG-liposomes was observed at a dosing interval of 10 days, but not at intervals of 5, 7, and 14 days (18). In line with the results of our group, Ishida et al. confirmed that the presence of the PEG-coating is not crucial for the induction of the phenomenon. Remarkably, it rather plays a role in preventing it. The use of a longer PEG-chain (Molecular weight [Mw] 5000 vs. Mw 2000) or a higher amount of PEG2000-lipids in the liposomal formulation attenuated the induction of the ABC effect (20). Furthermore, upon lowering the lipid dose of the first injection from 25 μmol/kg to 0.001 μmol/kg, the pharmacokinetic changes in mice became less pronounced (20).

Another striking observation reported by the group of Ishida et al. was that the injection of charged liposomes [by incorporating stearylamine (SA, positive) or dipalmitoylphosphatidylglycerol (negative) into the lipid bilayer] did not result in the ABC of a subsequent dose of PEG-liposomes in mice (20). A more recent study published by this group showed that in rats non-PEGylated hydrogenated egg phosphatidylcholine and positively or negatively charged liposomes (containing SA or dicetylphosphate, respectively) with a size of approximately 100 nm did induce a slightly enhanced clearance of subsequently injected (five days later) PEG-liposomes (21). Small (60 nm) liposomes, either PEGylated, negatively or positively charged induced a stronger ABC effect than their larger equivalents (100 nm). Remarkably, this size effect was not seen with non-PEGylated, noncharged liposomes (21). Based on these results, an additive effect of the PEG-coating and small liposome size on the induction of the ABC phenomenon was suggested (21). In contrast to the results obtained by our group, the Ishida et al. showed that the PEG-coating is an essential factor during the effectuation phase of the ABC effect in rats, no changes were observed in the circulation time of non-PEG-liposomes when injected five days after a dose of PEG-liposomes (21). In a recent study by Semple et al. no changes in pharmacokinetics of PEG-liposomes at one hour postinjection upon repeated (weekly) administration of PEGylated distearoylphosphatidylcholine liposomes in mice were observed (22). This lack of effect is probably related to the high dose of $70 \mu mol/kg$ that was administered. Interestingly, when these liposomes were loaded with antisense oligodeoxynucleotides (ODN), enhanced clearance was observed of the second and subsequent injections of either ODN-loaded or empty PEG-liposomes at the same high lipid dose. The accelerated clearance was accompanied by morbidity and in some cases even mortality. The effect was independent on the nucleic acid composition of the ODN, but the ratio of ODN-to-lipid ratio did play a role: No changes in pharmacokinetics were observed in case of ODN/lipid ratios lower than 0.04 w/w. Apparently, the presence of a critical amount of ODN evokes the induction of the ABC effect of PEG-liposomes at high lipid dose (22). In addition, no changes in pharmacokinetics were observed upon repeated injections of ODN-loaded liposomes without PEG-coating or with a rapidly exchangeable PEG-lipid in the bilayer (PEG-CerC14), which suggests that the presence of PEG-lipid in the membrane is essential for the ABC effect of ODN-loaded liposomes (22).

From the observations described above, it is clear that there are discrepancies between the findings reported in the different publications. These may be explained by the variation in experimental setup and in particular in the critical factors, such as lipid dose and dosing interval. Also, differences in the type of liposomes and animal species may have introduced variable results. An overview of the observations described in the different publications is given in Table 2.

Mechanistic Aspects

The exact mechanism that is responsible for the changes in the pharmacokinetics of PEG-liposomes upon repeated aministration has not yet been elucidated. The first article (17) that reported on the existence of the ABC phenomenon demonstrated the involvement of a serum factor: Transfusion of "pretreated" serum (collected from rats that received a dose of PEG-liposomes one week earlier) into nontreated rats caused the enhanced clearance of a first injection of PEG-liposomes in these rats (17). When the "pretreated" serum was heated to 56°C for 30 minutes before transfusion, the ABC effect was not observed. This suggested that a heat labile factor is involved. The effect of the transfusion appeared to be dose-dependent: When the amount of transfused serum was reduced, the changes in pharmacokinetics of PEG-liposomes became smaller. The heat-labile transfusable serum factor coeluted on a size-exclusion column with a 150-kD protein. However, studies with immunoglobulin (Ig)G- and IgM-depleted serum suggested that the factor was not an antibody molecule (17). This suggestion is not in line with recently published observations by other groups (22–24), suggesting the involvement of PEG-reactive IgG (23) or IgM (22,24).

Sroda et al. studied the effect of rabbit sera obtained after weekly injections of small liposomes containing 20 mol% PEG-phosphatidylethanolamine (PE) on the in vitro release of carboxyfluorescein (CF) from PEG-liposomes (23). An increased CF-release was observed upon incubation of the liposomes with the collected sera of rabbits that received one or more injections with PEG-liposomes ("pretreated" sera), as compared to "nontreated" serum, suggesting an "antiliposomal" activity of the "pretreated" sera. The strongest effect was seen with the serum collected after the second injection with liposomes and, in correspondence with the in vivo results of Dams et al. (17), the effect diminished upon preheating the serum to 56°C (23). To investigate the binding of serum proteins to the PEG-liposomes upon incubation with rabbit serum, the liposomes were isolated by gel filtration and analyzed with the immunoblot technique using goat antirabbit IgG. Upon incubation of PEG-liposomes with rabbit sera obtained after the second, fourth, and sixth (weekly) injection an increasing concentration of a 55-kD polypeptide was observed to bind to the liposome surface, whereas incubation with "nontreated" serum did not reveal such binding of the 55-kD polypeptide. The polypeptide was identified as an IgG heavy chain and appeared to have anti-PEG activity: Non-PEGylated liposomes did not bind the polypeptide and the presence of additional free PEG during the incubation of PEG-liposomes with rabbit sera inhibited the IgG-binding (23).

Semple et al. studied the presence of antiliposomal antibodies in sera of mice that were injected with ODN-loaded PEG-liposomes using enzyme-linked immunosorbent assay with biotinylated liposomes bound to streptavidin coated microplates. The presence of IgM was evaluated by incubation with rat antimouse IgM monoclonal antibodies. It was observed that the amount of IgM bound to the liposomes was increased in case of incubation of PEG-liposomes with "pretreated" mouse serum, as compared to incubation with "nontreated" serum and to incubation of non-PEGylated liposomes with "pretreated" serum. This suggests the involvement of PEG-reactive IgM in the ABC effect.

Comparison of the effect of repeated administration in immunocompetent mice with the effect in immunocompromised (athymic) mice (in which the ABC effect also occurred), and in SCID-Rag2 mice (in which no changes were observed), suggests an antibody response with a critical role for B-cells and not for T-cells in the ABC phenomenon for ODN-loaded PEG-liposomes (22). In line with these results, Ishida et al. demonstrated the presence of PEG-reactive proteins in serum collected from rats five days after an injection with PEG-liposomes by SDS-PAGE analysis of PEG-liposomes that were incubated with this serum. These proteins, which were not observed in control serum, were further identified using LC-MS/MS. The major protein identified was an IgM molecule (24). Though there is no consensus about the identity of the serum factor(s) involved in the ABC effect, most papers suggest the involvement of the complement system in the effectuation phase of the ABC effect (17,19,21–23). One of the mechanisms to initiate the complement cascade is the classical pathway, which is initiated by the interaction of the first component of the complement system (C1) with immunoglobulins of the IgM and IgG isotype (25). Wang et al. hypothesized that antiPEG IgM is secreted by macrophages during the induction phase of the ABC phenomenon, which binds to the subsequent dose of PEG-liposomes and activates the complement system (possibly via binding to C1), resulting in the accelerated clearance (21). In rats, total hemolytic complement was indeed significantly decreased one hour after the second injection, suggesting complement consumption by the injected PEG-liposomes (17). Ishida et al. did not observe any complement consumption at 24 hours after the second injection (19), which may be explained by the much later time point after the second injection taken for the complement determination.

Investigation of the intrahepatic distribution of [^{3}H]-labeled PEG-liposomes showed that 18 hours after the second injection (given one week after the first injection), the radiolabel was mainly present in the Kupffer cells (17). Macrophage depletion studies using intravenously injected clodronate-liposomes indicated that indeed macrophages play an essential role in both the induction and the effectuation phase of the ABC effect (13). Histopathological evaluation of liver and spleen by Ishida et al. did not show any changes induced by an injection of PEG-liposomes in both mice and rats. The amount of Kupffer cells remained unchanged as well (18,19). Recent in vitro studies by Kamps et al. with cultured Kupffer cells confirmed that decreased circulation times in case of repeated administration are the result of enhanced uptake by Kupffer cells (Kamps JAAM, Laverman, P, Morselt HWM, et al. manuscript in preparation). The presence of "pretreated" serum (collected from rats that received an injection of PEG-liposomes one week earlier) significantly increased the binding and uptake of PEG-liposomes by Kupffer cells in vitro as compared to "nontreated" serum. It was also demonstrated that in the presence of "nontreated" serum, Kupffer cells isolated from rats that received an injection of PEG-liposomes one week earlier did not show an enhanced capacity to internalize PEG-liposomes in comparison to control Kupffer cells. This indicates that the presence of one or more serum factors from liposome-treated rats are essential to enhance the uptake of PEG-liposomes by Kupffer cells (Kamps JAAM, Laverman P, Morselt HWM, et al. Manuscript in preparation). To study the role of the macrophages in the ABC effect in more detail, Moghimi, et al. used poloxamine-908, as this polymer has the capacity to enhance the clearance of subsequently injected colloidal particles by resident Kupffer cells and splenic macrophages in both mice and rats (14,26). Indeed, an in vivo study in which the enhanced macrophage phagocytic activity was mimicked by treating animals with poloxamine-908 did show accelerated clearance and higher hepatosplenic uptake of subsequently injected (two days later) PEG-lipsomes (27).

CLINICAL IMPLICATIONS

Long-circulating PEG-liposomes have become a clinical reality, with PEG-liposomal doxorubicin (Doxil®, also named Caelyx® in Europe) as prime example (2). It is therefore of great importance to know whether the reported changes in pharmacokinetics related to lipid dose and repeated administration also occur with clinical formulations. So far, such changes have never been described in literature for clinically applied liposomal products. We studied the in vivo fate of In[111] labeled Doxil® (13). No irregularities in the circulatory behavior upon repeated administration of the doxorubicin-containing PEG-liposomes was observed in rats. However, it was surprising to note that a second injection of Doxil® was cleared more rapidly upon a first injection of empty PEG-liposomes (Fig. 4) (13). This observation suggested that the presence of doxorubicin inside the

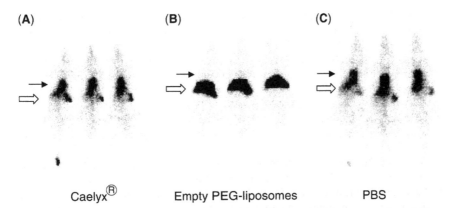

(A) **(B)** **(C)**

Caelyx®   Empty PEG-liposomes   PBS

Figure 4 Scintigraphic images of rats after injection with In[111] labeled Caelyx®, four hours post injection. Rats were injected with Caelyx® (**A**), empty PEG-liposomes (**B**) or phosphate buffered saline (**C**) one week earlier. →: indicates the heart region, ⇒: indicates the liver region. *Source*: From Ref. 13.

PEG-liposomes (Doxil®), given during the first injection, prevents the induction of the enhanced clearance of Doxil® given as second injection. This may be related to the reported toxicity of doxorubicin toward macrophages, resulting in reduced phagocytic activity (28,29). The study by Tardi et al. supports this hypothesis: An immune response toward ovalbumine-coated liposomes was prevented in mice when doxorubicin was encapsulated (30).

Though frequently suggested otherwise, it has become clear that PEG-liposomes are certainly not inert vehicles in vivo. It is obvious that unexpected changes in the pharmacokinetics and biodistribution of liposome-based drugs are highly undesirable. The disappearance of the stealth property of PEG-liposomes at low lipid dose or upon repeated administration may have important implications for their clinical use. Enhanced blood clearance could compromise the therapeutic efficacy and the increased uptake by liver and spleen could cause toxic effects toward these organs. Further research is mandatory to clarify the underlying mechanism(s) and to detail the clinical consequences.

REFERENCES

1. Allen TM, Cullis PR. Drug delivery systems: entering the mainstream. Science 2004; 303:1818.
2. Woodle MC, Storm G. Long Circulating Liposomes: Old Drugs, New Therapeutics. Berlin: Springer-Verlag, 1998.
3. Woodle MC, Lasic DD. Sterically stabilized liposomes. Biochim Biophys Acta 1992; 1113:171.
4. Storm G, Crommelin DJA. Liposomes: quo vadis? Pharmaceut Sci Technol Today 1998; 1:19.
5. Allen TM, Hansen C. Pharmacokinetics of stealth versus conventional liposomes: effect of dose. Biochim Biophys Acta 1991; 1068:133.
6. Utkhede DR, Tilcock CP. Effect of lipid dose on the bio distribution and blood pool clearance kinetics of PEG-modified technetium-labeled lipid vesicles. J Liposome Res 1998; 8:381.
7. Awasthi V, Goins B, Klipper, R, Loredo R, Korvick D, Phillips W. Imaging experimental osteomyelitis using radiolabeled liposomes. J Nucl Med 1998; 39:1089.
8. Goins B, Phillips W, Klipper R. Blood-pool imaging using technetium-99m-labeled liposomes. J Nucl Med 1996; 37:1374.
9. Laverman P, Boerman OC, Oyen WJ, Dams ET, Storm G, Corstens FH. Liposomes for scintigraphic detection of infection and inflammation. Adv Drug Deliv Rev 1999; 37:225.
10. Boerman OC, Laverman P, Oyen WJ, Corstens FH, Storm G. Radiolabeled liposomes for scintigraphic imaging. Prog Lipid Res 2000; 39:461.
11. Laverman P, Brouwers AH, Dams ET, et al. Preclinical and clinical evidence for disappearance of long-circulating characteristics of polyethylene glycol liposomes at low lipid dose. J Pharmacol Exp Ther 2000; 293:996.
12. Oja CD, Semple SC, Chonn A, Cullis PR. Influence of dose on liposome clearance: Critical role of blood proteins. Biochim Biophys Acta 1996; 1281:31.

13. Laverman P, Carstens MG, Boerman OC, et al. Factors affecting the accelerated blood clearance of polyethylene glycol-liposomes upon repeated injection. J Pharmacol Exp Ther 2001; 298:607.
14. Laverman P, Boerman OC, Oyen WJG, Corstens FHM, Storm G. In vivo applications of PEG liposomes: unexpected observations. Crit Rev Ther Drug Carrier Syst 2001; 18:551.
15. Goins B, Phillips WT, Klipper R. Repeat injection studies of technetium-99m-labeled PEG-liposomes in the same animal. J Liposome Res 1998; 265.
16. Oussoren C, Storm G. Effect of repeated i.v. administration on the circulation kinetics of poly(ethyleneglycol)-liposomes in rats. J Liposome Res 1999; 349.
17. Dams ET, Laverman P, Oyen WJ, et al. Accelerated blood clearance and altered biodistribution of repeated injections of sterically stabilized liposomes. J Pharmacol Exp Ther 2000; 292:1071.
18. Ishida T, Masuda K, Ichikawa T, Ichihara M, Irimura K, Kiwada H. Accelerated clearance of a second injection of PEGylated liposomes in mice. Int J Pharm 2003; 255:167.
19. Ishida T, Maeda R, Ichihara M, Irimura K, Kiwada H. Accelerated clearance of PEGylated liposomes in rats after repeated injections. J Cont Release 2003; 88:35.
20. Ishida T, Ichikawa T, Ichihara M, Sadzuka Y, Kiwada H. Effect of the physicochemical properties of initially injected liposomes on the clearance of subsequently injected PEGylated liposomes in mice. J Cont Release 2004; 95:403.
21. Wang XY, Ishida TI, chihara M, Kiwada H. Influence of the physicochemical properties of liposomes on the accelerated blood clearance phenomenon in ratsJ Cont Release 2005; 104:91.
22. Semple SC, Harasym TO, Clow KA, Ansell SM, Klimuk SK, Hope MJ. Immunogenicity and rapid blood clearance of liposomes containing polyethylene glycol-lipid conjugates and nucleic acid. J Pharmacol Exp Ther 2005; 312:1020.
23. Sroda K, Rydlewski J, Langner M, Kozubek A, Grzybek M, Sikorski AF. Repeated injections of PEG-PE liposomes generate antiPEG antibodies. Cell Mol Biol Lett 2005; 10:37.
24. Ishida, T, Ichihara M, Harada M, Kashima S, Wang X, Kiwada H. IgM secreted in response to the first injection of PEGylated liposomes involves in induction of the ABC phenomenon. 31 Annual Meeting and Exposition of the Controlled Release Society, Honolulu, Hawaii, June 13, 2004.
25. Bradley AJ, Devine, DV. The complement system in liposome clearance: Can complement deposition be inhibited? Adv Drug Deliv Rev 1998; 32:19.
26. Moghimi SM, Gray TA. A single dose of intravenously injected poloxamine-coated long-circulating particles triggers macrophage clearance of subsequent doses in rats. Clin Sci 1997; 93:371.
27. Laverman P, Carstens MG, Storm G, Moghimi SM. Recognition and clearance of methoxypoly(ethyleneglycol)2000-grafted liposomes by macrophages with enhanced phagocytic capacity. Implications in experimental and clinical oncology. Biochim Biophys Acta 2001; 1526:227.
28. Daemen T, Regts J, Meesters M, Ten Kate MT, Bakker-Woudenberg IAJM, Scherphof GL. Toxicity of doxorubicin entrapped within long-circulating liposomes. J Control Release 1997; 44:1.

29. Storm G, ten Kate MT, Working PK, Bakker-Woudenberg IA. Doxorubicin entrapped in sterically stabilized liposomes: effects on bacterial blood clearance capacity of the mononuclear phagocyte system. Clin Cancer Res 1998; 4:111.
30. Tardi PG, Swartz EN, Harasym TO, Cullis PR, Bally MB. An immune response to ovalbumin covalently coupled to liposomes is prevented when the liposomes used contain doxorubicin. J Immunol Methods 1997; 210:137.

6

Adenosine Triphosphate–Loaded Liposomes for Myocardium Preservation Under Ischemic Conditions

Daya D. Verma, Tatayana S. Levchenko, William C. Hartner, Eugene A. Bernstein, and Vladimir P. Torchilin

Department of Pharmaceutical Sciences, Northeastern University, Boston, Massachusetts, U.S.A.

INTRODUCTION

Cardiovascular diseases are the leading cause of morbidity and mortality, disability, and economic loss in industrialized states (1). Myocardial ischemia-reperfusion injury and acute myocardial infarction (MI) are significant consequences of clinical events during cardiac surgery, progression of coronary artery disease, and cardiac arrest (2). Acute MI leads to left ventricular (LV) remodeling, including expansion or aneurysm formation due to cardiomyocyte death in infarcted regions and LV dilation associated with hypertrophy and fibrosis of noninfarcted regions (3). Targeted delivery of cardioprotective drugs into the ischemic myocardium to prevent the MI/R damage may be one of the possible ways to reduce the number of heart failures.

Adenosine triphosphate (ATP) levels in the cardiomyocytes during cardiac ischemia drop to 20% of their initial value after approximately 15 minutes (4). In myocardial ischemia, a portion of the myocardium is deprived of nutrients and oxygen, and levels of ATP to drive the contractile process can only be maintained by breakdown of the myocardial glycogen to glucose units, which can be used to produce small amounts of ATP via glycolysis. When the myocytes needs immediate energy for contraction, the intracellular high-energy phosphate, ATP, breaks down as a source of this energy. When aerobic metabolism is unable to support the requirement for ATP, glycolysis will provide some ATP, though in much smaller amounts. The concentration of the metabolically active ATP in the extracellular space in well-perfused tissue is approximately 40 nM or 10^5 times lower than the intracellular ATP concentration of approximately 5 to 7 mM. However, in the continuing absence of an oxygen supply, these temporary metabolic pathways also become depleted and ATP-dependent ion pumps in the outer membranes of myocytes cease to function; the ion balance of the cells is lost and they swell and burst, thus releasing their contents into the circulation. A quick instigation of membrane disruption has been reported as early as 20 minutes after onset of ischemia in acute MI (6). Moreover, in myocardial ischemia, a toxic or inflammatory insult to the myocardium results in the loss of sarcolemmal integrity, resulting in the exposure of intracellular myosin to the extracellular milieu (7,8).

The key factor responsible for eliciting the decrease in the ATP supply/demand ratio during the myocardial ischemia is the relative lack of ATP. Therefore, we hypothesized that the delivery of exogenous ATP would help restore its normal cellular level in myocytes leading to cardioprotective effect. The first question to be addressed is how to deliver ATP into myocytes because it has a very short half-life in the blood being immediately hydrolyzed to adenosine diphosphate (ADP), adenosine monophosphate (AMP), and adenosine via a cascade of extracellular ecto-nucleotidases (9). Additionally, ATP, like other hydrophilic and strongly charged anions, cannot enter cells through the plasma membrane (9,10). These limitations confine the direct use of exogenous ATP as an efficient therapeutically significant bioenergic substrate. Some alternative ways for delivering ATP into ischemic cardiomyocytes have to be found.

The targeting of liposomes to the heart aims at diagnostic imaging cardiac diseases as well as delivering of pharmaceuticals to affected areas (11). The ischemic myocardium seems to be a possible good target for liposomes because the spontaneous accumulation of liposomes, especially positively charged ones, in the regions of experimental MI compared to the normal myocardium has been clearly demonstrated (12,13). Different components of the cardiovascular system have been used as targets for the delivery of liposomal drugs. Some of these targets include vessel wall, endothelial cells, atherosclerotic lesions, and infarcted myocardium (14–16). The accumulation of liposomes and other nanoparticular drug carriers (such as micelles) in ischemic tissues is a general phenomenon and might be explained, at least in part, by the impaired filtration in these areas, resulting in the trapping of drugs carriers within the ischemic zone (13,17) via the enhanced permeability and retention (EPR) effect (18,19).

A major obstacle to the drug therapy of MI is the limited access for drugs in active form to the ischemic myocardium. All currently available methods of delivering drugs to an ischemic zone are dependent on myocardial blood flow to that area, which is always impaired. Accumulation of positively charged liposomes in the region of experimental infarcted myocardium was reported in as early as 1977 by Caride (12). It was also suggested that liposomes may "plug" and "seal" the damaged myocyte membranes, thereby protecting the myocytes to a certain degree and possibly other structures against ischemic and reperfusion injury (20). These reports let us conclude that drug-loaded liposomes can be used for "passive" drug delivery to infarcted myocytes (13). Liposomes with a thrombolytic enzyme, streptokinase, were able to accelerate thrombolysis and reperfusion in a canine model of MI (21). Quercetin-filled liposomes provided reliable protection against peroxynitrite-induced myocardial injury in isolated cardiac tissues and anesthetized animals (22). An acute raise in the serum and myocardial levels of CoQ was observed following intravenous infusion of liposomal CoQ 10, resulting in the improved function, efficiency, and decrease in oxidant injury after ischemia/reperfusion (23).

Earlier, some encouraging results with the application of ATP-loaded liposomes (ATP-L) in certain in vitro and in vivo models were reported. Thus, liposomal ATP was shown to efficiently protect human endothelial cells from energy failure in a cell culture model of sepsis (24). In a brain ischemia model, it was observed that the use of ATP-L increased the number of ischemic episodes tolerated before brain electrical silence and death (25,26). In a hypovolemic shock-reperfusion model in rats, the administration of the ATP-L provided effective protection to the liver (27). The addition of the ATP-L during the cold storage preservation of the rat liver improved liver energy state and metabolism (28,29). Co-incubation of ATP-L with sperm cells showed the induction of the process of capacitating in vitro (30). Biodistribution studies with the ATP-L demonstrated their significant accumulation in the damaged myocardium (31).

In the following sections, we describe the ATP encapsulation in liposomes, attachment of antimyosin antibody on the surface of liposomes, their characterization, and effective cardioprotection ex vivo in isolated rat heart model and in vivo in rabbits with an induced MI.

ENCAPSULATION OF ATP IN LONG-CIRCULATING LIPOSOMES

We compared several methods for ATP encapsulation in liposomes (32) and found that relatively higher ATP encapsulation efficiency was achieved with both the reverse-phase evaporation (approximately 36 mol%) and freezing–thawing (around 38 mol%) methods (Table 1). We have chosen freezing–thawing as method for our further studies because it is simple, does

Table 1 Liposome Preparation Methods and Adenosine Triphosphate Encapsulation Efficiency

Methods of liposome preparation	ATP encapsulation efficiency (ATP per μmol of total lipid)
Lipid film hydration	5.1 ± 1.4
pH gradient	10.2 ± 2.1
Reverse phase evaporation	36 ± 3.5
Freezing–thawing	38 ± 2.4

Note: All liposomes were prepared from PC:Ch:PEG-DSPE (70:30:0.5 molar ratio). The encapsulation efficiency is represented as mean $\pm$ SD, $n = 3$.
Abbreviations: PEG-DSPE, 1,2-distearoyl-*sn*-glycero-3-phosphoethanolamine-*N*-[methoxyl(polyethyleneglycol)]; ATP, adenosine triphosphate; PC, phosphatidyl choline; Ch, cholesterol.

not require the presence of organic solvents or detergents, and avoids any possibility of vesicle damage associated with sonication (33).

In order to prepare long-circulating ATP-L, poly(ethylene glycol) (PEG)-DSPE was used as a component of the lipid mixture for liposome preparation for long circulation of liposomes. The data (Table 2) clearly demonstrated that the ATP encapsulation efficiency was significantly influenced by the quantity of the PEG-DSPE in liposomal membrane because of an "excluded" volume inside liposomes occupied by PEG chains. The maximum encapsulation of 0.38 μmol of ATP per μmol of total lipid (or 38 mol%) was achieved at 0.5 mol% of PEG-DSP in liposomal membrane.

Thus, ATP-L prepared by the freezing–thawing method and containing a quantity of PEG sufficient to make them long-circulating but not so high as to hinder the efficient ATP incorporation seems to be an optimal system for further in vitro and in vivo studies. For further experiments, ATP-L were

Table 2 The Effect of the PEG-DSPE Concentration in Liposomes on the Adenosine Triphosphate Encapsulation Efficiency

Amount of PEG$_{2000}$ DSPE (mol%)	Adenosine triphosphate (ATP) encapsulation efficiency (ATP per μmol of total lipid)
0.5	38.0 ± 2.4
2.5	25.2 ± 3.4
5.0	17.4 ± 2.6

Note: All liposomes were prepared by the freezing–thawing method. The encapsulation efficiency is represented as mean $\pm$ SD, $n = 3$.
Abbreviations: ATP, adenosine triphosphate; PEG, poly(ethylene glycol); DSPE, distearoylphosphatidylethanolamine.

prepared with the addition of 1,2-dioleoyl-3-trimethyl-ammonium-propane (3.3 mol%) to have slightly positive charge to facilitate their interaction with cardiac cells.

EFFECTIVE PROTECTION OF ISCHEMIC HEART BY ATP-L IN AN ISOLATED RAT HEART MODEL

Langendorff Apparatus, Experimental Ischemia, and Reperfusion Protocol

The Langendorff isolated rat heart model was used for measuring systolic and diastolic function of the left ventricle as indexes of cardiac function after global ischemia and reperfusion (34). Two pressure transducers measured the coronary perfusion pressure, left ventricular developed pressure (LVDP), and left ventricular end diastolic pressure (LVEDP). The liposomes were infused over a one-minute duration, prior to the ischemia onset. The calculated amount of the liposomal ATP administered into the isolated heart preparations was 9 ± 1 mmol/g wet weight of the heart (mean $\pm$ SE). A global ischemia was established for 25 minutes by decreasing the perfusion pressure to zero within 60 seconds, followed by reperfusion for 30 minutes.

Protection of the Systolic and Diastolic Functions of the Myocardium in Isolated Heart Model

The group treated with ATP-L at ATP concentration of 1 mg/mL clearly exhibited the markedly best recovery of LV contractile function of all groups. The recovery of systolic function (LVDP) at the end of reperfusion (30 minutes) in this ATP-L group was 72% as compared to 26% ($p < 0.005$), 40% ($p < 0.005$), and 51% ($p < 0.05$), in Krebs-Henseleit (KH) buffer, empty liposomes (EL), and free ATP in KH-buffer (F-ATP) group, respectively, (Fig. 1A).

ATP-L showed effective protection of diastolic function (LVEDP) after ischemia/reperfusion. At the end of reperfusion, the LVEDP was significantly reduced by 61% ($p = 0.0002$) in ATP-L group (1 mg/mL ATP), by 27% in EL ($p = 0.004$), and by 47% F-ATP groups ($p = 0.001$), as compared to KH buffer (no significant difference between KH buffer and EL groups) (Fig. 1B).

LVDP and LVEDP measurements demonstrated a substantial protective effect of the ATP-L at higher ATP concentration on the diastolic function after ischemia and reperfusion. It is worth mentioning here that the use of the physical mixture of "working concentrations" of EL and F-ATP produced the results close to that for F-ATP (the data not shown).

After the treatment with the ATPase, the protective effect of the F-ATP on the ischemic myocardium was completely eliminated, while ATP-L

(A)   **(B)**

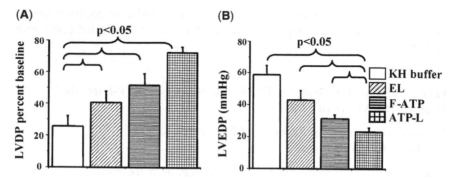

Figure 1 Protective effect of ATP-L infusion on LVDP **(A)** and LVEDP **(B)** after global ischemia and reperfusion in isolated rat heart (mean ± SE), $n = 7$–10. *Abbreviations*: LVDP, left ventricular developed pressure; LVEDP, left ventricular end diastolic pressure; KH, Krebs-Henseleit; EL, empty liposomes; ATP-L, adenosine triphosphate-loaded liposomes; F-ATP, free-adenosine triphosphate.

(1 mg/mL ATP) still provided significant recovery of LV contractile function (Fig. 2A). The LVDP recovery at the end of reperfusion (30 minutes) in the ATP-L group was 70% as compared to 28% ($p = 0.01$) and 27% ($p = 0.002$) in KH buffer and F-ATP groups, respectively. Similarly, at the end of the reperfusion, LVEDP was significantly reduced by 61% in the ATP-L group and by 40% in the F-ATP group ($p = 0.024$) as compared to the KH buffer group ($p = 0.003$) (Fig. 2B).

(A)   **(B)**

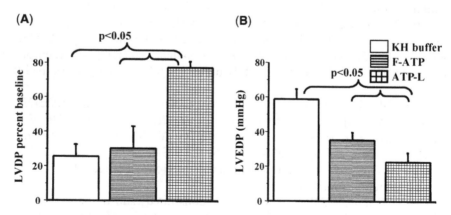

Figure 2 Protective effect of ATP-L (incubated with ATPase for 60 minutes at 37°C prior to infusion) on LVDP **(A)** and LVEDP **(B)** after global ischemia and reperfusion in isolated rat heart (mean ± SE), $n = 4$. *Abbreviations*: LVDP, left ventricular developed pressure; LVEDP, left ventricular end diastolic pressure; KH, Krebs-Henseleit; ATP-L, adenosine triphosphate-loaded liposomes; F-ATP, free-adenosine triphosphate.

Both the F-ATP and ATP-L (1 mg/mL ATP) provided significant protection to the ischemic myocardium, the maximum effect still being observed with the ATP-L. However, F-ATP can hardly protect the ischemic myocardium in vivo because it is immediately hydrolyzed by extracellular enzymes (10,35). We clearly proved it by simulating in vivo conditions by preincubating F-ATP and ATP-L with ATPase at 37°C. The encapsulation into liposomes prevented the hydrolysis of ATP by ATPase. After two minutes incubation, only 21% of the F-ATP remained nonhydrolyzed as compared to 90% in the case of ATP-L. After 25 minutes incubation, less than 0.5% of the F-ATP remained as compared to 85% in the case of ATP-L (36).

The beneficial (though smaller) effects in the case of EL and F-ATP (Fig. 1A and B) could follow different protective mechanisms. Thus, the beneficial effect of the EL may be explained by the earlier demonstrated nonspecific "plugging and sealing" of the damaged cell membranes (20); the cardioprotection of the isolated heart by the F-ATP that cannot enter cardiomyocytes may involve endogenous physiological mechanisms because the role of the ATP in the heart is not restricted to energy-requiring processes inside cells. Extracellular ATP is an important signal molecule (5,37) and F-ATP could provide a certain kind of protection to the ischemic myocardium by known mechanisms involving the chain of interstitial adenine nucleotides, interstitial enzymes, myocyte purinoreceptors, and potassium adenosine triphosphate (KATP) channels as a final common pathway for endogenous cardioprotection (5,9,37,38). For instance, the intravascular administration of ATP has been shown to induce a significant coronary vasodilation in the isolated heart (39,40). Still, whatever mechanisms of the protective action of F-ATP might be, it can hardly be considered as an in vivo cardioprotector because of rapid ATP dephosphorylation in the circulation by ecto-nucleotidases located on the surface of myocytes (5,9,41) via an efficient and "almost instantaneous" process (38).

Although the amount of liposomal ATP actually delivered into the ischemic myocytes may not be very high, it nevertheless may "tip the balance" of the ischemic injury and provide cardioprotection.

Time- and Dose-Dependent Protection of Ischemic Myocardium by ATP-L

Here, we infused 4 mg lipids/1 mg ATP/mL of liposomes at different time intervals of the myocardial ischemia, i.e., 0, 5, 10, and 15 minutes. A delay in the treatment resulted in the decrease in the recovery of LVDP in hearts treated with ATP-L at 0 (72%), 5 (55%), 10 (35%), and 15 (31%) minutes of ischemia, respectively. Also, delayed treatment followed in the increase in the LVEDP with ATP-L 0 (23 mmHg), 5 (25 mmHg), 10 (37 mmHg), and 15 (45 mmHg) minutes of ischemia, respectively.

In the dose-dependent studies, the ATP-L were infused just before ischemia, with the dose of 0.4 mg lipids/0.1 mg ATP/mL and 4 mg lipids/1 mg ATP/mL of ATP-L. The phenomenon of LVDP recovery was ATP concentration-dependent because a 10-fold decrease in the dose of ATP delivered by ATP-L (perfusate with 0.1 mg ATP/mL was used) resulted in decreased LVDP recovery after 30 minutes reperfusion to only 47%. Also, the 10-fold decrease in ATP concentration resulted in 1.7-fold increase in LVEDP at the end of reperfusion.

Visualization of Liposome Accumulation in the Ischemic Tissue and Increase in the ATP Level in the Ischemic Myocardium After ATP-L Delivery

The hypothesis that the selective accumulation of ATP-L in the ischemic tissue and direct ATP delivery to ischemic cells are responsible for myocardial protection was confirmed by fluorescence microscopy of heart cryosections after the perfusion with liposomes labeled with rhodamine (Rh)-phosphatidylethanol-amine (PE) and loaded with fluoroscein-isothiocyanate (FITC)-dextran. Figure 3A shows an extensive association of fluorescently labeled liposomes (Fig. 3A, 3) and intraliposomal load (Fig.3A, 2) with ischemic areas (Fig 3A, 4-superimposition of 2 and 3), but not with the normal myocardium (Fig. 3B, 2-3-4). The presence of FITC-dextran in the infarcted zone indicates that liposomes can deliver their hydrophilic load, such as ATP, into the ischemic myocardium.

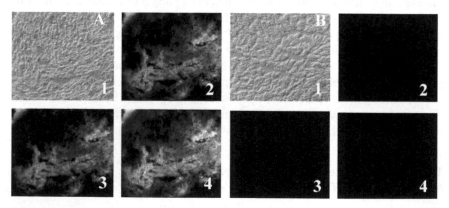

Figure 3 Microscopy of 7-μm thick heart cryosections fixed with 4% formaldehyde, washed with phosphate buffered saline, and mounted with flour mounting media (Trevigen). **(A)** Extensive association of Rh-PE (3) and fluoroscein-isothiocyanate (FITC) fluorescence (2) with infarcted tissue, 4, superposition of 2 and 3; **(B)** lack of fluorescence associated with normal tissue. 1, Transmission microscopy; 2, fluorescence microscopy with FITC filter; 3, fluorescence microscopy with Rh filter; 4, superposition of 2 and 3.

ATP in the heart tissue was determined by high performance liquid chromatography (HPLC). After 30 minutes reperfusion, the heart sample was washed with liposome-free buffer to remove ATP-L from the vasculature. Tissue samples obtained after perfusion were quickly frozen in liquid nitrogen and stored at $-80°C$. To extract ATP, the heart tissue was homogenized (Yamato Scientific) with 0.9 M perchloric acid, vortexed for one minute, and centrifuged at 3000 rpm for five minutes at $0°C$ (42). From this mixture, 1 mL of supernatant was mixed with 0.5 mL of 0.66 M potassium phosphate, and again vortexed and centrifuged. Then 50 μL of the supernatant was injected into the HPLC system to measure the amount of ATP. The ATP level in ventricular tissue (cardiomyocytes) at the end of reperfusion in the ATP-L (1 mg/mL ATP) group was 43.2 ± 2.4 μg/100 mg wet weight as compared to 32.4 ± 1.8 μg in the KH buffer group $(p = 0.02)$. This statistically significant increase by 30% in the levels of ATP found in ATP-L-treated hearts (ATP in liposomes accumulated in the interstitium via the EPR effect and in ischemic cardiomyocytes) supports our hypothesis. The mechanism involved in cardioprotection by ATP-L seems to be "nonphysiological" and may directly deliver ATP into the myocytes, thus recovering heart mechanical functions after the ischemia/reperfusion. Assuming that the use of ATP-L eventually resulted in an increased level of ATP in cardiomyocytes, 25 minutes of ischemia in the isolated heart model should have provided a sufficient time for ATP-L to accumulate in the ischemic area via the EPR effect and "unload" ATP in concentrations sufficient to protect ischemic cells. This is in agreement with earlier reports that ATP-L are rapidly taken up by hearts during a Langendorff perfusion (43) or protect the rat liver during 30 minutes of hypovolumic shock (27).

The ability of liposomes to cross the biological barriers, such as capillary endothelium, and deliver ATP directly into the cell by transporting themselves through endothelial tight junction opening and increased endothelial endocytosis was clearly shown (26,44). The opening of the endothelial tight junctions is the primary mechanism through which liposomes reach the tissue in the ischemic brain (10,45) and in the liver (27). It was also reported that liposomes could cross continuous walls of myocardial capillaries in the isolated heart through endocytosis (46,47). In the ischemic myocardium, liposomes were found in the cytoplasm of both cardiomyocytes and endothelial cells (12).

In summary, in the isolated heart model, ATP-L effectively protects ischemic myocardium from the ischemia/reperfusion damage and significantly improves both systolic and diastolic functions after ischemia and reperfusion.

ENHANCED CARDIOPROTECTION BY ATP-IL IN ISOLATED RAT HEART MODEL

In order to further reduce myocardial ischemic injury, we attempted a specific antibody-mediated targeting of ischemic myocardium. Cardiac myosin, which is a highly insoluble contractile protein that is not washed away following cell disintegration, is a convenient target antigen for the infarcted myocardium (7,8). This way of targeting is based on the observation that normal myocardial cells with intact membranes do not permit extracellular macromolecules, such as antimyosin antibody, to traverse the cell membrane. However, injured cardiomyocytes with disrupted membranes will allow interaction of exposed cytoskeleton component myosin with the antibody (7).

Preparation of ATP-IL

p-Nitrophenylcarbonyl-PEG-phosphatidylethanol-amine (*p*NP-PEG-PE), used to modify antibodies for their attachment to the liposomal membrane, was synthesized according to a previously reported method (48). Because the PEG-PE-modified 2G4 antibody becomes amphiphilic and forms micelles, these micelles were incubated with ATP-L for two hours at 37°C in phosphate buffered saline (PBS) (49). In good agreement with all these earlier data, we observed almost complete (86%) incorporation of 2G4-PEG-PE into ATP-containing liposomes after two hours at 37°C. The antibody incorporation was between 50 and 100 molecules per single liposome with a diameter of 200 nm.

In Vitro ATP Release Studies

ATP release from various fresh liposomal preparations was determined over a 24-hour period at 37°C in KH buffer both in the absence and of 2G4-PEG-PE micelles. The percent release of ATP was measured by the HPLC (50). The 24-hour long incubation of ATP-L in the KH buffer at 37°C resulted in the leakage of approximately 12% of the total liposomal ATP. Interestingly, the presence of 2G4-PEG-PE micelles did not influence the ATP leakage rate and extent: the ATP release pattern was approximately the same with and without 2G4-PEG-PE. In other words, the gradual incorporation of the PEG-PE-modified 2G4 antibody into the liposomal membrane via its PE moiety did not provoke any noticeable membrane destabilization.

Specific Binding of ATP–Containing 2G4-IL to the Myosin Substrate

The results (Fig. 4) of in vitro assay of the specific binding ability of the PEG-PE-modified 2G4 antibody and 2G4-PEG-PE-containing ATP-loaded immunoliposomes (IL) indicated that 2G4-PEG-PE and 2G4-bearing ATP-containing PEG-liposomes retained their myosin-binding activity.

During and after antibody attachment, liposomes do not lose the entrapped ATP, while the liposome-attached 2G4 antibody preserves its specific binding ability and can successfully target ATP-L to a myosin monolayer

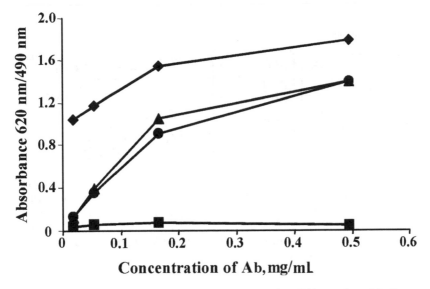

Figure 4   Enzyme-linked immunosorbent assay (ELISA) results with the myosin monolayer for the native 2G4 antimyosin antibody (♦); 2G4-poly(ethylene glycol) (PEG)-phosphatidylethanol-amine conjugate (●); 2G4- PEG-phosphatidylethanol-amine-bearing adenosine triphosphate (ATP)-containing PEG-liposomes (▲); and 2G4-free ATP-L containing PEG-liposomes (■).

in vitro. Such liposomes, due to their specificity (20) and prolonged circulation, may effectively deliver ATP to ischemically damaged cells and so provide a better chance for cell salvage.

Protection of the Systolic and Diastolic Functions of the Myocardium by ATP-IL in an Isolated Rat Heart Model

In the Langendroff perfused ischemic rat heart, ATP concentrations decrease rapidly to 60% in the first minute, with a rapid secondary decrease by 13 minutes due to contracture (51). In the previous section, we showed that the ATP-L delivered for one minute before induction of global ischemia can effectively protect the rat heart from ischemia-reperfusion damages. Recently, it was reported that cytoskeleton-specific immunoliposome resulted in preservation of myocardial viability (52). Therefore, our IL loaded with ATP may result in further recovery of the mechanical functions after 25 minutes of global ischemia and 30 minutes of reperfusion.

ATP for these experiments were prepared by the freezing–thawing method (as described in earlier section) with addition of 1,2-dioleoyl-3-trimethyl-ammonium-propane (3.3 mol%) with a diameter of approximately 200 nm. Here, we present the data on the significant cardioprotective effect

(A)   **(B)**

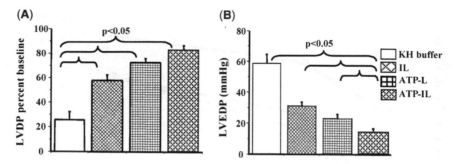

Figure 5 Protective effect of ATP-IL, IL, and ATP-L infusion on LVDP (A), and LVEDP (B) after global ischemia and reperfusion in isolated rat heart (mean ± SE), $n = 6$–11. *Abbreviations*: LVDP, left ventricular developed pressure; LVEDP, left ventricular end diastolic pressure; ATP-IL, adenosine triphosphate-loaded immunoliposomes; ATP-L, adenosine triphosphate-loaded liposomes; KH, Krebs-Henseleit; IL, immunoliposomes.

of the ATP-IL on isolated rat hearts subjected to global ischemia. The same protocol was used as in the case of ATP-L mentioned earlier.

The ATP-L-treated isolated rat heart group exhibited pronounced recovery of LV contractile function of all groups. The LVDP at the end of reperfusion (30 minutes) in the ATP-IL group was 83% as compared to 26% in the KH buffer group ($p = 0.0006$), 60% in the IL group ($p = 0.002$), and 72% in the ATP-L group ($p = 0.046$) (Fig. 5A). At the end of reperfusion, the LVEDP was significantly reduced by 74% in ATP-IL group ($p = 0.00002$), by 47% in the IL ($p = 0.004$), and by 61% in the ATP-L ($p = 0.001$) groups as compared to the KH buffer. Thus, the ATP-IL group had a much smaller increase in LVEDP after ischemia and reperfusion (Fig. 5B). The results of LVDP and LVEDP measurements demonstrate a substantial protective effect of the ATP-IL on the systolic and diastolic functions after ischemia and reperfusion.

Interestingly, the protective mechanisms among these cases might be rather different. Our results also revealed that this protective effect is a combination of the protection conferred separately by the IL and by the loaded ATP. Experiments demonstrated that IL by itself have some protective effect on the ischemic myocardium and showed 60% of postischemic recovery, which may be explained by the "plug and seal" phenomenon of the damaged cell membranes of the myocytes (20). Moreover, encapsulation of ATP in IL resulted in the further dramatic increase of recovery of mechanical functions upon reperfusion by a factor of 1.4. These results can be considered as a significant step towards the antibody-mediated targeted delivery of pharmaceuticals to the ischemic myocardium and its protection against damage resulting from the ischemia and reperfusion.

The Effect of the Amount of the Liposome Surface-Attached Antibody on Preparation Efficacy

We observed an antibody dose-dependent increase in the myocardial preservation. The protective effect of the ATP-IL on the ischemic myocardium was comparable to ATP-L ($p = \mathrm{NS}$) and by a 10-fold decrease in the amount of antibody (0.0216 mg/mL) on the surface of ATP-IL, whereas a 10-fold increase in amount of antibody (2.16 mg/mL) showed an increase in the recovery of LVDP ($p = 0.028$) after 15 minutes of reperfusion as compared to the ATP-L. However, a 10-fold increase in the amount of antibody on the surface of ATP-IL resulted in a significantly higher recovery of LVDP (94%) as compared to the ATP-IL (83%) after 15 minutes of reperfusion ($p = 0.018$) (Fig. 6A). After 30 minutes of reperfusion, a 10-fold decrease in the amount of antibody on the surface of ATP-IL showed significantly elevated LVEDP value (23 ± 3 mmHg) ($p = 0.033$) (data not shown), while a 10-fold increase in the amount of antibody showed decreased LVEDP (Fig. 6B). However, a 20-fold increase in the amount of antibody on the surface of liposome did not lead to any further significant reduction in the LVEDP as compared to the ATP-IL after 30 minutes of reperfusion because the diastolic pressure in the case of ATP-IL (15 ± 2 mmHg) was already very close to the baseline value of 10 mmHg (date not shown). To summarize, an increase in the amount of antibody on the surface of liposomes to 0.0216 mg/mL resulted in the preparation allowing for almost

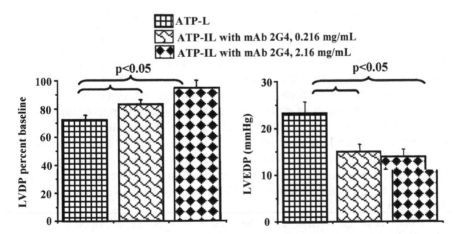

Figure 6 Dependence of the degree of cardioprotection of the systolic (*left*) and diastolic (*right*) functions of the myocardium on the quantity of mAb 2G4 attached to the ATP-liposomes. *Abbreviations*: LVDP, left ventricular developed pressure; LVEDP, left ventricular end diastolic pressure; ATP-IL, adenosine triphosphate–loaded immunoliposomes; ATP-L, adenosine triphosphate–loaded liposomes.

complete normalization of the mechanical function (94%) after 25 minutes of global ischemia and 30 minutes of reperfusion.

The cardioprotection with IL and ATP-IL is essentially related to the potentiation of two or more cardioprotective mechanisms. Our data supports the conclusion that cardioprotection by liposomes loaded with ATP is an integration of physiological functions through the phenomenon of membrane repair and ATP delivery by the IL and by cardioprotection by the effect of exogenous ATP. Prevention of cell death in myocardial ischemia by delivering bioenergetics molecules like ATP encapsulated in IL specific for damaged myocytes could have significant clinical utility and might become an interesting strategy to protect myocardium under ischemia.

SIGNIFICANT REDUCTION IN THE SIZE OF THE IRREVERSIBLY DAMAGED FRACTION OF THE "AREA AT RISK" BY ATP-L IN RABBITS WITH EXPERIMENTAL MI

ATP-L were also found to be capable of providing significant myocardial preservation in vivo in rabbits with experimental MI. The procedure for an experimental MI was previously described by Narula et al. (53). Briefly, New Zealand (NZW) rabbits (2.5–3.5 kg) were anesthetized with ketamine and xylazine, intubated via a tracheostomy, and ventilated. The heart was exposed through a parasternal thoracotomy and a flexible plastic catheter was inserted into the left atrium for rapid coronary infusions. An anterior branch of the left coronary artery was isolated with a 3–0 suture for control of flow with an occlusive snare. Approximately 3 mL of ATP-L (45 mg lipid/12 mg ATP) or KH buffer, pH 7.4, were infused during brief clamping of the aorta as the occluding snare was tightened. After 30 minutes, the snare was released and reperfusion established for three hours. The coronary artery was reoccluded and 3 mL of a 1:5 diluted Unispearse was infused via the atrial catheter to demarcate the area at risk. The anesthetized animal was immediately sacrificed and the left ventricle excised, sliced between apex and base into five slices of equal thickness, and digitally photographed on both sides to estimate the occlusion-induced area at risk. Slices were stained with preheated nitroblue tetrazolium in PBS, pH 7.4 at 45°C for 20 minutes to outline the infarcted fraction of the area at risk, rephotographed and weighed. The area at risk and the fraction of infarcted area at risk were determined from planimetry of both sides of all slices using Adobe Photoshop 7.0.

The size of the "area at risk" (the net area of hypoxia developed as a result of occlusion) was between approximately 20% and 30% of the total heart tissue in all infarcted animals as shown by Unispearse Blue dye staining. The effect of the experimental and control treatments was estimated by measuring the fraction of the area at risk, which at the end of the experiment demonstrated irreversible damage according to the nitro blue tetrazolium (NBT) staining. The intracoronary administration of the ATP-L (36 mg of ATP total)

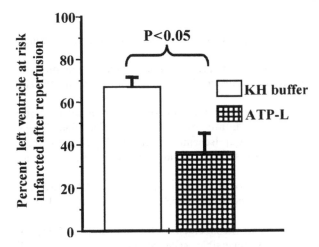

Figure 7 Cardioprotective effect of ATP-L during 30 minutes of coronary occlusion and following three hours of reperfusion in rabbits with experimental MI. Summary graph showing the fraction of infarcted area as a percentage of the total area at risk in a KH buffer-treated control group ($n = 3$) and an ATP-L-treated group ($n = 5$). *Abbreviations*: KH, Krebs-Henseleit; ATP-L, adenosine triphosphate-loaded liposomes; MI, myocardial infarction.

effectively protected the ischemic heart muscle in rabbits with experimental MI, which follows from a significantly decreased (by approximately 45% according to NBT staining, $p < 0.05$) fraction of the irreversibly damaged heart within the total area at risk (which remained essentially the same in all experiments) in ATP-L-treated animals as compared to the animals receiving the KH buffer (Fig. 7). The KH buffer was used here as the only control for the ATP-L treatment, similar to how it was done by Khudairi and Khaw (52) because using the F-ATP as an additional control does not make any sense because of its instant degradation in vitro and in vivo (9,38). EL liposomes did not yield a statistically significant difference relative to the KH buffer even under more favorable conditions of the Langendorff model (Fig. 1); in addition, in studies with IL, the effect of ELs in similar systems was always at the level of the free buffer (20).

CONCLUSIONS

In conclusion, in the isolated heart model, ATP-L and ATP-IL effectively protect ischemic myocardium from the ischemia/reperfusion damage and significantly improve (almost complete normalization) both systolic and diastolic functions after ischemia and reperfusion. In vivo, in rabbits with experimental MI, ATP-L significantly diminishes the size of the irreversibly damaged zone in the heart. ATP-L may provide an effective exogenous source of the ATP in vivo and serve as a tool for cardiomyocyte protection in ischemically damaged hearts.

REFERENCES

1. Carden DL, Granger DN. Pathophysiology of ischaemia-reperfusion injury. J Pathol 2000; 190:255.
2. Karmazyn M. The 1990 Merck Frosst Award. Ischemic and reperfusion injury in the heart, cellular mechanisms and pharmacological interventions. Can J Physiol Pharmacol 1991; 69:719.
3. McKay RG, Pfeffer MA, Pasternak RC, et al. Left ventricular remodeling after myocardial infarction: a corollary to infarct expansion. Circulation 1986; 74:693.
4. Kingsley PB, Sako EY, Yang MQ, et al. Ischemic contracture begins when anaerobic glycolysis stops: a 31P-NMR study of isolated rat hearts. Am J Physiol 1991; 261:H469.
5. Kuzmin AI, Lakomkin VL, Kapelko VI, Vassort G. Interstitial ATP level and degradation in control and postmyocardial infarcted rats. Am J Physiol 1998; 275:C766.
6. Jennings RB, Schaper J, Hill ML, Steenbergen C Jr., Reimer KA. Effect of reperfusion late in the phase of reversible ischemic injury. Changes in cell volume, electrolytes, metabolites, and ultrastructure. Circ Res 1985; 56:262.
7. Khaw BA, Beller GA, Haber E, Smith TW. Localization of cardiac myosin-specific antibody in myocardial infarction. J Clin Invest 1976; 58:439.
8. Khaw BA, Scott J, Fallon JT, Cahill SL, Haber E, Homcy C. Myocardial injury: quantitation by cell sorting initiated with antimyosin fluorescent spheres. Science 1982; 217:1050.
9. Gordon JL. Extracellular ATP: effects, sources and fate. Biochem J 1986; 233:309.
10. Puisieux F, Fattal E, Lahiani M, et al. Liposomes, an interesting tool to deliver a bioenergetic substrate (ATP) in vitro and in vivo studies. J Drug Target 1994; 2:443.
11. Torchilin VP, Trubetskoy VS. Which polymers can make nanoparticulate drug carriers long-circulating. Adv Drug Delivery Rev 1995; 16:141.
12. Caride VJ, Zaret BL. Liposome accumulation in regions of experimental myocardial infarction. Science 1977; 198:735.
13. Palmer TN, Caride VJ, Caldecourt MA, Twickler J, Abdullah V. The mechanism of liposome accumulation in infarction. Biochim Biophys Acta 1984; 797:363.
14. Torchilin VP, Khaw BA, Smirnov VN, Haber E. Preservation of antimyosin antibody activity after covalent coupling to liposomes. Biochem Biophys Res Commun 1979; 89:1114.
15. Torchilin VP, Narula J, Halpern E, Khaw BA. Poly(ethylene glycol)-coated anti-cardiac myosin immunoliposomes: factors influencing targeted accumulation in the infarcted myocardium. Biochim Biophys Acta 1996; 1279:75.
16. Trubetskaya OV, Trubetskoy VS, Domogatsky SP, et al. Monoclonal antibody to human endothelial cell surface internalization and liposome delivery in cell culture. FEBS Lett 1988; 228:131.
17. Lukyanov AN, Hartner WC, Torchilin VP. Increased accumulation of PEG-PE micelles in the area of experimental myocardial infarction in rabbits. J Control Release 2004; 94:187.
18. Maeda H, Wu J, Sawa T, Matsumura Y, Hori K. Tumor vascular permeability and the EPR effect in macromolecular therapeutics: a review. J Contr Rel 2000; 65:271.

19. Maeda H. The enhanced permeability and retention (EPR) effect in tumor vasculature: the key role of tumor-selective macromolecular drug targeting. Adv Enzyme Regul 2001; 41:189.
20. Khaw BA, Torchilin VP, Vural I, Narula J. Plug and seal: prevention of hypoxic cardiocyte death by sealing membrane lesions with antimyosin-liposomes. Nat Med 1995; 1:1195.
21. Nguyen PD, O'Rear EA, Johnson AE, et al. Accelerated thrombolysis and reperfusion in a canine model of myocardial infarction by liposomal encapsulation of streptokinase. Circ Res 1990; 66:875.
22. Soloviev A, Stefanov A, Parshikov A, et al. Arrhythmogenic peroxynitrite-induced alterations in mammalian heart contractility and its prevention with quercetin-filled liposomes. Cardiovasc Toxicol 2002; 2:129.
23. Li S, Huang L. In vivo gene transfer via intravenous administration of cationic lipid-protamine-DNA (LPD) complexes. Gene Ther 1997; 4:891.
24. Han YY, Huang L, Jackson EK, et al. Liposomal ATP or NAD+ protects human endothelial cells from energy failure in a cell culture model of sepsis. Res Commun Mol Pathol Pharmacol 2001; 110:107.
25. Laham A, Claperon N, Durussel JJ, et al. Intracarotidal administration of liposomally-entrapped ATP: improved efficiency against experimental brain ischemia. Pharmacol Res Commun 1988; 20:699.
26. Laham A, Claperon N, Durussel JJ, et al. Liposomally entrapped adenosine triphosphate. Improved efficiency against experimental brain ischaemia in the rat. J Chromatogr 1988; 440:455.
27. Konno H, Matin AF, Maruo Y, Nakamura S, Baba S. Liposomal ATP protects the liver from injury during shock. Eur Surg Res 1996; 28:140.
28. Neveux N, De Bandt JP, Chaumeil JC, Cynober L. Hepatic preservation, liposomally entrapped adenosine triphosphate and nitric oxide production: a study of energy state and protein metabolism in the cold-stored rat liver. Scand J Gastroenterol 2002; 37:1057.
29. Neveux N, De Bandt JP, Fattal E, et al. Cold preservation injury in rat liver: effect of liposomally-entrapped adenosine triphosphate. J Hepatol 2000; 33:68.
30. Skiba-Lahiani M, Auger J, Terribile J, et al. Stimulation of movement and acrosome reaction of human spermatozoa by PC12 liposomes encapsulating ATP. Int J Androl 1995; 18:287.
31. Xu GX, Xie XH, Liu FY, et al. Adenosine triphosphate liposomes: encapsulation and distribution studies. Pharm Res 1990; 7:553.
32. Liang W, Levchenko TS, Torchilin VP. Encapsulation of ATP into liposomes by different methods: optimization of the procedure. J Microencapsul 2004; 21:251.
33. Pick U. Liposomes with a large trapping capacity prepared by freezing and thawing of sonicated phospholipid mixtures. Arch Biochem Biophys 1981; 212:186.
34. Galinanes M, Hearse DJ. Assessment of ischemic injury and protective interventions: the Langendorff versus the working rat heart preparation. Can J Cardiol 1990; 6:83.
35. Grobben B, Anciaux K, Roymans D, et al. An ecto-nucleotide pyrophosphatase is one of the main enzymes involved in the extracellular metabolism of ATP in rat C6 glioma. J Neurochem 1999; 72:826.

36. Verma DD, Levchenko TS, Bernstein EA, Torchilin VP. ATP-Loaded liposomes effectively protect mechanical functions of the myocardium from global ischemia in an isolated rat heart model. In: Transactions of the 31st Annual Meeting of the Controlled Release Society. Hawaii: Controlled Release Society, 2004.
37. O'Rourke B, Myocardial K. (ATP) Channels in preconditioning. Circ Res 2000; 87:845.
38. Korchazhkina O, Wright G, Exley C. No effect of aluminium upon the hydrolysis of ATP in the coronary circulation of the isolated working rat heart. J Inorg Biochem 1999; 76:121.
39. Hopwood AM, Burnstock G. ATP mediates coronary vasoconstriction via P2x-purinoceptors and coronary vasodilatation via P2y-purinoceptors in the isolated perfused rat heart. Eur J Pharmacol 1987; 136:49.
40. Burnstock G. The past, present and future of purine nuclcotides as signalling molecules. Neuropharmacology 1997; 36:1127.
41. Yagi K, Shinbo M, Hashizume M, et al. ATP diphosphohydrolase is responsible for ecto-ATPase and ecto-ADPase activities in bovine aorta endothelial and smooth muscle cells. Biochem Biophys Res Commun 1991; 180:1200.
42. Bedford GK, Chiong MA. High-performance liquid chromatographic method for the simultaneous determination of myocardial creatine phosphate and adenosine nucleotides. J Chromatogr 1984; 305:183.
43. Kayawake S, Kako KJ. Association of liposomes with the isolated perfused rabbit heart. Basic Res Cardiol 1982; 77:668.
44. Chapat S, Frey V, Claperon N, et al. Efficiency of liposomal ATP in cerebral ischemia: bioavailability features. Brain Res Bull 1991; 26:339.
45. Kuroiwa T, Shibutani M, Okeda R. Blood-brain barrier disruption and exacerbation of ischemic brain edema after restoration of blood flow in experimental focal cerebral ischemia. Acta Neuropathol (Berl.) 1988; 76:62.
46. Mixich F, Mihailescu S. In: deBoer, B, Sutanto W, eds. Drug Transport Across the Blood-Brain Barrier. Amsterdam: Harwood Acad Publ, GMBH, 1997:201.
47. Zipper J, Sarrach D, Halle W. Interaction of liposomes with vascular endothelial cells. Biomed Biochim Acta 1988; 47:713.
48. Torchilin VP, Levchenko TS, Lukyanov AN, et al. *p*-Nitrophenylcarbonyl-PEG-PE-liposomes: fast and simple attachment of specific ligands, including monoclonal antibodies, to distal ends of PEG chains via *p*-nitrophenylcarbonyl groups. Biochim Biophys Acta 2001; 1511:397.
49. Liang W, Levchenko T, Khaw B, Torchilin V. ATP-containing immunoliposomes specific for cardiac myosin. Curr Drug Deliv 2004; 1:1.
50. Leoncini G, Buzzi E, Maresca M, Mazzei M, Balbi A. Alkaline extraction and reverse-phase high-performance liquid chromatography of adenine and pyridine nucleotides in human platelets. Anal Biochem 1987; 165:379.
51. Hearse DJ, Garlick PB, Humphrey SM. Ischemic contracture of the myocardium: mechanisms and prevention. Am J Cardiol 1977; 39:986.
52. Khudairi T, Khaw BA. Preservation of ischemic myocardial function and integrity with targeted cytoskeleton-specific immunoliposomes. J Am Coll Cardiol 2004; 43:1683.
53. Narula J, Petrov A, Pak KY, Lister BC, Khaw BA. Very early noninvasive detection of acute experimental nonreperfused myocardial infarction with 99mTc-labeled glucarate. Circulation 1997; 95:1577.

7

Targeting Tumor Angiogenesis Using Liposomes

Raymond M. Schiffelers, Evisa Gjini, Marcel H. A. M. Fens, and Gert Storm

Department of Pharmaceutics, Utrecht Institute of Pharmaceutical Sciences, Utrecht University, Utrecht, The Netherlands

INTRODUCTION

The vasculature is involved in many (patho)physiological processes. Apart from the role of the endothelium as passive barrier between blood and tissues, endothelial cells (ECs) can actively perform complex functions to regulate pathological and physiological processes such as vascular homeostasis, thrombosis, inflammation, and angiogenesis. During these pathological processes, the vascular system undergoes phenotypical changes. Often, the vasculature is characterized by an increased capillary permeability during disease progression, a characteristic that has been used for passive targeting of drug carriers to diseased sites (1).

Alternatively, to actively target the vasculature at diseased sites, the changes that occur in expression profiles of membrane proteins on ECs can be used for site-specific delivery. As the endothelium plays a crucial role in the disease process, targeting the upregulated surface proteins of ECs may be used to modulate EC function, which may result in therapeutic effects in several pathologies (2–6). The strong therapeutic effects that can be generated by targeting the endothelium have stimulated a strong interest in active targeting of the vasculature at pathological sites (7–10). This chapter gives an overview of the literature regarding the targeting of liposomes for delivery of drugs to the endothelium during tumor angiogenesis.

ANGIOGENESIS

The generation of new capillaries from pre-existing blood vessels is called angiogenesis. The angiogenesis process takes place during embryogenesis and in the adult, for example, in the female reproductive system and wound healing. Additionally, angiogenesis occurs in pathological conditions such as cancer, macular degeneration, psoriasis, and rheumatoid arthritis (5).

Angiogenesis and tumor progression are very closely linked with each other. Tumor cells are dependent on angiogenesis because their growth and expansion requires oxygen and nutrients, which are made available through the angiogenic vasculature. Investigational studies on tumor development have shown that an alteration in the blood supply can noticeably affect tumor growth and metastasis formation. Based on previous research, the shutdown of blood vessels resulted in tumor regression. Therefore, tumor angiogenesis has become an important area in cancer research (8,11).

Different cells and stimulating factors are involved in angiogenesis. Some of the cells engaged are the ECs, lymphocytes, macrophages, and mast cells. Vascular endothelial growth factor (VEGF) and fibroblast growth factor (FGF) are two of the major among many proangiogenic factors. Both cells and stimulating factors play different roles during tumor angiogenesis. This process includes two major phases: the activation phase and the formation phase.

During the activation phase, growth factors are produced. When a condition such as hypoxia is present in the tumor tissue, the tumor cells experiencing hypoxia promote the angiogenic switch, which is mediated by hypoxia-inducible factor-1 (HIF-1) and induces angiogenesis (12–14).

HIF-1 binds to hypoxia-response elements and activates a number of hypoxia-response genes such as VEGF. Additionally, the tumor cells export FGF, VEGF, and FGF bind to their receptors on the ECs. Both VEGF and FGF activate signal transduction pathways, activating in this way the ECs (15). In the first phase, the adventitial cells and pericytes retract, while the basal membrane of the pre-existing vessels is degraded by proteases, for example, by members of the matrix metalloproteinase (MMP) family (16). MMPs are produced by the activated ECs. After the basement membrane barrier is disrupted, the ECs, which cover the internal wall of a blood vessel, are able to migrate from pre-existing vessels toward the angiogenic stimuli and proliferate. The migration of the ECs is based on cell-extracellular interaction that is mediated by vascular cell-adhesion molecules as, for example, integrin $\alpha v\beta 3$ (17,18). Research has shown that this molecule, which mediates cell adhesion, plays an important role in angiogenesis.

During the formation phase, the ECs, after migrating, are structured into tubes to form capillary-like structures, which mature into functional capillaries, and then the blood flow is initiated. The formation phase is thought to be dependent on E-selectin, which is a transmembrane cell-adhesion glycoprotein (19). Moreover, mesenchymal cells play a role in the formation of mature blood vessels. These cells express angiopoetin-1, which binds to Tie-2 receptors expressed on the EC. This binding is thought to help in pericyte recruitment, vessel sprouting, and vessel stabilization. Tie receptors (Tie-1 and Tie-2) are tyrosine kinases and their expression follows vascular endothelial growth factor receptor (VEGFR) expression. The ligand for Tie-2 is angiopoetin-1, which upon binding to Tie-2 induces tyrosine phosphorylation of Tie-2. Angiopoetin-1 has showed induction of capillary sprouts formation and EC survival support (20).

After recruitment, the mesenchymal cells differentiate into smooth-muscle cell-like pericytes, which cover the vascular tree. A special role in pericyte recruitment plays the platelet-derived growth factor (PDGF). PDGF is excreted by the EC and it functions as a chemoattractant for pericyte precursors, which after associating with EC differentiate into pericytes. The role of pericytes is not yet completely understood, but it is believed that they play a role in stabilizing the newly formed blood vessels (21).

TARGETING TUMOR ANGIOGENESIS

Targeting tumor angiogenesis rather than tumor cells has several advantages:

- Quiescent ECs are not affected by the treatment; therefore, side effects to nontarget endothelium are expected to be limited.
- Proliferating ECs share similar phenotypes among different tumors, making vascular targeting treatment applicable to many tumor types.
- ECs are genetically stable (unlike tumor cells), reducing the risk of development of drug resistance.
- Many tumor cells are dependent on a single blood vessel leading to large amplification of the effect.
- ECs are easily accessible from the bloodstream upon intravenous (IV) injection.

To target liposomes specifically to activated angiogenic ECs in the tumor region, the overexpression of several specific cell surface proteins can be exploited (10,22–25).

ANTIBODY-BASED TARGETING OF LIPOSOMES

Antibodies are Y-shaped molecules with a molecular weight of approximately 150 kDa consisting two light chains and two heavy chains. Both heavy and light chains have a constant region and a unique variable part that contains the antigen-binding site. By treatment with papain, the antigen-binding fragment (Fab) can be cleaved from the crystalline fragment. The

Fab fragments contain the variable domains, consisting of three hypervariable amino acid sequences responsible for the antibody specificity. The high affinity of antibodies for specific antigens and the possibility to raise antibodies to virtually any target molecule has made them popular targeting ligands for site-specific drug delivery.

Antibody to Vascular Cell Adhesion Molecule 1(VCAM1)

Chiu et al. coupled an anti-VCAM-1 antibody to the distal end of poly(ethylene glycol) (PEG) chains on the surface of sterically stabilized liposomes. The liposomes contained phosphatidylserine and were designed to target tumor vasculature using the antibody and subsequently to induce local thrombogenesis by the phosphatidylserine (26,27). The target for the antibody, VCAM-1, was induced to be overexpressed on the surface of ECs by activation of EC with cytokines. However, VCAM-1 is also expressed on other cell types, such as macrophages, myoblasts, dendritic cells, and tumor cells (9). VCAM-1 interacts with integrin $\alpha 4\beta 1$ (very late antigen 4), leading to signal transduction in ECs, resulting in a change of cell morphology and promoting leukocyte extravasation by enabling adhesion. VCAM-1 expression on tumor endothelium has been clearly demonstrated, together with expression of other adhesion molecules such as intercellular adhesion molecule-1 and E-selectin (26).

The study by Chiu et al. demonstrates that VCAM-1-targeted liposomes are bound to interferon-stimulated ECs in vitro. Binding was positively correlated with antibody concentration. The authors demonstrated in a previous study that the use of a sheddable PEG-coating allowed unmasking of the phosphatidylserine, which can induce blood coagulation (28). On the basis of these in vitro data, it was suggested that premature coagulation could be prevented in vivo by modifying the length of the PEG-lipid anchor (and thereby modifying the kinetics of PEG-shedding) and by modulating phosphatidylserine concentration. However, this strategy has, up to now, not been followed up in vivo.

Antibody to CD105

CD105, also known as endoglin or transforming growth factor-beta (TGFB)-receptor, is expressed as transmembrane glycoprotein dimmer mainly on the vascular EC surface (29,30). However, expression has also been demonstrated on vascular smooth muscle cells, fibroblasts, macrophages, and erythroid precursors, although densities on those cell types are generally lower. During angiogenesis, expression of CD105 is strongly upregulated, as shown by immunohistochemical staining in a variety of human tumors derived from breast, brain, lung, prostate, and cervix tissue. CD105 (-/-) mice have a multitude of vascular and cardiac abnormalities that cause death at an early embryonic stage, underlining the importance of the protein in neovascularization.

Immunoliposomes equipped with a single-chain monoclonal antibody Fv fragment directed against CD105 showed enhanced binding to CD105-positive ECs in vitro, while showing limited interaction with CD105 negative cell types (31). At 37°C, cell binding was followed by internalization. As a consequence, immunoliposomes loaded with doxorubicin could effectuate strong cytotoxicity in CD105-positive ECs in vitro. In vivo studies were only focused on a pharmacokinetic analysis in healthy animals and showed that the immunoliposomes were cleared at an extremely fast rate with an approximate half-life of three minutes. These findings could considerably hamper the use of these CD105-targeted liposomes as there may be insufficient time for interaction with the target cells.

PEPTIDE-BASED TARGETING OF LIPOSOMES

Biopanning experiments using phage-display has yielded a number of peptide motifs that show preferential binding to angiogenic tumor vasculature. Coupling of such peptides has been shown to result in preferential accumulation of liposomes at sites of angiogenesis.

Arg-Leu-Pro-Leu-Pro-Gly (RLPLPG)-Mediated Targeting of Membrane Type 1-Matrix Metalloproteinase

Kondo et al. used a peptide substrate of membrane type 1-matrix metalloproteinase (MT1-MMP) with the sequence RLPLPG. This peptide was coupled to a stearoyl moiety inserted into the liposomal membrane (32). The target, MT1-MMP, is a metalloproteinase consisting of three domains, which are exposed at the EC surface. A small domain, composed of 20 hydrophobic amino acids, crosses the plasma membrane, and is completed by a short cytoplasmic tail (33). The enzyme can degrade various extracellular membrane components and activate other pro-MMP, such as (MMP)-2, pro-MMP-13, and, indirectly, pro-MMP-9. Therefore, its expression is crucial for remodeling connective tissue during development and as such it has an important role during physiological and pathological angiogenesis (34,35). The expression of MT1-MMP is upregulated in proangiogenic milieu (36–38).

In vitro, RLPLPG-modified liposomes displayed high binding ability to human umbilical vein ECs compared to unmodified liposomes, although the involvement of some specific interactions cannot be ruled out. The arginine residue in the targeting peptide exerted a positive surface charge to the liposomes, which may have contributed to cell binding as a result of an electrostatic interaction with the negatively charged cell membrane. In vivo, RLPLPG-targeted liposomes showed a four-fold higher degree of localization in tumor tissue than liposomes without the targeting peptide. Encapsulation of 5'-O-dipalmitoylphosphatidyl 2'-C-cyano-2'-deoxy-1-beta-D-arabino-pentofuranosylcytosine in RLPLPG-targeted liposomes inhibited tumor growth rate at a higher degree than unmodified liposomes loaded with this drug.

Arg-Gly-Asp–Mediated Targeting of αvβ3-Integrin

Integrins belong to the family of heterodimeric transmembrane glycoproteins and mediate the interaction between cells and the extracellular matrix (39,40). They consist of an α and β subunit that are noncovalently bound. Different combinations of the various α and β subunits can be formed. As a result, over 20 distinct integrins can be recognized. Two integrins, αvβ3- and αvβ5-integrin, are expressed preferentially on the angiogenic vascular endothelium. Both have high affinity for Arg-Gly-Asp (RGD) RGD-motif containing proteins and peptides. In situ, αvβ5-integrin binds only to vitronectin, whereas αvβ3 binds vitronectin, fibronectin, von Willebrand factor, thrombospondin, osteopontin, laminin, denatured collagen, and developmentally regulated endothelial locus 1.

The importance of these integrins in angiogenesis is supported by the observation that antibodies or cyclic RGD-peptides that can block binding of vitronectin to ECs can inhibit angiogenesis in vitro and in vivo. However, contrary to expectations, αv-knockout mice still can mount a pronounced angiogenic response and β3/β5 knockouts also appear to exhibit normal angiogenesis. The apparent discrepancy may be explained by the role integrins play in regulating pro- and antiangiogenic factors. Knockdown of an integrin may, thereby, actually upregulate expression of other proangiogenic receptors and factors.

Articles by Koning et al. (41), Schiffelers et al. (42), and Janssen et al. (43) demonstrated that PEG liposomes bearing cyclic RGD-peptides on their surface could bind to and were internalized by activated ECs in vitro. Binding was positively correlated with RGD-density on the liposome surface. PEG-liposomes bearing RAD-peptide did not induce EC binding when coupled to the PEG-liposomes. This shows the high specificity of the RGD-motif, as the RAD-peptide differs from the RGD-peptide by only one methyl group. When RGD-liposomes where loaded with doxorubicin or ^{10}B they displayed cytotoxicity toward ECs in vitro (41,42). In a doxorubicin-resistant murine colon carcinoma model, RGD-targeted PEG-liposomes encapsulating doxorubicin were able to reduce tumor growth, whereas PEG-liposomes and RAD-liposomes failed, indicating that RGD-targeted doxorubicin liposomes effectuated antitumor effects through their action on the angiogenic endothelium (42).

Publications by Fahr et al. and Muller et al. dealt with RGD-mediated targeting of poly(ethylene imine)-complexed DNA. The particles demonstrated more efficient transduction of ECs in vitro when the RGD-peptide was present in the complex (44,45).

As the RGD-targeted integrins are not exclusively expressed on the neovasculature, liposomes equipped with RGD-targeting peptides may interact with other cell types, most notably tumor cells and activated platelets (46–51). Obviously, this is an important consideration for the design of targeted drug-delivery systems.

Asn-Gly-Arg–Mediated Targeting of Aminopeptidase N

Aminopeptidase N is a protease that is anchored to the cell surface (52). Its expression is confined to angiogenic endothelium. Absence of expression on quiescent vasculature makes it an attractive target for tumor vasculature targeting (53). Panning experiments using phage-display revealed that peptides containing an Asn-Gly-Arg (NGR) NGR-motif display binding to aminopeptidase N. It has been demonstrated that NGR-peptides also induce vasculature-specific cytotoxic effects when coupled to doxorubicin (54).

Pastorino et al. coupled NGR-peptides to the surface of PEG-liposomes loaded with doxorubicin (55). These liposomes were studied in an orthotopic neuroblastoma xenograft model in severe combined immunodeficient (SCID) mice. Pharmacokinetic analysis revealed that coupling of the peptide did not have major effects on the blood circulation time of the PEG-liposomes, in contrast to observations made in our laboratory for RGD-PEG-liposomes (42,43). This is probably related to a rather selective expression of the aminopeptidase N by angiogenic ECs as compared to the αv-integrin, which is also expressed on macrophages. Interestingly, NGR-peptide induced targeting of PEG-liposomes resulted in 10-fold increased tumor accumulation as compared to nontargeted PEG-liposomes, while, remarkably, tumor accumulation of control Ala-Arg-Ala-peptide-PEG-liposomes was absent. The specificity of the liposomes was confirmed by coinjection of the liposomes with an excess of free NGR-peptide, which had a complete inhibition of binding to tumor tissue as a consequence. Antitumor efficacy was studied in a different tumor model of adrenal carcinoma. High weekly doses yielded increased efficacy of the targeted formulation in combination with a decreased vascularization. Frequent low dosing was even shown to result in a complete eradication of tumors. The authors postulate that the targeted liposome formulation has a dual mechanism of action: destruction of tumor vasculature by EC-delivered doxorubicin and killing of tumor cells by doxorubicin released in the tumor interstitium. This attack on two fronts has the advantage that it also attacks viable tumor cells that can survive after successful collapse of tumor vasculature.

Ala-Pro-Arg-Pro-Gly–Mediated Targeting

The use of phages to find peptide motifs that can accumulate at sites of angiogenesis has yielded a number of peptides for which it is unclear to which target receptor the peptide motifs bind. For Ala-Pro-Arg-Pro-Gly (APRPG), the target receptor has not been identified as yet (56). Nevertheless, APRPG-peptide modified liposomes showed binding to human umbilical vein ECs in vitro and to angiogenic ECs in murine tumor models in vivo. Overall, tumor accumulations of targeted and untargeted liposomes were not different, but the intratumoral distribution had markedly changed to a preferential uptake by endothelium for APRPG-liposomes. In addition, APRPG-modified liposomes containing doxorubicin or 5'-O-dipalmitoyl-phosphatidyl 2'-C-cyano-2'-deoxy-1-beta-D-arabino-pentofuranosylcytosine effectively inhibited tumor growth in several murine tumor models (57,58).

CHARGE-BASED TARGETING OF LIPOSOMES

Inclusion of lipids that bear a net positive charge into the liposome lipid composition has been shown to result in preferential accumulation of liposomes in neovasculature (59,60). One of the first studies, performed by McLean et al., focused on liposomes composed of cationic 1-[2-[9-(Z)-octadecenoyloxy]]-2-[8](Z)-heptadecenyl]-3-[hydroxyethyl] imidazolinium chloride:cholesterol complexed to DNA in healthy mice (61). After IV administration, the cationic lipid complexes showed aggregation in the circulation through interactions with blood components leading to a rapid clearance by the mononuclear phagocyte system. In addition to this rapid uptake by macrophages in liver and spleen, certain ECs exhibited pronounced uptake of the cationic liposomes as well. It was shown that the endothelium in capillaries of lungs, ovaries, anterior pituitary, and in the high venules of lymph nodes were primarily responsible for liposome uptake, whereas ECs in other organs only took up few liposomes. Binding to ECs was already observed at five minutes postinjection. Bound liposomes were subsequently internalized and processed through the endosomal/lysosomal pathway within four hours. The authors concluded from the specific distribution pattern of liposomal uptake by the endothelium that a heterogeneously distributed endothelial membrane receptor was responsible for the uptake of cationic complexes. Nevertheless, earlier studies have demonstrated that the EC can be an important cell type for clearance of foreign particles, like polystyrene beads and bacteria, from the blood stream. As such, the distribution pattern of liposome uptake may reflect activity of the endothelium in the clearance of foreign material rather than distribution of a theoretical receptor (62). These studies showed that normal endothelium could bind and internalize liposomes bearing a cationic charge. This is an important finding when employing cationic liposomes for targeting neovasculature, as specific uptake of these liposomes by normal endothelium can occur and can produce side effects.

Cationic albumin-functionalized sterically stabilized liposomes have been studied for targeting brain ECs. The presence of cationic albumin indeed induced binding and internalization by these cells; however, for in vitro systems, cationic charge usually promotes interaction with any cell type. As in vivo observations are lacking, it is uncertain whether targeting of brain endothelium would occur in animals (63).

In murine models of pancreatic islet cell carcinoma or chronic *Mycoplasma pulmonis*-induced airway inflammation, cationic liposomes were shown to be preferentially taken up by the activated (angiogenic) endothelium. Degree of uptake in these areas was approximately 15- to 30-fold higher than by quiescent endothelium in disease-free animals (64). The majority of EC-associated liposomes were already internalized at 20 minutes postinjection. Within the angiogenic endothelium, uptake was not homogenous. Certain areas displayed pronounced uptake, whereas uptake in other regions was much lower. This heterogeneity in angiogenic EC uptake may reflect differences in phase of angiogenesis and EC activity.

Campbell et al. investigated intratumoral distribution of EC-targeted cationically charged PEG-coated liposomes (65). Tumor uptake was similar for liposomes with or without the cationic charge. However, intravital microscopy revealed that by increasing the cationic lipid content from 10 to 50 mol% the degree of liposomal uptake by the tumor vasculature increased two-fold (66). The authors suggest that cationic liposomes will interact preferentially with tumor endothelium because of the slow and irregular tumor blood flow. In addition, tumor vessels are tortuous and leaky (65,67). As a result, cationic liposomes have more opportunities to interact with anionic structures, like proteoglycans, on the angiogenic endothelium than with these structures in normal blood vessels where blood velocity is higher.

Cationic liposomes loaded with drugs have been investigated for therapeutic efficacy in preclinical models. In a model of amelanotic hamster melanoma grown in a dorsal skinfold window chamber, cationic liposome-encapsulated paclitaxel effectuated strong inhibition of tumor growth (66). At a dose of 5 mg/kg body weight of paclitaxel in cationic liposomes, tumor volume was approximately 6-fold, 6-fold, and 10-fold lower as compared to Taxol, empty cationic liposomes, or buffer-treated animals, respectively, and significantly delayed local lymph node metastasis. The observation that control treatments (paclitaxel and empty cationic liposomes) provide a modest therapeutic effect on their own may suggest that the antitumor effects for paclitaxel-loaded cationic liposomes represent a merely additive effect. Nevertheless, it may also be the result of angiogenic EC delivery of paclitaxel by cationic liposomes. The focus of a subsequent study was the mechanism of action of paclitaxel-loaded cationic liposomes in this model. Analysis of microvessels in the tumor showed that functional vessel density was reduced by liposomal paclitaxel. After treatment, vessel diameters were smaller, leading to reduced blood flow resulting in a reduced microcirculatory perfusion index. Staining for apoptosis revealed that the lower index was mainly associated with microvessels in treated tumors, indicating that the changes in microcirculation are the result of cytotoxic effects on tumor endothelium (68).

Kunstfeld et al. used a humanized SCID mouse melanoma model, in which human melanoma cells grow on human dermis and are partly supported by human microvasculature (69). In this model, cationic liposomal paclitaxel reduced tumor growth and tumor invasiveness and improved the lifespan of the mice. Interestingly, by measuring the mitotic index of endothelium in vivo, it was demonstrated that cationic liposome-encapsulated paclitaxel particularly reduced EC proliferation.

Similar observations were made in a model of Meth-A sarcoma where porphyrins were delivered by cationic liposomes to the mouse tumor vasculature. After laser irradiation of the tumor, neovascular destruction was seen with concomitant reduced tumor growth along with a prolonged survival time of the mice. Immunohistochemistry was used to confirm that antitumor effects were related to the destruction of angiogenic endothelium resulting in tumor cell apoptosis (70).

CONCLUSIONS

Targeting liposomes to tumor vasculature has been shown to be a viable strategy for antiangiogenic treatment in preclinical models. Other diseases where angiogenesis plays a role and where targeting liposomes to the vasculature may be beneficial, such as rheumatoid arthritis or neovascularization of the eye, have not been investigated to the same extent as tumor angiogenesis. Clearly, there are still many opportunities to explore liposomal targeting of pathological vasculature for other diseases as well.

An important point to address in the development of angiogenesis-targeted liposomes for clinical application is the heterogeneity of tumor vasculature. Preclinical studies are usually performed in rodents bearing tumors that are generally synchronized, strongly dependent on angiogenesis, and rapidly proliferating. In patients, however, different regions of the tumor can go through different phases the angiogenic process. At the same time, the proliferation rate and, thereby, dependency on angiogenesis may be lower. Therefore, liposomes need to be administered at the right time and equipped with a variety of ligands to attack various phases of angiogenesis at the same time. In this perspective, gene profiling of tumor vasculature has identified a number of targets that have not been explored for angiogenesis-targeted liposomes as yet (71–74). Together with previously investigated target proteins, liposomes may be tailored to attack different phases of the angiogenic process, to destroy or inhibit tumor vasculature, or, ultimately, to achieve vessel normalization (75).

REFERENCES

1. Tanaka T, Shiramoto S, Miyashita M, Fujishima Y, Kaneo Y. Tumor targeting based on the effect of enhanced permeability and retention (EPR) and the mechanism of receptor-mediated endocytosis (RME). Int J Pharm 2004; 277(1–2): 39–61.

2. Caballero AE. Endothelial dysfunction in obesity and insulin resistance: a road to diabetes and heart disease. Obes Res 2003; 11(11):1278–1289.
3. Middleton J, Americh L, Gayon R, et al. Endothelial cell phenotypes in the rheumatoid synovium: activated, angiogenic, apoptotic and leaky. Arthritis Res Ther 2004; 6(2):60–72.
4. Jain RK. Molecular regulation of vessel maturation. Nat Med 2003; 9(6):685–693.
5. Carmeliet P. Angiogenesis in health and disease. Nat Med 2003; 9(6):653–660.
6. Wick G, Knoflach M, Xu Q. Autoimmune and inflammatory mechanisms in atherosclerosis. Annu Rev Immunol 2004; 22:361–403.
7. Morishita Y, Sakube Y, Sasaki S, Ishibashi K. Molecular mechanisms and drug development in aquaporin water channel diseases: aquaporin superfamily (superaquaporins): expansion of aquaporins restricted to multicellular organisms. J Pharmacol Sci 2004; 96(3):276–279.
8. Siemann DW, Chaplin DJ, Horsman MR. Vascular-targeting therapies for treatment of malignant disease. Cancer 2004; 100(12):2491–2499.
9. Szekanecz Z, Koch AE. Vascular endothelium and immune responses: implications for inflammation and angiogenesis. Rheum Dis Clin North Am 2004; 30(1):97–114.
10. Alessi P, Ebbinghaus C, Neri D. Molecular targeting of angiogenesis. Biochim Biophys Acta 2004; 1654(1):39–49.
11. Bergers G, Benjamin LE. Tumorigenesis and the angiogenic switch. Nat Rev Cancer 2003; 3(6):401–410.
12. Vaupel P. The role of hypoxia-induced factors in tumor progression. Oncologist 2004; 9(suppl 5):10–17.
13. Covello KL, Simon MC. HIFs, hypoxia, and vascular development. Curr Top Dev Biol 2004; 62:37–54.
14. Brahimi-Horn MC, Pouyssegur J. The hypoxia-inducible factor and tumor progression along the angiogenic pathway. Int Rev Cytol 2005; 242:157–213.
15. Cross MJ, Claesson-Welsh L. FGF and VEGF function in angiogenesis: signalling pathways, biological responses and therapeutic inhibition. Trends Pharmacol Sci 2001; 22(4):201–207.
16. Handsley MM, Edwards DR. Metalloproteinases and their inhibitors in tumor angiogenesis. Int J Cancer 2005; 115(6):849–860.
17. Hwang R, Varner J. The role of integrins in tumor angiogenesis. Hematol Oncol Clin North Am 2004; 18(5):991–1006, vii.
18. Stupack DG, Cheresh DA. Integrins and angiogenesis. Curr Top Dev Biol 2004; 64:207–238.
19. Kannagi R, Izawa M, Koike T, Miyazaki K, Kimura N. Carbohydrate-mediated cell adhesion in cancer metastasis and angiogenesis. Cancer Sci 2004; 95(5):377–384.
20. Thurston G. Role of Angiopoietins and Tie receptor tyrosine kinases in angiogenesis and lymphangiogenesis. Cell Tissue Res 2003; 314(1):61–68.
21. Yu J, Ustach C, Kim HR. Platelet-derived growth factor signaling and human cancer. J Biochem Mol Biol 2003; 36(1):49–59.
22. Ran S, Huang X, Downes A, Thorpe PE. Evaluation of novel antimouse VEGFR2 antibodies as potential antiangiogenic or vascular targeting agents for tumor therapy. Neoplasia 2003; 5(4):297–307.

23. Ruoslahti E. Specialization of tumour vasculature. Nat Rev Cancer 2002; 2(2):83–90.
24. Zurita AJ, Arap W, Pasqualini R. Mapping tumor vascular diversity by screening phage display libraries. J Control Release 2003; 91(1–2):183–186.
25. Carson-Walter EB, Watkins DN, Nanda A, Vogelstein B, Kinzler KW, St Croix B. Cell surface tumor endothelial markers are conserved in mice and humans. Cancer Res 2001; 61(18):6649–6655.
26. Gulubova MV. Expression of cell adhesion molecules, their ligands and tumour necrosis factor alpha in the liver of patients with metastatic gastrointestinal carcinomas. Histochem J 2002; 34(1–2):67–77.
27. Chiu GN, Bally MB, Mayer LD. Targeting of antibody conjugated, phosphatidylserine-containing liposomes to vascular cell adhesion molecule 1 for controlled thrombogenesis. Biochim Biophys Acta 2003; 1613(1–2):115–121.
28. Chiu GN, Bally MB, Mayer LD. Effects of phosphatidylserine on membrane incorporation and surface protection properties of exchangeable poly(ethylene glycol)-conjugated lipids. Biochim Biophys Acta 2002; 1560(1–2):37–50.
29. Fonsatti E, Altomonte M, Nicotra MR, Natali PG, Maio M. Endoglin (CD105): a powerful therapeutic target on tumor-associated angiogenetic blood vessels. Oncogene 2003; 22(42):6557–6563.
30. Fonsatti E, Maio M. Highlights on endoglin (CD105): from basic findings towards clinical applications in human cancer. J Transl Med 2004; 2(1):18.
31. Volkel T, Holig P, Merdan T, Muller R, Kontermann RE. Targeting of immunoliposomes to endothelial cells using a single-chain Fv fragment directed against human endoglin (CD105). Biochim Biophys Acta 2004; 1663(1–2):158–166.
32. Kondo M, Asai T, Katanasaka Y, et al. Anti-neovascular therapy by liposomal drug targeted to membrane type-1 matrix metalloproteinase. Int J Cancer 2004; 108(2):301–306.
33. Osenkowski P, Toth M, Fridman R. Processing, shedding, and endocytosis of membrane type 1-matrix metalloproteinase (MT1-MMP). J Cell Physiol 2004; 200(1):2–10.
34. Sounni NE, Baramova EN, Munaut C, et al. Expression of membrane type 1 matrix metalloproteinase (MT1-MMP) in A2058 melanoma cells is associated with MMP-2 activation and increased tumor growth and vascularization. Int J Cancer 2002; 98(1):23–28.
35. Holmbeck K, Bianco P, Caterina J, et al. MT1-MMP-deficient mice develop dwarfism, osteopenia, arthritis, and connective tissue disease due to inadequate collagen turnover. Cell 1999; 99(1):81–92.
36. Chiarugi V, Magnelli L, Dello Sbarba P, Ruggiero M. Tumor angiogenesis: thrombin and metalloproteinases in focus. Exp Mol Pathol 2000; 69(1):63–66.
37. Lafleur MA, Handsley MM, Edwards DR. Metalloproteinases and their inhibitors in angiogenesis. Expert Rev Mol Med 2003; 2003:1–39.
38. Ottino P, Finley J, Rojo E, et al. Hypoxia activates matrix metalloproteinase expression and the VEGF system in monkey choroid-retinal endothelial cells: involvement of cytosolic phospholipase A2 activity. Mol Vis 2004; 10:341–350.
39. Hodivala-Dilke KM, Reynolds AR, Reynolds LE. Integrins in angiogenesis: multitalented molecules in a balancing act. Cell Tissue Res 2003; 314(1): 131–144.

40. Jin H, Varner J. Integrins: roles in cancer development and as treatment targets. Br J Cancer 2004; 90(3):561–565.
41. Koning GA, Fretz MM, Woroniecka U, Storm G, Krijger GC. Targeting liposomes to tumor endothelial cells for neutron capture therapy. Appl Radiat Isot 2004; 61(5):963–967.
42. Schiffelers RM, Koning GA, ten Hagen TL, et al. Anti-tumor efficacy of tumor vasculature-targeted liposomal doxorubicin. J Control Release 2003; 91(1–2): 115–122.
43. Janssen AP, Schiffelers RM, ten Hagen TL, et al. Peptide-targeted PEG-liposomes in anti-angiogenic therapy. Int J Pharm 2003; 254(1):55–58.
44. Fahr A, Muller K, Nahde T, Muller R, Brusselbach S. A new colloidal lipidic system for gene therapy. J Liposome Res 2002; 12(1–2):37–44.
45. Muller K, Nahde T, Fahr A, Muller R, Brusselbach S. Highly efficient transduction of endothelial cells by targeted artificial virus-like particles. Cancer Gene Ther 2001; 8(2):107–117.
46. Scott ES, Wiseman JW, Evans MJ, Colledge WH. Enhanced gene delivery to human airway epithelial cells using an integrin-targeting lipoplex. J Gene Med 2001; 3(2):125–134.
47. Nahde T, Muller K, Fahr A, Muller R, Brusselbach S. Combined transductional and transcriptional targeting of melanoma cells by artificial virus-like particles. J Gene Med 2001; 3(4):353–361.
48. Lestini BJ, Sagnella SM, Xu Z, et al. Surface modification of liposomes for selective cell targeting in cardiovascular drug delivery. J Control Release 2002; 78(1–3):235–247.
49. Jain S, Mishra V, Singh P, Dubey PK, Saraf DK, Vyas SP. RGD-anchored magnetic liposomes for monocytes/neutrophils-mediated brain targeting. Int J Pharm 2003; 261(1–2):43–55.
50. Kurohane K, Namba Y, Oku N. Liposomes modified with a synthetic Arg-Gly-Asp mimetic inhibit lung metastasis of B16BL6 melanoma cells. Life Sci 2000; 68(3):273–281.
51. Harvie P, Dutzar B, Galbraith T, et al. Targeting of lipid-protamine-DNA (LPD) lipopolyplexes using RGD motifs. J Liposome Res 2003; 13(3–4): 231–247.
52. Sato Y. Role of aminopeptidase in angiogenesis. Biol Pharm Bull 2004; 27(6):772–776.
53. Pasqualini R, Koivunen E, Kain R, et al. Aminopeptidase N is a receptor for tumor-homing peptides and a target for inhibiting angiogenesis. Cancer Res 2000; 60(3):722–727.
54. Arap W, Pasqualini R, Ruoslahti E. Cancer treatment by targeted drug delivery to tumor vasculature in a mouse model. Science 1998; 279(5349):377–380.
55. Pastorino F, Brignole C, Marimpietri D, et al. Vascular damage and anti-angiogenic effects of tumor vessel-targeted liposomal chemotherapy. Cancer Res 2003; 63(21):7400–7409.
56. Oku N, Asai T, Watanabe K, et al. Anti-neovascular therapy using novel peptides homing to angiogenic vessels. Oncogene 2002; 21(17):2662–2669.
57. Asai T, Shimizu K, Kondo M, et al. Anti-neovascular therapy by liposomal DPP-CNDAC targeted to angiogenic vessels. FEBS Lett 2002; 520(1–3):167–170.

58. Asai T, Oku N. Liposomalized oligopeptides in cancer therapy. Methods Enzymol 2005; 391:163–176.
59. Simberg D, Weisman S, Talmon Y, Faerman A, Shoshani T, Barenholz Y. The role of organ vascularization and lipoplex-serum initial contact in intravenous murine lipofection. J Biol Chem 2003; 278(41):39858–39865.
60. Dass CR. Improving anti-angiogenic therapy via selective delivery of cationic liposomes to tumour vasculature. Int J Pharm 2003; 267(1–2):1–12.
61. McLean JW, Fox EA, Baluk P, et al. Organ-specific endothelial cell uptake of cationic liposome-DNA complexes in mice. Am J Physiol 1997; 273(1 Pt 2): H387–H404.
62. Ryan U. Phagocytic properties of endothelial cells. In: Ryan U, ed. Endothelial Cells. Vol. 3. Boca Raton: CRC Press Inc., 1988:33–49.
63. Thole M, Nobmanna S, Huwyler J, Bartmann A, Fricker G. Uptake of cationised albumin coupled liposomes by cultured porcine brain microvessel endothelial cells and intact brain capillaries. J Drug Target 2002; 10(4):337–344.
64. Thurston G, McLean JW, Rizen M, et al. Cationic liposomes target angiogenic endothelial cells in tumors and chronic inflammation in mice. J Clin Invest 1998; 101(7):1401–1413.
65. Camspbell RB, Fukumura D, Brown EB, et al. Cationic charge determines the distribution of liposomes between the vascular and extravascular compartments of tumors. Cancer Res 2002; 62(23):6831–6836.
66. Krasnici S, Werner A, Eichhorn ME, et al. Effect of the surface charge of liposomes on their uptake by angiogenic tumor vessels. Int J Cancer 2003; 105(4):561–567.
67. Jain RK. The next frontier of molecular medicine: delivery of therapeutics. Nat Med 1998; 4(6):655–657.
68. Strieth S, Eichhorn ME, Sauer B, et al. Neovascular targeting chemotherapy: encapsulation of paclitaxel in cationic liposomes impairs functional tumor microvasculature. Int J Cancer 2004; 110(1):117–124.
69. Kunstfeld R, Wickenhauser G, Michaelis U, et al. Paclitaxel encapsulated in cationic liposomes diminishes tumor angiogenesis and melanoma growth in a "humanized" SCID mouse model. J Invest Dermatol 2003; 120(3):476–482.
70. Takeuchi Y, Kurohane K, Ichikawa K, Yonezawa S, Nango M, Oku N. Induction of intensive tumor suppression by antiangiogenic photodynamic therapy using polycation-modified liposomal photosensitizer. Cancer 2003; 97(8): 2027–2034.
71. Nanda A, St Croix B. Tumor endothelial markers: new targets for cancer therapy. Curr Opin Oncol 2004; 16(1):44–49.
72. Nanda A, Carson-Walter EB, Seaman S, et al. TEM8 interacts with the cleaved C5 domain of collagen alpha 3(VI). Cancer Res 2004; 64(3):817–820.
73. Brekken RA, Thorpe PE. VEGF-VEGF receptor complexes as markers of tumor vascular endothelium. J Control Release 2001; 74(1–3):173–181.
74. Donate F, Juarez JC, Guan X, et al. Peptides derived from the histidine-proline domain of the histidine-proline-rich glycoprotein bind to tropomyosin and have antiangiogenic and antitumor activities. Cancer Res 2004; 64(16):5812–5817.
75. Jain RK. Normalization of tumor vasculature: an emerging concept in antiangiogenic therapy. Science 2005; 307(5706):58–62.

8

Targeting Liposomes to Endothelial Cells in Inflammatory Diseases

Jan A. A. M. Kamps and Grietje Molema
*Medical Biology Section, Department of Pathology and
Laboratory Medicine, Endothelial Biomedicine and Vascular Drug Targeting,
Groningen University Medical Center, University of Groningen,
Groningen, The Netherlands*

INTRODUCTION

The endothelium covers the vascular wall of all blood vessels in the body and comprises 10^{13} endothelial cells (ECs) in an adult (1). Their excellent accessibility for drugs present in the systemic circulation and their involvement in a large variety of physiological and pathophysiological processes make ECs ideal targets for targeted liposome-mediated drug delivery. The heterogeneity of the endothelium with respect to appearance and function allows for drug delivery approaches that are either organ and/or disease specific. Despite these features, research on liposome-mediated drug delivery to ECs is limited. This is undoubtedly related to the fact that ECs are generally refractory to liposome uptake (2) and that the use of specific targeting devices are a prerequisite for liposome uptake by selected cell subsets.

In this chapter, we discuss the role of ECs in health and disease, and summarize the strategies that have been applied so far to selectively deliver liposomes and their contents into different EC populations. In the methodology section, we address liposome technology related to EC targeting systems generally applied for in vitro and in vivo studies, and advanced methods to establish EC localization and (therapeutic) effects of targeted liposomal formulations.

Endothelial Cell Function in Health and Disease

The blood vessels in our body are highly heterogeneous with regard to architecture and function. The larger blood vessels consist of ECs, smooth muscle cells, connective tissue, and elastic elements, and are mainly responsible for transport of blood through the body and blood pressure control (3). In the microvascular capillary bed of organs, the ECs reside on a basal lamina and are only supported by sparsely distributed pericytes that facilitate vascular integrity (4,5).

ECs exert four main functions (Fig. 1). They serve as a (semipermeable) barrier for transport of soluble molecules from the blood into the underlying tissues, and maintain hemostatic balances via the production of anticoagulants such as thrombomodulin and tissue factor pathway inhibitor, and procoagulants including von Willebrand factor and tissue factor (3).

Coordinated recruitment of neutrophils, lymphocytes, and monocytes into tissues is another essential task of ECs to facilitate efficient immune responses during an inflammatory insult (6). This process takes place in the postcapillary venules in most microvascular beds except those of spleen, lungs, and liver. Activation of ECs by proinflammatory cytokines such as tumor necrosis factor (TNF)α and interleukin (IL)-1β lead to the production of a series of cell adhesion molecules, chemokines, and cytokines. As a result, leukocytes are guided into the underlying tissue and costimulated during transmigration to become fully activated.

The fourth function of ECs is to actively participate in neovascularization. In chronic inflammatory diseases, the ongoing leukocyte recruitment, cellular activation, and cell death induction all put a heavy demand on local oxygen supply. Consequently, new blood vessel formation or angiogenesis takes place. Similarly, hypoxic conditions in growing tumors and/or the unbalance in production of pro- and antiangiogenic factors via oncogenic transformation of the tumor cells can activate capillary bed ECs to become proangiogenic.

For the development of effective therapeutic entities aimed at selectively interfering with ECs located in inflammatory or angiogenic sites, detailed knowledge regarding the molecular control of their behavior during disease initiation and progression is required. By this means, both suitable intracellular pharmacological targets and transmembrane expressed targets for the drug-delivery systems can be identified. Due to space limitations, a short summary of the state-of-the-art knowledge on drug targets and cellular targets for immunoliposomes is given. For reviews, the reader is referred to references (7–12).

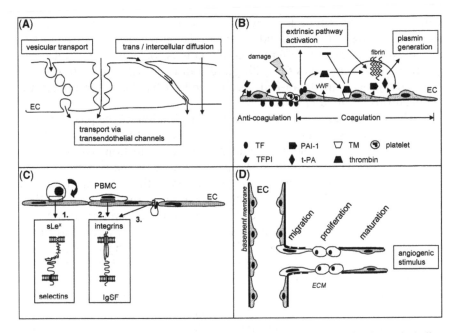

Figure 1 ECs exert several important functions in the body. (**A**) The endothelium forms a semipermeable barrier for the transport of blood-borne peptides, proteins, and other soluble molecules to underlying tissue; (**B**) via the regulated expression of pro- and anticoagulative activities endothelium actively participates in the hemostatic balance in the body; (**C**) under the influence of proinflammatory cytokines, ECs upregulate a variety of cellular adhesion molecules, cytokines, chemokines, and coagulation factors (the latter groups are not depicted) to tether and activate leukocytes and facilitate leukocyte adhesion and transmigration from the blood into the tissue; (**D**) during wound healing and tumor growth, among others, angiogenesis takes place. In this process, an active role exists for ECs. *Abbreviations*: EC, endothelial cell; ECM, extracellular matrix; IgSF, Ig superfamily; PAI-1, plasminogen activator inhibitor-I; PBMC, peripheral blood mononuclear cells; sLe^x, sialyl Lewis X; TFPI, tissue factor pathway inhibitor; TM, thrombomodulin; t-PA, tissue-type plasminogen activator. *Source*: From Ref. 6.

Endothelial Cell Activation in Inflammatory Diseases

One of the first steps in the immune response during inflammation is cytokine-mediated activation of ECs. As a result, P-selectin, E-selectin, vascular cell adhesion molecule-1 (VCAM-1), and intercellular adhesion molecules (ICAM-1), -2, and -3, are coordinately expressed to interact with their counterligands on the leukocytes. These ligands include sialyl Lewis X for the selectins, and the integrins very late antigen-4 and leukocyte function antigen-1, respectively, for VCAM-1 and ICAM-1. Together with cytokines and chemokines such as IL-6, IL-8, and Fraktalkine, the cell adhesion molecules actively initiate leukocyte–EC contact and facilitate subsequent leukocyte rolling, activation, and transendothelial migration. Different leukocyte subsets employ different combinations of adhesion molecules, cytokines, and chemokines for rolling and transendothelial migration, while also heterogeneity in organ-bed–specific endothelial behavior affects the combinations required (13,14).

The above processes are often initiated by binding of the proinflammatory cytokines TNFα and/or IL-1β to their respective receptors. These cytokines activate multiple intracellular signal transduction pathways that lead to an inflammatory state of the EC (15). One of the common downstream effects of both cytokines is the activation of the transcription factor nuclear factor κB (NFκB). As a result of binding of the cytokines to their receptors, the inhibitor of κB (IκB) kinase complex phosphorylates NFκB complexed IκB. Phosphorylation labels IκB for ubiquitinylation and subsequent proteasome degradation, leading to the unmasking of the nuclear localization sequence of NFκB (8,16). In the nucleus, NFκB binds to NFκB B consensus sites in the promoter sequence of selected genes, allowing subsequent interaction with coactivators and other components of the gene transcription machinery. As a result, ECs start to express an array of functionally related genes, including the proinflammatory adhesion molecules, cytokines, cyclooxygenase-2 (COX-2), and inducible nitric oxide synthase (iNOS) (17–19).

Another signal transduction pathway common to both TNFα and IL-1β receptor activation in ECs involves the mitogen-activated protein kinase (MAPK) pathway. p38 MAPK is the main family member associated with EC activation in inflammation (20). It is involved in TNFα-driven NFκB activation and controls mRNA stability of various inducible cytokines that are short-lived. Also IL-1–induced IL-6, COX-2, collagenase-1, and stomelysin-1 were shown to be p38 MAPK controlled in human umbilical vein ECs (HUVEC) (21).

In inflamed sites in rheumatoid arthritis patients, especially the p38 MAPK family member was present in ECs (22), while also activated NFκB could be immunohistochemically detected in these cells (23). Similarly, in human glomerulonephritis, activated p38 MAPK was present in glomerular ECs (24). In inflammatory bowel disease, the pharmacological effects of glucocorticoid therapy were paralleled by diminished NFκB activity in, among others, ECs (25).

The sinusoidal ECs in the liver (representing approximately 20% of the total number of liver cells) are characterized by specialized functions based on their location in the liver. Sinusoidal ECs play a major role in maintaining metabolic homeostasis between the liver and other organs. Via endothelial fenestrae, the size and number of which vary under the influence of different agents (26), they control diffusion of endogenous macromolecules and particulate matter to the space of Disse and therefore to the parenchymal cells. For instance, the access of lipoproteins carrying cholesterol and vitamin A to hepatocytes and stellate cells is controlled by the sinusoidal ECs. In liver cirrhosis, loss of fenestrae and production of a fibrous collagen matrix occur. This may unfavorably influence the exchange of nutrients (27). Liver ECs have a high endocytic and digestive capacity. By virtue of a variety of receptors present on these cells, they are heavily involved in the clearance of various substances from the blood circulation. Receptors reported on theses cells include scavenger receptors AI, AII, and BI (28), collagen α-chain receptor, Fc receptor, and mannose receptor (29). The fenestrated endothelial lining of the sinusoid allows the free exchange of fluids and solutes, and small particles from the blood to hepatocytes. In addition, transferrin, insulin, and ceruloplasmin are taken up by ECs and transcytosed to the parenchymal cells.

Similar to the microvascular capillary bed endothelium elsewhere in the body, sinusoidal liver ECs express ICAM-1 and VCAM-1 for the recruitment of neutrophils and lymphocytes to the liver sinusoids during inflammatory processes (30). In contrast to the situation in microvascular beds in other organs, however, in the liver E-selectin and P-selectin are often not expressed to participate in this process (30). Among the cytokines produced by liver ECs upon activation are prostaglandin I_2, prostaglandin E_2, IL-1, IL-6, and interferon (29). Together with the Kupffer cells, liver ECs are assumed to be the most important resident cell population involved in the local production of inflammatory mediators.

Understanding the molecular control of EC activation during inflammatory conditions, and the effects thereof on the expression of membrane markers, is a prerequisite in the design of vascular-directed drug-targeting strategies. Both the choice of the drug to be included in the delivery vehicle and the target epitopes aimed at on the cell membrane of the inflamed endothelium are dependent on this knowledge. Furthermore, the complex profile of gene expression in the endothelium during inflammation can provide us with disease activity markers as read-out parameters for assessment of therapeutic success of the drug-delivery strategy.

Molecular Control of Angiogenesis During Inflammation

Active neovascularization is a hallmark of chronic inflammation (31–33). Although the molecular mechanisms of angiogenesis have been most extensively studied in neoplastic growth, the general molecular and cell biological concepts of angiogenesis are thought to be comparable irrespective of the nature of the disease.

During angiogenic sprouting EC activation, migration and proliferation takes place in a highly orchestrated manner. Often, local hypoxia is the initiator of vascular endothelial growth factor (VEGF) production, although leukocytes can also deposit considerable amounts of this growth factor in the inflamed tissue. VEGF signals mainly through VEGF receptor-2 to activate a network of kinases and other downstream effectors, leading to EC migration, proliferation, and survival (34). Induction of iNOS activity is an essential step in early stage vessel dilation, acting in conjunction with the release of vascular endothelial (VE)-cadherin from its actin anchor. As a consequence, endothelial permeability will increase (35). Subsequently, the basement membrane of the capillary bed is degraded by matrix metalloproteinases, allowing serum components to leak out of the vessel and to form a provisional matrix onto which ECs can migrate into the tissue. Upon hypoxia, local angiopoietin (Ang)-2 production is also augmented, as a result of which it can compete with Ang-1 for their mutual Tie-2 receptor. While Ang-1 stabilizes capillaries, Ang-2–Tie-2 interaction facilitates ECs to become responsive to growth factors (36). During the course of neovessel formation, prevention of endothelial apoptosis is effected by VEGF-R and αvβ3 integrin–mediated signal transduction. αvβ3 integrin ligation furthermore upregulates matrix metalloproteinase-2 to permit EC invasion (37).

After cellular proliferation and new vessel formation, functional blood vessels should form. This maturation process is controlled by the recruitment of pericytes and the formation of new extracellular matrix components. Growth factors intricately associated with vessel maturation include platelet-derived growth factor, transforming growth factor-β, Ang-1, and sphingosine-phosphate (38).

Drugs Interfering with Endothelial Cell Function in Inflammation

Various types of drugs have been investigated for their activity in interfering with EC function in inflammation. They include antibodies and small chemical entities that block leukocyte–EC interaction, and neutralize cytokines and drugs that interfere with intracellular signalling events associated with EC activation. Among the latter are NFκB inhibitors, p38 MAPK inhibitors, glucocorticosteroids, nonsteroidal anti-inflammatory drugs, and tyrosine kinase inhibitors (7–9,39). Other experimental approaches include the development of antisense oligonucleotides (ODN), small interference RNAs, and plasmid DNA to modulate EC function (40,41).

It is now well recognized that besides the development of new drugs with improved selectivity for their molecular targets, proper delivery of the drugs has become a prerequisite for treatment success (42). While biotech-derived drugs often need protection from degradation in biological conditions (e.g., antisense ODN and plasmids encoding therapeutic genes are rapidly degraded in the systemic circulation if not protected from the responsible enzymes), small chemical inhibitors of signal transduction pathways often interfere with their targets in nondiseased cells (43,44). The resulting toxicity is unacceptable and hampers further development of highly potent new chemical entities into clinically applicable drugs. Incorporation of the drugs into targeted liposomes specifically delivering their cargo in the (endothelial) cells offers important opportunities in this respect. Although not yet explored in great detail for EC-specific delivery of inhibitors of cell activation in inflammatory diseases, prospects for further development in this field of application are excellent.

LIPOSOMES TARGETED TO ENDOTHELIAL CELLS

Macrophages in liver (Kupffer cells) and in spleen are the most important cells in the uptake of nontargeted liposomes from the blood. Depending on liposome size and composition, hepatocytes also can contribute substantially to liposome uptake from the blood (45). Despite their obvious accessibility from the blood circulation, ECs seem to be refractory for uptake of nontargeted liposomes. However, this is not an intrinsic feature of ECs. We demonstrated that liver ECs cultured in serum-free conditions are highly capable of endocytosing considerable amounts of negatively charged liposomes containing phosphatidylserine via a scavenger receptor–mediated pathway (46). Also in the intact liver in a serum-free perfused rat liver set-up, liposomes containing phosphatidylserine were readily cleared from the perfusate by liver ECs (47). In vivo plasma-derived proteins effectively shield the negative charges of the phosphatidylserine headgroup, thereby preventing proper interaction of the liposomes with the scavenger receptors on the ECs.

For the delivery of drugs into ECs in an inflamed tissue, modifications of the liposomes are necessary to circumvent uptake by cells of the mononuclear phagocyte system and to create specificity for the activated ECs. One of the main determinants for therapeutic success of drug targeting is the selectivity of the cellular target molecule in combination with the selectivity of the homing ligand in the targeting construct. In theory, any protein expressed in the membrane of the ECs can serve as a target provided it is absent from other (endothelial) cells in the body. Essential in this respect is knowledge on the expression patterns of the epitopes by other vascular beds and by other cells in the body. By this means, one can either ensure cell selectivity of the target or identify the location in the body where potential side effects could occur. Furthermore, knowledge regarding cellular handling after ligand binding is essential, as it determines the choice of effector molecules to be delivered. In the case of using, e.g., bacterial toxins, toxic drugs, or plasmids encoding therapeutic proteins, intracellular delivery is a prerequisite, as their effects are exerted in the cells' interior. In the following paragraphs, we will briefly address the different approaches studied to selectively target liposomes to specific EC (sub)populations.

Liposomes Targeted to Inflammatory Endothelial Cells

As indicated above, various adhesion molecules are specifically (over)expressed by ECs in inflammatory sites. They may, therefore, act as selective targets for liposome-based drug delivery (48). So far, E-selectin has been most widely studied as a target for EC-specific delivery of lipososomes. Specific antibodies against E-selectin as well as sialyl Lewis X mimetics were also employed for this purpose (49–52). Anti-E-selectin–based immunoliposomes are efficiently internalized by activated ECs, followed by release of the entrapped liposomal content (53). In a murine delayed-type hypersensitivity skin-inflammation model, we demonstrated that anti-E-selectin immunoliposomes selectively accumulated in the activated endothelium of the inflamed skin (54).

In a similar approach, anti-VCAM-1 antibodies were employed in immunoliposomes composed of thrombogenic phosphatidylserine to induce blood coagulation (55). The anti-VCAM-1 immunoliposomes showed up to 16-fold higher binding to activated HUVEC as compared to control liposomes. Upon binding to VCAM-1 coated plates, the immunoliposomes were capable of inducing blood coagulation. Application of these types of immunoliposomes may be considered for therapy of solid tumors to selectively induce thrombosis in the tumor vasculature. Whether inflammatory conditions would favor from inhibition of blood flow needs to be established.

Although ICAM-1 is constitutively expressed on ECs, its expression is often upregulated in inflammation-related diseases. In one of the first studies where liposomes were targeted to adhesion molecules on ECs, it was demonstrated that anti-ICAM-1 immunoliposomes bind to ICAM-1 expressing cells in amounts that correlated with the actual ICAM-1 expression (56). Later on, bronchial epithelial cells were shown to internalize anti-ICAM-1 immunoliposomes after binding (57). Recently, it was demonstrated that ECs could also internalize anti-ICAM-1 immunoliposomes. The uptake mechanism involved multimeric interaction of the antibody-targeted particles with membrane expressed ICAM-1, thereby activating a specific uptake route that is different from general receptor-mediated endocytosis (58).

Liposomes Targeted to Angiogenic Endothelial Cells

Delivering the contents of liposomes into angiogenic ECs offers a challenging perspective to interfere with neovascularization processes. Angiogenic ECs (over)express a variety of epitopes that allow differentiation between resting and neovasculature and that can serve as targets for the liposomes (59,60). Although until now focus has been primarily on angiogenic ECs in tumors, the pronounced neovascularization that is taking place in chronic inflammatory diseases justifies investigations on application in these pathologies as well.

Arg-Gly-Asp (RGD)-liposomes targeted to $\alpha v \beta 3$ integrins overexpressed on angiogenic tumor endothelium are among the best described liposomal targeting approaches studied so far (61,62). Synthetic cyclic RGD-peptides have a high affinity for $\alpha v \beta 3$ integrin and can be relatively easily coupled to the distal end of poly(ethylene glycol) (PEG)-grafted liposomes. Thus prepared, RGD-PEG-liposomes loaded with the chemotherapeutic drug doxorubicin were more efficacious over nontargeted liposomes in inhibiting C26 tumor outgrowth in mice. The effects were thought to be a result of direct effects on the tumor endothelium because C26 tumor cells are doxorubicin insensitive (63). Other approaches studied to target liposomes to angiogenic ECs were: in vivo studies using Asn-Gly-Arg (NGR) peptide coupled to the distal end of liposome surface grafted PEG targeted to aminopeptidase N (64), Gly-Pro-Leu-Pro-Leu-Arg (GPLPLR) peptide coupled to the lipid bilayer targeted to membrane type-1 metalloproteinase (65), in vitro studies applying nonpegylated immunoliposomes targeted to the kinase insert domain containing receptor [kinase domain region (KDR or VEGF receptor)] (66), and liposomes conjugated with a single-chain Fv fragment directed to human endoglin (CD105) (67). Cationic liposomes have been shown to preferentially target angiogenic ECs in tumors and chronic inflammation in mice as compared to nonangiogenic ECs (68). Application of this selective delivery is, however, hampered by fast clearance of these liposomes from the blood by liver, spleen, and lung, and by the toxicity of cationic liposomes (69).

Liposomes Targeted to Liver Sinusoidal Endothelial Cells

When human serum albumin (HSA) derivatized with *cis*-aconitic anhydride (Aco-HSA) was covalently coupled to liposomes, it mediated efficient in vivo uptake of these liposomes by scavenger receptors present on liver ECs (70). Within 30 minutes of injection of these targeted liposomes, the liver accounted for more than 80% of the uptake from the blood. Liver ECs contributed for about two-thirds of this liver uptake, and more than 80% of the EC population participated in the uptake of the Aco-HSA liposomes. No uptake of liposomes was observed in ECs elsewhere in the body. The capacity of liver ECs to internalize Aco-HSA liposomes in vivo was dependent on the particle size. When the liposome diameter exceeded 100 nm, EC uptake decreased.

Because liver ECs play an important role in pathophysiological conditions, several attempts were undertaken to modulate their function to interfere with disease-related processes. Therefore, we designed several lipid-based particles based on the scavenger receptor–mediated, Aco-HSA targeting concept (71–73). As effector molecules, antisense oligonucleotides inhibiting ICAM-I expression were incorporated (74–76), after which the Aco-HSA was covalently coupled to the particles. In vivo, up to 66% of the injected dose of these Aco-HSA-lipid-based particles were efficiently taken up by liver ECs. Despite the capability of these targeted particles to exert antisense activity in cultured J774 cells that express scavenger receptors and have an inducible ICAM-1 expression, we were however not able to generate biological activity in vivo (72,73). Apparently, the transfection efficiency of antisense containing particles in vivo is a delicate balance between serum stability of the particle and release of the antisense from the endocytic compartment. This will especially hold true for scavenger receptor–mediated uptake mechanisms because their physiological role is the efficient uptake and intralysosomal destruction of macromolecules.

Besides the targeting of liposomes to ECs in inflammation, several studies investigated liposomal drug targeting to ECs of the blood-brain barrier, in most cases to enable drugs to cross the blood-brain barrier. We will not address this topic here as it is not in the scope of this chapter; the reader is referred to Refs. (77–79) for further reading on this subject.

METHODOLOGY

Many methods involved in the development of liposome-based drug-delivery systems for drug targeting to ECs are derived from general methodology in liposome research. By incorporating ligands specific for activated, pro-inflammatory or proangiogenic endothelium, the desired specificity can be created. There is, however, an important difference between the ECs as targets, and targets such as tumor cells and organ parenchymal cells. The ECs aimed at are lining microvascular capillary beds in a selected part of the body, and present themselves in a single cell layer organized in a complex cell environment. Moreover, in disease conditions, the integrity of the endothelial monolayer is often compromised, leading to increased vascular permeability and enhanced permeability retention–based accumulation of the targeted liposomes. Specialized techniques to visualize the targeted liposomes and/or liposomal contents are, therefore, essential tools to demonstrate proof of localization of the constructs in the endothelium. Determination of EC-specific pharmacological effects of the targeted liposomes is another challenge to demonstrate cell-specific efficacy of the delivered drug. Below we will discuss how to experimentally deal with some of these issues and new advances in tool development for these purposes.

Coupling of Endothelial Cell–Specific Targeting Devices to Liposomes

Modification of the liposome surface is a necessity to selectively deliver liposomal contents into ECs. There are a wide variety of chemical protocols available enabling surface modifications with proteins, peptides, sugars, or polymers. For reviews on liposome surface modifications the reader is referred to more general publications (80,81). The techniques employed, include:

1. Insertion of galactose-terminated PEG chains coupled to long-chain diacyl glyceride for anchoring in the liposome bilayer (82),
2. Coupling of proteins either directly to the lipid surface (83,84) or to the distal end of PEG chains (73,85), using a sulfhydryl-maleimide coupling method,

3. Coupling of antibodies to a PEG-terminal cyanuric chloride (86,87), and
4. Hydrazone linkage of the hydrazide moiety at the distal end of PEG chains to the oxidized carbohydrates in antibodies (88,89).

Although all of these methods have their advantages and disadvantages, the sulfhydryl maleimido coupling methods are convenient methods to conjugate endothelial targeting ligands in our experience. In this procedure, free sulfhydryl groups are introduced in the protein to be coupled to the liposome, using *N*-succinimidyl-*S*-acethylthioacetate (SATA) as a heterobifunctional reagent. After separation of the free SATA from the protein by gel permeation chromatography, the acetylthioacetate-protein conjugate is deacetylated by addition of a freshly prepared solution of 0.5 M hydroxylamine-HCl, 0.5 M Hepes, 25 M ethylenediaminetetraacetate (EDTA), pH 7.0. After deacetylation, the thioacetyl-protein conjugate is allowed to react for four hours at room temperature or overnight at 4°C with liposomes containing either maleimido-4-(*p*-phenylbutyryl) phosphatidylethanolamine as a functionalized lipid or 1,2-distearoyl-sn-glycero-3-phosphoethanolamine-*N*-[methoxy(PEG)-2000]-maleimide as functionalized, bilayer anchored PEG chain. The coupling reaction is stopped by the addition of excess *N*-ethylmaleimide to cap unreacted sulfhydryl groups. Liposomes can be separated from unconjugated protein by metrizamide or OptiPrep® gradient ultracentrifugation (84). This method produces stable and reproducible homing ligand modified liposomes and lipid-based particles. We recently showed that coupling of liver EC-specific Aco-HSA either to the surface of PEG stabilized particles or to the distal end of the PEG chain dramatically influenced the particle's targeting capacity (73). Upon intravenous injection, bilayer coupled Aco-HSA particles were cleared from the blood at the same low rate of untargeted particles, while PEG distal-end coupled particles were rapidly cleared from the blood and taken up by liver ECs (55% of the injected dose after 30 minutes) (Fig. 2). In contrast to bilayer coupled Aco-HSA PEG-stabilized particles, classical liposomes with bilayer coupled Aco-HSA were readily taken up by liver ECs (70). It is likely that this apparent discrepancy is caused by the presence of positively charged lipids in the bilayer of PEG stabilized particles that may (partly) neutralize the negative charges of the anionized albumin.

Studying the Fate of Liposomes: Liposomal Markers

To investigate the in vivo fate of liposomes, numerous liposomal markers are available. Radiolabels, either encapsulated water-soluble compounds or bilayer-incorporated lipid labels, provide a sensitive and powerful tool to determine liposome biodistribution. Most lipids can be purchased in one or more radiolabeled forms. The choice of the radiolabeled compound primarily depends on the nature of the experiment, e.g., long-term or short-term experiments, determination of binding and uptake, or degradation of

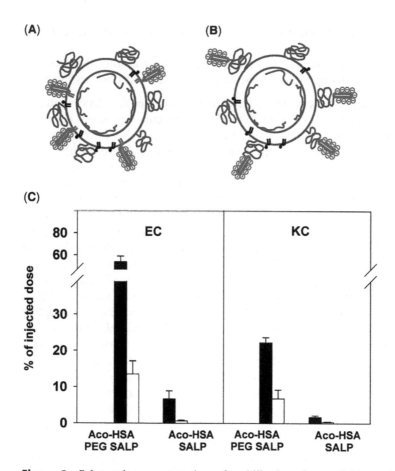

Figure 2 Schematic representation of stabilized antisense lipid particles (SALP) with Aco-HSA coupled to the particle's surface (Aco-HSA SALP) (**A**) and stabilized antisense particles with Aco-HSA coupled to the distal end of surface grafted PEG chains (Aco-HSA PEG SALP) (**B**). The intrahepatic distribution of Aco-HSA SALP and Aco-HSA PEG SALP is shown 30 minutes after intravenous injection into rats (*filled bars*) (**C**). Pretreatment of the rats with polyinosinic acid, an inhibitor of scavenger receptor–mediated uptake, inhibits uptake of Aco-HSA PEG SALP by ECs and KC. *Abbreviations*: SALP, stabilized antisense lipid particles; EC, endothelial cell; KC, Kupffer cell; Aco-HSA, human serum albumin derivatized with *cis*-aconitic anhydride; PEG, poly(ethylene glycol). *Source*: From Ref. 73.

liposomes. Radioactive lipid labels can be incorporated into the liposomal bilayer when mixed with the other lipids during liposome preparation, irrespective of the method of liposome preparation. It is important that a lipid marker is stably incorporated in the liposomal bilayer. This will avoid transfer of the marker from the liposomal membrane to cell membranes or to endosomal membranes, or to serum components such as lipoproteins. In addition to stability requirements, the radioactive marker also has to be metabolically inert during the time interval of the experiment to avoid unjustified interpretations of the experimental data generated. Examples of radioactive markers that fulfill these characteristics are [^{3}H]cholesteryl-hexadecyl ether and [^{3}H]cholesteryloleyl ether. The ether bond in these markers ascertains that, following internalization by cells, the markers are not significantly metabolized within a period of days. These markers provide a convenient way for rapid and quantitative assessment of tissue distribution after parenteral administration of liposomes.

There are also fluorescent lipid labels on the market, which are valuable tools for studying the in vivo behavior of liposomes (90). In addition to labeling of the liposome itself, liposome-associated compounds such as encapsulated material and homing devices can be labeled. Fluorescence methods in liposome research are increasingly important to determine processes such as liposome fusion, liposome release, and intracellular trafficking of liposomes. These methods, including appropriate fluorescence markers, are reviewed in Ref. (91).

The usefulness of labels as markers of the fate of liposomes will largely depend on the specifications of the labeled compound and the nature of the label itself. Using liposomes that were (triple) labeled with [^{3}H]cholesteryloleyl ether, 1,2-di[1-^{14}C]palmitoylcholine and, *N*-(lissamine rhodamine-b sulfonyl)-phosphatidylethanolamine, we observed that the rhodamine label was eliminated from the blood twice as fast as the radiolabels. It appeared that the *N*-(lissamine rhodamine-B sulfonyl)-phosphatidylethanolamine was selectively removed from the liposomes by scavenger B-1 receptors expressed in the liver (92). This example indicates that to ensure whether the fate of a liposomal marker is representative of the fate of the entire liposome, use of proper double-labeled liposomes is recommendable.

For morphological studies on the intracellular fate of liposomes and/or their components, fluorescent markers are powerful tools at the light microscopic level. They can be visualized either with conventional fluorescence microscopy or with confocal laser scanning microscopy (93–95). For studies at the electron microscopic level, liposome-encapsulated colloidal gold or the reaction products of liposome-encapsulated horseradish peroxidase are convenient and readily detectable markers (96–99).

Endothelial Cell Systems for In Vitro Studies

In vivo experiments are essential to unequivocally show proof of targeting specificity and pharmacological effectiveness in relation to disease progression. Yet, in vitro studies using EC systems allow execution of detailed studies on cell-binding capacity, cellular handling, and kinetics of these processes. They provide a means to generate valuable data on drug-delivery efficacy and pharmacological effects induced. Besides the availability of a wide variety of EC lines (100), in our laboratory we most frequently use cultures of primary ECs isolated from human umbilical cord veins and of primary sinusoidal ECs isolated from rat liver. The former cells can be cultured for four passages without losing typical features, whereas the latter cells are only used as nonpassaged isolates. Here we briefly present the isolation methods for both cell types as they are routinely executed in our laboratory. For a more detailed descriptions, the reader is referred to Ref. (101–103).

Isolation, Purification, and Culturing of Human Umbilical Vein Endothelium

HUVEC are generally isolated from two umbilical cords to improve reproducibility of the data generated. The human umbilical cords are kept in cold sterile buffer containing 140 mM NaCl, 4 mM KCl, 11 mM D-glucose, 10 mM Hepes, pH 7.4 (buffer A) until the start of the isolation procedure. The vein is cannulated and rinsed with buffer A. The vein is then filled with 0.2 mg/mL chymotrypsin in buffer containing 0.15 M NaCl, 10 mM NaPi, pH 7.4 (37°C). The blood vessels are closed and ECs are allowed to detach during 15- to 20-minute incubation at 37°C. The cells are collected by perfusion of the vessels with culture medium. The perfusate is centrifuged for eight minutes at 200 g and resuspended in culture medium containing RPMI 1640 supplemented with 20% heat-inactivated fetal calf serum (FCS), 2 mM L-glutamine, 5 U/mL heparine, 50 µg/mL EC growth factor either harvested from calf brain according to Maciag et al. (104) or commercially obtained, 100 µg/mL streptomycin, and 100 IU/mL penicillin. Cells are cultured to confluency (i.e., $\sim$60,000 cells/cm^2) in two 25-cm^2 flasks coated with 1% gelatin, which takes four to seven days. At confluency, cells are passaged by trypsinization followed by resuspending into culture medium and reseeding at 1 : 3 density in gelatin-coated flasks. Typically, passages 2 to 4 are used for binding, uptake, and pharmacological effect studies. HUVEC passage 1 can be frozen and stored in liquid nitrogen for several months. After thawing and putting into culture at cell densities at which they were harvested for freezing, they require five to seven days before they can be used in experiments.

Isolation, Purification, and Culturing of Liver Endothelial Cells

Before starting the liver cell isolation, the following solutions have to be prepared:

A. Preperfusion buffer [142 mM NaCl, 6.7 mM KCl, 10 mM *N*-2-hydroxyethylpiperazine-*N'*-2-ethanesulfonic acid (Hepes), pH 7.6]: oxygenate preperfusion buffer by carbogen bubbling for at least 20 minutes at 37°C, adjust pH to 7.6 immediately before the use of buffer.

B. Collagenase buffer (66.7 mM NaCl, 6.7 mM KCl, 4.8 mM CaCl$_2$. 2H$_2$O, 10 mM Hepes, pH 7.4): oxygenate collagenase buffer for 20 minutes at 37°C, dissolve bovine serum albumin (BSA) in 10 mL of this buffer [2% (w/v) BSA final concentration], add collagenase (collagenase A, Roche Diagnostics) [0.05% (w/v) collagenase final concentration], mix gently with rest of the buffer to avoid air bubbles, and adjust pH to 7.4.

C. Postperfusion buffer: oxygenate 0.1 L of the Hanks' solution (137 mM NaCl, 5.4 mM KCl, 0.8 mM MgSO$_4$. 7H$_2$O, 0.33 mM Na$_2$HPO$_4$. 2H$_2$O, 0.44 mM KH$_2$PO$_4$, 10 mM Hepes, 5 mM glucose pH 7.4) for 20 minutes at 37°C, dissolve BSA in 10 mL of this buffer (2% BSA final concentration), adjust pH to 7.4.

D. Dissolve BSA (0.3% BSA final concentration) in 1.5 L of Hanks' solution and adjust pH to 7.4, keep at 4°C.

The perfusion system, consisting of a peristaltic pump temperature-controlled by a water bath (37°C), is first rinsed with 70% ethanol followed by rinsing with excess water. Then the tubing is filled with preperfusion buffer. An anesthetized rat is placed on a warmed surface. The abdomen is opened and the portal vein is cannulated using a 20-gauge braunule. The chest is opened by cutting the rib cage up each side of the sternum. Leave the diaphragm and the lowest ribs in place. A 16-gauge braunule is carefully inserted into the vena cava between the heart and the diaphragm pointing away from the heart. The tubing of the peristaltic pump is connected to the portal vein braunule and also the vena cava braunule is connected to the tubing, which in turn is connected to a reservoir for waste perfusate. The liver is then perfused with 400 mL of preperfusion buffer at 20 mL/min. As the blood is displaced from the liver by the perfusate, the liver will become paler. At this point the pump speed can be increased to 28 mL/min. After the preperfusion medium has passed, the tubing is changed to the collagenase buffer reservoir and the collagenase buffer is allowed to recirculate for 10 minutes at 28 mL/min. The tubing is changed to postperfusion buffer and 100 mL is passed through the liver at 28 mL/min. The liver is taken out by cutting around the diaphragm and subsequently cutting the liver free from the remnants of diaphragm. At this stage, the liver should be a fragile soft bag of cells. The liver capsule is opened and the released cells are gently filtered through a sterile nylon mesh (100 μm). The filtered cells are pooled in 50-mL tubes and centrifuged for 45 s at 50 g (no brake). The supernatant containing the nonparenchymal Kupffer and ECs is saved. The parenchymal cell (hepatocyte) pellet is washed one more time and the supernatant is pooled with the first supernatant.

For preparation of liver ECs, the pooled supernatants are centrifuged for 10 minutes at 500 g at 4°C to sediment the nonparenchymal cells. The pellet is resuspended in two plastic 14-mL tubes with 8.5 mL Hanks' BSA solution each and mixed with 3.2 mL of Optiprep™ gradient solution. The cell mixtures are layered with 1 mL of Hanks' solution and centrifuged for 15 minutes at 1350 g at 4°C. The nonparenchymal cells are now separated from red blood cells and cell debris and can be collected as a cell layer in the buffer phase just on top of the Optiprep solution. The cells are resuspended in 10 mL of Hanks' BSA solution and centrifuged for 10 minutes, at 500 g, 4°C. Then the cell pellet is resuspended in 5 mL Hanks' BSA solution and the cells are flushed into the Beckman, type JE-6 elutriation rotor at 4°C, at a flow rate of 13°mL/min, and a rotor speed of 2500 rpm (750 g). At this flow rate, cell debris is flushed out in 200 mL of Hanks' solution containing 0.3% BSA. Liver ECs are collected in 150 mL at a flow rate of 23 mL/min, an intermediate cell fraction containing large endothelial and small Kupffer cells is collected in 150 mL at a flow rate of 25 mL/min, and Kupffer cells can be collected in 150 mL at 46 mL/min. The EC fraction is concentrated by centrifugation for 10 minutes at 500 g, 4°C. The cells are resuspended in 10 mL of Hanks' solution or culture medium and the number of cells is determined by microscopic examination. Liver ECs in RPMI 1640 supplemented with 20% heat-inactivated FCS, 2 mM L-glutamine, 100 IU/ mL penicillin, 100 μg/mL streptomycin, and 10 ng/mL endothelial growth factor can now be plated in collagen-I coated plates (105) after careful resuspension of the cells, at the desired density. Typical liver EC densities in 24-well plates are 5×10^5 cells per 500 μL per well. The cells are allowed to adhere to the substrate for at least four hours, but preferably overnight. Adhered cells are washed to remove nonattached cells and further cultured in medium containing 10% FCS. Liposome–cell interaction experiments are performed between the first and third day after culturing of the cells.

In Vitro Liposome-Endothelial Cell Binding, Uptake, and Metabolism Studies

Primary cultures of ECs and EC lines are useful tools to study the interaction of liposomes with ECs, including effect studies of the delivered liposomal drugs. Experimental conditions such as the presence or absence of serum in the culture medium may seriously influence liposome–cell interactions in vitro. To study liposome–cell interactions and to compare the advantages of targeted liposome over nontargeted ones, it is therefore important to standardize the experimental setups used. The general protocol presented below to quantitatively determine binding and uptake of liposomes by ECs can be a starting point for the development of more advanced in vitro systems if required.

Using either confluent or subconfluent monolayers of ECs, one should first replace the culture medium by serum-free medium and incubate the cells for one hour at 37°C in a humidified 5% CO_2/95% air atmosphere. This step may be omitted if the incubation cannot be performed in serum-free conditions. Remove the medium and add new serum-free cultured medium containing the appropriate amounts of labeled liposomes and if necessary other additions. The label can be an inert radiolabel such as [^{3}H]cholesteryloleyl ether or a label that is metabolically degraded, such as cholesteryl [^{14}C]oleate or a combination of the two. In the latter case, the rapid release from the cell of [^{14}C]oleate derived from degraded cholesteryl[^{14}C]oleate and the cellular retention of the metabolically inert [^{3}H]cholesteryloleyl ether leads to a change in cellular ^{3}H/^{14}C ratio, which is a sensitive measure for intracellular degradation. Incubate the cells for three hours at 37°C to determine cell association (binding and uptake) or for three hours at 4°C to determine exclusively cell binding. After the incubation, the plates are placed on ice and washed five to seven times with ice-cold buffer containing 0.15 M NaCl, 10 mM NaPi, pH 7.4. The cells are then lysed in 0.1 M NaOH for one hour at 37°C. The cell-associated radioactivity is determined by liquid scintillation counting of aliquots of the lysed cell suspension. The radioactivity is normalized to the amount of cellular protein determined according to Lowry et al. (106). A similar protocol can be followed when using liposomes that are labeled with fluorescent markers, yet the readout systems, fluorescent microscopy, confocal laser scanning microscopy, or flow cytometry will only allow for a qualitative measure of liposome binding and/or internalization. The fluorescence microscopy analyses have the advantage of enabling the localization of the liposomes within the cells' interior, as described above.

In Vitro Drug Effector Read-Out Systems

The majority of studies employing (immuno)liposomes for the selective delivery of drugs into ECs dealt with the delivery of toxic drugs. In vitro read-out of cell death induction can be performed using general cell counts or mitochondrial activity studies in which mitochondrial activity is related to cell number. To establish whether the mechanism of cell death is either necrosis or apoptosis, analysis of cellular apoptosis features can nowadays be easily performed using commercially available reagents. These include fluorescently labeled Annexin-V that interacts with phosphatidylserine molecules exposed toward the cells' exterior upon apoptosis induction, and bio or fluorescent substrates that are enzymatically cleaved by activated caspases that control execution of apoptosis. Also, loss of integrity of the mitochondria via dihexaoxacarbocyanine iodide (DiOC6) staining as a marker of apoptosis induction can be easily established. Although several of these parameters can be visualized by fluorescence microsopy, the more quantitative read-out protocols require a flow cytometer or bioluminescence/fluorescence plate-readers.

In case of delivery of drugs interfering, e.g., with endothelial kinase activity in TNFα or IL-β–induced intracellular signaling pathways, often the downstream effects on gene expression levels are established. Although complex gene expression profiles can be determined using microarray analysis (107), real-time reverse transcriptase (RT)-polymerase chain reaction (PCR) is the technique to quantitatively determine the pharmacological effects of the intracellularly delivered drugs. Until recently, conventional RT-PCR was applied for this purpose. The disadvantage of this technique is, however, that it merely gives a qualitative measure of mRNA levels. As such, downregulation of mRNA levels to an almost zero level, compared to untreated control, can be easily detected. More delicate changes can, however, not be analyzed this way. The advent of real-time RT-PCR that monitors the formation of the PCR product in an online mode provides a sensitive new way to quantitatively determine gene expression profiles in a short span of time.

In Vivo Pharmacological Effects of Endothelial Cell–Targeted Liposomes

Caution should be taken when interpreting in vitro results for the purpose of extrapolation to the in vivo situation. Cultured cells generally have a life cycle and cellular makeup that is different from the cells in their natural environment in vivo. In the organs in the body, ECs are strongly influenced by their local environment. Furthermore, blood flow–induced shear stress and continuous interaction with blood-derived cellular and soluble components are continuously sensed by the capillary endothelium. These conditions strongly affect EC behavior (108–110), yet cannot be easily mimicked in in vitro cell culture systems. For this reason, the final experiments studying the pharmacological effects of the delivered drugs should be performed in animal models of disease.

Various animal models for inflammatory diseases are available to study the pharmacological effects of liposomally delivered drugs (111–114). In our quest for a suitable animal model to investigate the effects of E-selectin–based targeted delivery of anti-inflammatory drugs into ECs, we recently made an inventory of E-selectin expression in different murine models of inflammation (50). Although in some disease models E-selectin expression could be detected, in others no evidence could be generated on its presence during disease-initiation and/or progression. Furthermore, using RT-PCR and immunohistochemistry, we found a discrepancy between the presence of E-selectin protein in the ECs in the inflamed site and the presence of mRNA encoding for this protein. An explanation for this observation might be that the antibodies used for immunohistochemical staining identified protein epitopes that were either not present in the lesions or not accessible for the antibodies during the staining procedure. Only a careful account of the kinetics of expression of the target epitope within the target molecule at different stages of the disease allows for a proper choice of the model.

In absolute cell numbers, the endothelium forms a minor part of an organ. In whole tissue homogenates measurements of pharmacological effects at a molecular level in ECs are often masked by the signal caused by the other cells present. Only analysis of EC-specific marker genes will confirm EC-specific effects of the delivered drugs. This severely limits the pharmacological read-out options because true EC-specific genes and proteins are rare. Immunohistochemical analysis of proteins allows detection of disease-associated markers to be combined with cellular localization. This analysis is, however, a qualitative measure of protein expression that does not enable quantitation of effects. Similarly, in situ hybridization combines gene expression data with cellular localization within a tissue. Yet true quantitation of expression levels in presence and absence of targeted drugs cannot be made unless the effects represent a clear difference in gene expression. For the further development of liposomal drug carriers targeted to ECs, it is essential to have means to determine EC-specific effects and to relate these effects to disease progression. In the case of studying liver ECs, analysis of gene and protein expression levels can be relatively easily performed after isolation of the ECs after treatment of an experimental animal with targeted liposomes. For other capillary endothelium, no such specific isolation methods that facilitate fast and reproducible EC isolation at high cell yield are on hand. We, therefore, developed protocols to isolate microvascular ECs from their pathophysiological environment by laser dissection microscopy (115). The use of snap-frozen tissue biopsies ensured that levels of mRNA expressed by the cells represent their levels at the time of harvesting. No artifacts are introduced because time-consuming enzymatic digestion and selection procedures are not required. Using this method, we demonstrated that it is feasible to isolate intact RNA from the dissected ECs at sufficient amounts and integrity to perform RT-PCR analysis of several genes at a time (Fig. 3) (115, Asgeirsdottir SA, Werner N, Kuldo JM, et al., submitted.). At present, protocols for linear amplification of genes in combination with quantitative RT-PCR are being applied to analyze complex gene expression profiles of endothelium in vivo. An additional aim is to isolate protein from the dissected samples for future broad kinase activity screens (116). By this means, e.g., endothelial effects of immunoliposome-delivered MAPK inhibitors can be studied and related to disease outcome. The micro- to nanoscale conditions enable application of these technologies to needle biopsies from patients as well.

PERSPECTIVES OF ENDOTHELIAL CELL–TARGETED LIPOSOMES

Over 40 years of experience with liposomes as vehicles for different classes of classical and biotech-derived drugs has revealed clear potentials and limitations of these macromolecular carrier systems for therapeutic purposes. By aiming at the vascular ECs in diseased sites, one of the main hurdles experienced, the vascular wall, is eliminated. Although homing ligands specific for activated ECs were highly capable of endowing PEG liposomes

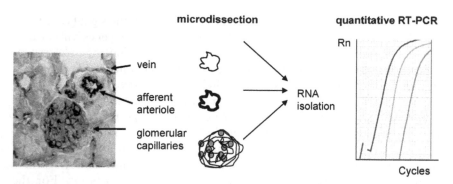

Figure 3 Measurement of in vivo pharmacological effects of drugs selectively delivered into activated ECs using immunoliposomes can be performed by laser microdissection followed by real-time (RT)-polymerase chain reaction (PCR). By this means microvascular ECs from different origins can be isolated from their pathophysiological (snap frozen) tissue biopsies without disturbing their original gene expression makeup. Endothelial RNA isolated from the dissected material is subsequently subjected to quantitative RT-PCR analysis for relevant genes to enable gene expression profiling studies.

with selectivity for the inflamed endothelium, the occurrence of endothelial heterogeneity during disease progression rationalizes the search for multiple targets. Only by this means, delivery of the drugs into all diseased ECs is ensured. In addition, methods for stable incorporation of drugs into either the water phase or the lipid bilayer of the liposomes without compromising drug release upon internalization are still not widely available. Systematic studies on the physicochemical requirements for liposomal incorporation of inhibitors of inflammatory signal transduction pathways will become critical in transforming these drugs of the future into selective, effective, and clinically applicable targeted drugs of today.

REFERENCES

1. Galley HF, Webster NR. Br J Anaesth 2004; 93(1):105–113.
2. Spanjer HH, van Galen M, Roerdink FH, Regts J, Scherphof GL. Biochim Biophys Acta 1986; 863(2):224–230.
3. The Endothelial Cell in Health and Disease. Stuttgart: Schattauer Verlagsgesellschaft, 1995.
4. Ball HJ, McParland B, Driussi C, Hunt NH. Brain Res Brain Res Protoc 2002; 9(3):206–213.
5. Hellstrom M, Gerhardt H, Kalen M, et al. J Cell Biol 2001;153(3):543–553.
6. Griffioen AW, Molema G. Pharmacol Rev 2000; 52(2):237–268.
7. Kuldo JM, Ogawara KI, Werner N, et al. Curr Vasc Pharmacol 2005; 3(1):11–39.
8. Karin M, Yamamoto Y, Wang QM. Nat Rev Drug Discov 2004; 3(1):17–26.

9. English JM, Cobb MH. Trends Pharmacol Sci 2002; 23(1):40–45.
10. Pober JS. Arthritis Res 2002; 4(suppl 3):S109–S116.
11. Harlan JM, Winn RK. Crit Care Med 2002; 30(suppl 5):S214–S219.
12. Madge LA, Pober JS. Exp Mol Pathol 2001; 70(3):317–325.
13. Lim YC, Garcia-Cardena G, Allport JR, et al. Am J Pathol 2003; 162(5) 1591–1601.
14. Von Andrian UH, Mackay CR. N Engl J Med 2000; 343(14):1020–1034.
15. Itoh T, Arai KI. Mechanisms of signal transduction. In: Jacques T, ed. The Cytokine Network and Immune Function. Oxford: Oxford University Press, 1999:169–190.
16. O'Neill LA, Dinarello CA. Immunol Today 2000; 21(5):206–209.
17. Viemann D, Goebeler M, Schmid S, et al. Blood 2004; 103(9):3365–3373.
18. Zhao B, Stavchansky SA, Bowden RA, Bowman PD. Am J Physiol Cell Physiol 2003; 284(6):C1577–C1583.
19. Raab M, Daxecker H, Markovic S, Karimi A, Griesmacher A, Mueller MM. Clin Chim Acta 2002; 321(1–2):11–16.
20. Kumar S, Boehm J, Lee JC. Nat Rev Drug Discov 2003; 2(9):717–726.
21. Salven P, Hattori K, Heissig B, Rafii S. FASEB J 2002; 16(11):1471–1473.
22. Schett G, Tohidast-Akrad M, Smolen JS, et al. Arthritis Rheum 2000; 43(11): 2501–2512.
23. Marok R, Winyard PG, Coumbe A, et al. Arthritis Rheum 1996; 39(4):583–591.
24. Stambe C, Nikolic-Paterson DJ, Hill PA, Dowling J, Atkins RC. J Am Soc Nephrol 2004; 15(2):326–336.
25. Thiele K, Bierhaus A, Autschbach F, et al. Gut 1999; 45(5), 693–704.
26. Braet F, Wisse E. Comp Hepatol 2002; 1(1):1.
27. Beljaars L, Melgert BN, Meijer DKF, Molema G, Poelstra K. Neoglyco- and neopeptide albumins for cell-specific delivery of drugs to chronically diseased livers. In: Schreier H, ed. Drug Targeting Technology: A Critical Analysis of Physical, Chemical, and Biological Methods. New York: Marcel Dekker, Inc, 2001:189–240.
28. Malerod L, Juvet K, Gjoen T, Berg T. Cell Tissue Res 2002; 307(2):173–180.
29. Wisse E, Braet F, Luo D, et al. Toxicol Pathol 1996; 24(1):100–111.
30. Lalor PF, Adams DH. Mol Pathol 1999; 52(4):214–219.
31. Paleolog EM. Arthritis Res 2002; 4(suppl 3):S81–S90.
32. Fearon U, Griosios K, Fraser A, et al. J Rheumatol 2003; 30(2):260–268.
33. Canete JD, Pablos JL, Sanmarti R, et al. Arthritis Rheum 2004; 50(5): 1636–1641.
34. Ferrara N, Gerber HP, LeCouter J. Nat Med 2003; 9(6):669–676.
35. Weis S, Shintani S, Weber A, et al. J Clin Invest 2004; 113(6):885–894.
36. Ramsauer M, D'Amore PA. J Clin Invest 2002; 110(11):1615–1617.
37. Silletti S, Kessler T, Goldberg J, Boger DL, Cheresh DA. Proc Natl Acad Sci USA 2001; 98(1):119–124.
38. Jain RK. Nat Med 2003; 9(6):685–693.
39. Fabian MA, Biggs WH III, Treiber DK, et al. Nat Biotechnol 2005; 23(3): 329–336.
40. Dykxhoorn DM, Novina CD, Sharp PA. Nat Rev Mol Cell Biol 2003; 4(6):457–467.

41. Rijcken E, Krieglstein CF, Anthoni C, et al. Gut 2002:51(4):529–535.
42. Basu P. Nat Med 2003; 9(9):1100–1101.
43. Chen LW, Egan L, Li ZW, Greten FR, Kagnoff MF, Karin M. Nat Med 2003; 9(5):575–581.
44. Blink Bvan den, Juffermans NP, ten Hove T, et al. J Immunol 2001; 166(1):582–587.
45. Scherphof GL, Kamps JA. Prog Lipid Res 2001; 40(3):149–166.
46. Kamps JA, Morselt HW, Scherphof GL. Biochem Biophys Res Commun 1999; 256(1):57–62.
47. Rothkopf C, Fahr A, Fricker G, Scherphof GL, Kamps JAAM. Biochim Biophys Acta-Biomembr 2005; 1668(1):10–16.
48. Koning GA, Schiffelers RM, Storm G. Endothelium 2002; 9(3):161–171.
49. Spragg DD, Alford DR, Greferath R, et al. Proc Natl Acad Sci USA 1997; 94(16):8795–8800.
50. Everts M, Asgeirsdottir SA, Kok RJ, et al. Inflamm Res 2003; 52(12):512–518.
51. Ehrhardt C, Kneuer C, Bakowsky U. Adv Drug Del Rev 2004; 56(4): 527–549.
52. Stahn R, Grittner C, Zeisig R, Karsten, U, Felix SB, Wenzel K. Cell Mol Life Sci 2001; 58(1):141–147.
53. Kessner S, Krause A, Rothe U, Bendas G. Biochim Biophys Acta-Biomembr 2001; 1514(2):177–190.
54. Everts M, Koning GA, Kok RJ, et al. Pharm Res 2003; 20(1):64–72.
55. Chiu GN, Bally MB, Mayer LD. Biochim Biophys Acta 2003; 1613(1–2): 115–121.
56. Bloemen PG, Henricks PA, van Bloois L, et al. FEBS Lett 1995; 357(2): 140–144.
57. Mastrobattista E, Storm G, van Bloois L, et al. Biochim Biophys Acta 1999; 1419(2):353–363.
58. Muro S, Wiewrodt R, Thomas A, et al. J Cell Sci 2003; 116(Pt 8):1599–1609.
59. Schaft DWJ van der, Ramakrishnan S, Molema G, Griffioen AW. Tumor vasculature targeting. In: Molema G, Meijer DKF, eds. Drug Targeting— Organ-Specific Strategies. Weinheim, New York: Wiley-VCH, 2001:233–254.
60. Schraa AJ, Everts M, Kok RJ, Asgeirsdottir SA, Molema G. Biotechnol Ann Rev 2002; 8:133–165.
61. Janssen AP, Schiffelers RM, ten Hagen TL, et al. Int J Pharm 2003; 254(1): 55–58.
62. Dubey PK, Mishra V, Jain S, Mahor S, Vyas SP. J Drug Target 2004; 12(5):257–264.
63. Schiffelers RM, Koning GA, ten Hagen TL, et al. J Control Release 2003; 91(1–2):115–122.
64. Pastorino F, Brignole C, Marimpietri D, et al. Cancer Res 2003; 63(21): 7400–7409.
65. Kondo M, Asai T, Katanasaka Y, et al. Int J Cancer 2004; 108(2):301–306.
66. Benzinger P, Martiny-Baron G, Reusch P, et al. Biochim Biophys Acta-Biomembr 2000; 1466(1–2):71–78.
67. Volkel T, Holig P, Merdan T, Muller R, Kontermann RE. Biochim Biophys Acta 2004; 1663(1–2):158–166.

68. Thurston G, McLean JW, Rizen M, et al. J Clin Invest 1998; 101(7):1401–1413.
69. Dass CR. J Mol Med 2004; 82(9):579–591.
70. Kamps JA, Morselt HW, Swart PJ, Meijer DK, Scherphof GL. Proc Natl Acad Sci USA 1997; 94(21):11681–11685.
71. Bartsch M, Weeke-Klimp AH, Meijer DK, Scherphof GL, Kamps JA. Pharm Res 2002; 19(5):676–680.
72. Bartsch M, Weeke-Klimp AH, Hoenselaar EP, et al. J Drug Target 2004; 12 (9–10):613–621.
73. Bartsch M, Weeke-Klimp AH, Morselt HW, et al. Mol Pharmacol 2005; 67(3):883–890.
74. Semple SC, Klimuk SK, Harasym TO, et al. Biochim Biophys Acta 2001; 1510(1–2):152–166.
75. Stuart DD, Allen TM. Biochim Biophys Acta 2000; 1463(2):219–229.
76. Stuart DD, Semple SC, Allen TM. Methods Enzymol 2004; 387:171–188.
77. Pardridge WM. Mol Interv 2003; 3(2):90–105, 51.
78. Cornford EM, Cornford ME. Lancet Neurol 2002; 1(5):306–315.
79. Begley DJ. Pharmacol Ther 2004; 104(1):29–45.
80. Torchilin VP, Weissig V, Matin FJ, Heath TD, New RRC. Surface modification of liposomes. In: Torchilin VP, Weissig V, eds. Liposomes a Practical Approach. 2nd ed. Oxford: Oxford University Press, 2003:193–229.
81. Hermanson GT. Bioconjugate Techniques. San Diego: Academic Press Inc., 1996.
82. Shimada K, Kamps JA, Regts J, et al. Biochim Biophys Acta 1997; 1326(2):329–341.
83. Derksen J, Scherphof G. Biochim Biophys Acta-Biomembr 1985; 814(1): 151–155.
84. Kamps JA, Swart PJ, Morselt HW, et al. Biochim Biophys Acta 1996; 1278(2):183–190.
85. Bartsch M, Weeke-Klimp A, Morselt HW, et al. Mol Pharmacol 2004.
86. Bendas G, Krause A, Bakowsky U, Vogel J, Rothe U. Int J Pharm 1999; 181(1):79–93.
87. Bendas G, Rothe U, Scherphof GL, Kamps JA. Biophys Acta 2003; 1609(1):63–70.
88. Hansen CB, Kao GY, Moase EH, Zalipsky S, Allen TM. Biochim Biophys Acta 1995; 1239(2):133–144.
89. Koning GA, Morselt HW, Gorter A, et al. Pharm Res 2003; 20(8):1249–1257.
90. Kamps JA, Koning GA, Velinova MJ, et al. J. Drug Target 2000; 8(4):235–245.
91. Düzgünes N, Bagatolli LA, Meers P, Oh YK, Straubinger RM. Fluorescence methods in liposome research. In: Torchilin VP, Weissig V, eds. Lipososmes a Practical Approach. 2nd ed. Oxford: Oxford University press, 2003:105–147.
92. Yan X, Poelstra K, Scherphof GL, Kamps JA. Biochem Biophys Res Commun 2004; 325(3):908–914.
93. Cerletti A, Drewe J, Fricker G, Eberle AN, Huwyler J. J Drug Target 2000; 8(6):435–446.
94. Daleke DL, Hong K, Papahadjopoulos D. Biochim Biophys Acta 1990; 1024(2):352–366.
95. Papadimitriou E, Antimisiaris SG. J Drug Target 2000; 8(5):335–351.

96. Koning GA, Morselt HW, Velinova MJ, et al. Biochim Biophys Acta 1999; 1420(1–2):153–167.
97. Ellens H, Morselt HW, Dontje BH, Kalicharan D, Hulstaert CE, Scherphof GL. Cancer Res 1983; 43(6):2927–2934.
98. Hong K, Friend DS, Glabe CG, Papahadjopoulos D. Biochim Biophys Acta 1983; 732(1):320–323.
99. Huang SK, Hong K, Lee KD, Papahadjopoulos D, Friend DS. Biochim Biophys Acta 1991; 1069(1):117–121.
100. Bouis D, Hospers G, Meijer C, Molema G, Mulder N. Angiogenesis 2001; 4(91):102.
101. Casteleijn E, Van Rooij HC, Van Berkel TJ, Koster JF. FEBS Lett 1986; 201(2):193–197.
102. Kamps JAAM, Scherphof GL. Liposomes in biological systems. In: Torchilin VP, Weissig V, eds. Liposomes a Practical Approach. 2nd ed. Oxford: Oxford University Press, 2003:267–288.
103. Mulder AB, Blom NR, Smit JW, et al. Thromb Res 1995; 80(5):399–411.
104. Maciag T, Cerundolo J, Ilsley S, Kelley PR, Forand R. Proc Natl Acad Sci USA 1979; 76(11):5674–5678.
105. Braet F, De Zanger R, Sasaoki T, et al. Lab Invest 1994; 70(6):944–952.
106. Lowry OH, Rosebrough NJ, Farr AL, Randall RJ. J Biol Chem 1951; 193: 265–275.
107. Asgeirsdottir SA, Kok RJ, Everts M, Meijer DKF, Molema G. Biochem Pharmacol 2003; 65:1729–1739.
108. Ohura N, Yamamoto K, Ichioka S, et al. J Atheroscler Thromb 2003; 10(5):304–313.
109. Brooks AR, Lelkes PI, Rubanyi GM. Endothelium 2004; 11(1):45–57.
110. Wasserman SM, Topper JN. Vasc Med 2004; 9(1):35–45.
111. Borza DB, Hudson BG. Kidney Int 2002; 61:1905–1906.
112. Sun Y, Chen HM, Subudhi SK, et al. Nat Med 2002; 8(12):1405–1413.
113. Veihelmann A, Harris AG, Krombach F, Schutze E, Refior HJ, Messmer K. Microcirculation 1999; 6(4):281–290.
114. Farkas S, Herfarth H, Rossle M, et al. Clin Exp Immunol 2001; 126(2): 250–258.
115. Asgeirsdottir SA, Werner N, Harms G, Van Den Berg A, Molema G. Ann NY Acad Sci 2002; 973:586–589.
116. Diks SH, Kok K, O'Toole T, et al. J Biol Chem 2004; 279(47):49,206–49,213.

9

Targeting of Cationic Liposomes to Endothelial Tissue

Uwe Michaelis and Heinrich Haas

MediGene AG, Martinsried, Germany

INTRODUCTION

This review focuses on cationic liposomes and lipid complexes as neovascular targeting agents or so-called vascular-disrupting agents (VDAs) (1) for tumor treatment. Attacking the already established tumor blood vessel system with VDAs is an emerging concept in the treatment of cancer. In contrast to conventional chemotherapy, which targets the tumor tissue compartment, the VDA action is directed against the endothelial cells lining the newly formed tumor vasculature. Destruction of the tumor endothelial cells should ultimately lead to thrombus formation with subsequent occlusion of the tumor blood vessels. The following reduction or even collapse of tumor blood flow may cause a reduction or complete remission of the tumor cell mass (2). Both the vascular-disrupting approach and the anti-angiogenic therapy are antivascular treatments, but there are important conceptional differences between the respective drug classes. VDAs target the existing tumor vasculature, thereby allowing the treatment of already established tumors. Inhibitors of angiogenesis interfere with the formation of new blood vessels and should prevent further tumor growth. Although new reports show that antiangiogenic agents do more than merely stop growth of new blood vessels (3–5), the key differences between the two anti-vascular treatment strategies still remain substantial.

Cationic colloidal carriers are a new option in the field of vascular targeting. They have high potential for the development of therapeutic agents for a wide range of indications in tumor therapy. In this contribution, the concept of neovascular targeting by cationic liposomal carriers is discussed in the context of other antivascular approaches in tumor therapy and of classical drug delivery by liposomes. At Medigene AG, new pharmaceutical products for tumor therapy are being developed on this basis. With the principle of cationic targeting the technology platform EndoTAG®[a] has been established, enabling to set up a wide range of new products for tumor therapy and other indications. An overview over the prospects and current products in the development pipeline on the basis of this platform technology is given.

VASCULAR-DISRUPTING AGENTS FOR TUMOR THERAPY

Conventional VDAs can be grouped to one of the two following drug classes. The low-molecular-weight VDAs include several microtubulin destabilizers and the flavonoid, 5,6-dimethylxanthenone-4-acetic acid (DMXXA). These drugs do not specifically target proliferating tumor endothelial cells and distribute throughout the whole body. These VDAs can induce a significantly higher acute reduction of blood flow in tumor tissue compared to other organs but it is still speculative why the tumor endothelial cells are more susceptible to treatment with these drugs compared to quiescent endothelial cells. An unsolved problem of this class of VDAs is that tumor cells in the periphery of the tumor always survive treatment and are responsible for rapid regrowth after end of treatment. Also, in most patients participating in phase I and phase II trials, all vascular parameters returned to baseline within 24 hours (2,6). A second group are ligand-based VDAs which target specific molecules that are overexpressed on the luminal or abluminal surface of tumor endothelial cells. Ligand-based VDAs show a modular structure by combining antibody fragments, peptides, or soluble receptors as targeting moities with toxins, procoagulant factors, or other effector moities (6,7). For an overview on VDAs and antivascular therapy approaches in general, please refer to the following reviews (2,6–10).

Cationic liposomes loaded with therapeutically active drugs define a third class of VDAs with several distinct key features. They are designed to preferentially target and destroy the proliferating tumor endothelial cells by delivering cytotoxic agents specifically into those activated endothelial cells (Fig. 1). Because endothelial cells do not vary in their properties as much as the cells of different tumors, the treatment is less dependent on tumor type and tumor stage, and the risk of developing resistance is lowered. Therefore, cationic liposomes have the potential for treatment of many

---

[a] EndoTAG® is a registered trademark of Medigene AG in Germany.

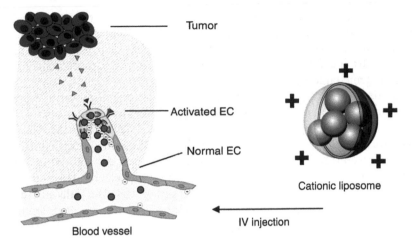

Figure 1 The activated and dividing tumor endothelial cell (EC) as a novel cellular target in oncology. Cationic liposomes carrying cytotoxic drugs directly target and destroy the tumor blood vessels. The EndoTAG® targeting concept could be superior to antiangiogenic approaches, which interfere only with one out of several angiogenic factors responsible for tumor vessel growth.

different solid tumors. The cationic targeting approach could be superior to indirect-acting antiangiogenic therapies, which interfere only with one out of several angiogenic factors responsible for tumor blood vessel proliferation (e.g., vascular endothelial growth factor), or only target specific tumor endothelial cell surface antigens. The efficacy of these drugs can strongly depend on the tumor stage and several possible mechanisms for development of resistance against treatment exist (11,12).

The preferential binding and uptake of cationic liposomes by angiogenic blood vessels in tumors and sites of chronic inflammation after intravenous (IV) administration was initially described by Thurston et al. They investigated the accumulation of fluorescently labeled cationic liposomes in blood vessels of pancreatic islet tumors and in airway blood vessels of a chronic airway inflammation model (13). In both the models, the angiogenic endothelial cells showed a more than 30-fold higher uptake of the fluorescence dye compared to the corresponding normal vessels. Despite this impressive accumulation in the tumor vessels, the normal pattern of uptake of the cationic liposomes by macrophages in the liver and spleen and by endothelial cells in the lung and various other organs was observed. It was demonstrated that binding and uptake of the cationic liposomes is a fast process: already 20 minutes after IV application 85% of the cationic liposomes, which were associated with the tumor vessels, were either bound to the luminal surface of the endothelium or taken up by the endothelial cells. Fluorescently labeled anionic, neutral, or sterically stabilized neutral liposomes served as control and did not show any binding and internalization. These authors were the first who speculated that charge interactions may be the driving force for the preferential binding of cationic carriers to the angiogenic endothelial cells.

The described targeting characteristics of cationic liposomes were confirmed in a variety of species, different tumor types, and tumor microenvironments (14–16). The cationic liposomes tightly associated with the tumor vascular area and did not extravasate to the tumor tissue compartment. In contrast to conventional liposomes, the cationic carriers showed a fast interaction with the tumor vasculature and were rapidly cleared from circulation. Importantly, the tumor blood vessel targeting was essentially proven for a broad range of cationic liposome formulations composed of different cationic lipids (CLs) and neutral colipids.

MEMBRANE CHARACTERISTICS OF DIVIDING ENDOTHELIAL CELLS

The molecular structures responsible for enhanced binding and uptake of CL complexes by tumor endothelial cells are not known, but electrostatic interactions are thought to be a paramount driving force. Therefore, negatively charged molecules and structures overexpressed during the process of angiogenesis are presumably the target and are responsible for the accumulation of the cationic carriers (Fig. 2). It is known that tumor endothelial cells divide much more rapidly compared to quiescent endothelial cells in normal tissue. Up to 100- to 1000-fold higher proliferation rates have been reported (17,18).

During the last years, anionic membrane phospholipids, principally phosphatidylserine, were shown to be highly selective markers for the tumor vasculature. These anionic lipids would represent excellent binding partners for the cationic liposomes and could be responsible for their accumulation in the tumor vascular bed. Under normal conditions, anionic phospholipids are found on the inner leaflet of the plasma membrane in most cell types. Recent studies have demonstrated that anionic phospholipids—most likely phosphatidylserine—become exposed on the luminal surface of tumor endothelial cells, probably due to oxidative stress and the action of inflammatory cytokines (19,20). Exposure of phosphatidylserine in a single tumor vessel can show some heterogeneity (21); however, phosphatidylserine is absent from endothelium of normal tissues.

In vitro, the exposure of endothelial cells to the standard chemotherapeutic agents docetaxel and/or gemcitabine triggers externalization of phosphatidylserine from the inner to the outer leaflet of the cell membrane—without inducing apoptosis (22,23). For docetaxel, it could already be demonstrated that systemic treatment of tumor-bearing mice strongly

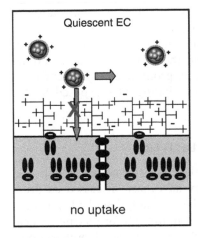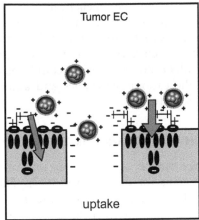

Figure 2 Schematic graph depicting membrane characteristics of quiescent and activated/dividing endothelial cells (ECs). Increased exposure of negatively charged surface molecular moieties is the basis for preferential binding and internalization of cationic liposomes to tumor vasculature.

increased the percentage of tumor blood vessel cells that expose anionic phospholipids—but not in quiescent endothelial cells in normal tissue (23). This observation allows the speculation that preadministration of gemcitabine or docetaxel might enhance the effectiveness of cationic liposome binding and uptake to the tumor vasculature, an important fact to be considered in the context of combination therapy studies.

Glycosaminoglycans and proteoglycans have a central role in regulation of biological processes as, for example, cellular proliferation and differentiation including angiogenesis (24,25). The expression pattern of negatively charged cell surface proteoglycans can change during angiogenesis (26–28). The enormous heterogeneity and complexity of the proteoglycan structures and expression pattern makes it, however, difficult to judge to which extend extracellular proteoglycans contribute to an increased exposure of anionic sites at the luminal surface of endothelial cells.

The contribution of proteins that are expressed on tumor endothelial cells to the targeting selectivity of cationic liposomes is a field for future research. It is known that the tumor microenvironment induces distinct protein expression on the tumor endothelial cell surface. Expression of negatively charged hyperglycosylated and hypersialylated membrane proteins on activated endothelial cells has been described in vitro and in vivo (29–31). Endosialin (TEM1), carrying abundantly sialylated, O-linked oligosaccharides, was identified as a tumor endothelial marker (32,33), although the extent of expression selectivity still needs more clarification (7,34). For an overview about the currently known proteins that are preferentially expressed on the proliferating tumor endothelial cells and the techniques for identification of new targets, see the recently published review (7).

Also, characteristic structural features and abnormalities of tumor vessels, e.g., their tortuous architecture and their sluggish and sometimes irregular blood flow (11,35,36), could possibly promote charge interactions between the cationic liposomes and anionic binding sites exposed by the tumor vasculature.

CATIONIC LIPOSOMES FOR TARGETED DELIVERY

Liposomes can be considered as one of the longest-established colloidal carrier systems for drug delivery (37). They have been discussed as drug-delivery vehicles already soon after they have been initially described more than 40 years ago (38), and the field of research on liposomes for pharmaceutical use is an active and growing area (39). Several liposomal preparations, including products for tumor therapy, have reached the market.

In a general perspective, the aim of liposomal drug formulation is to improve parameters like pharmacokinetics, pharmacodynamics, or bioavailability of a given compound (37,39). For example, liposomes can be useful to extend circulation times of small hydrophilic molecules, which are otherwise rapidly excreted by the kidneys. Liposomal encapsulation may also protect the active compound from undesired serum interactions and rapid degradation. A key application for liposomes is targeted delivery to a selected type of tissue (39). Such targeted delivery would be particularly desirable in tumor chemotherapy, to enhance delivery of the cytotoxic agent to the tumor and to reduce the cytotoxic burden to healthy tissue (40). Passive targeting of liposomes to a tumor can be obtained through the so-called enhanced permeability and retention effect, which makes use of the observation that tumor tissue displays an elevated permeability for small particles. In this case, it is important that the liposomes are small enough to permeate across the leaky tumor endothelial layer, and that their circulation time is long enough to achieve sufficient accumulation at the tumor site. The classical example for active targeting is immunoliposomes. These are functionalized with a specific binding agent, like antibodies or antibody fragments, for selective binding to a specific molecular ligand motif at the cellular membranes of the target tissue.

The innovative concept for targeted drug delivery of the EndoTAG® platform differs from these established and well-known approaches in several ways. In analogy to immunoliposomes, cationic colloidal carriers enable targeted delivery to a specific type of tissue, in this case activated (angiogenic) endothelial cells. However, specific binding is achieved not by an antigen–antibody type of interaction: tissue selectivity is based on certain changes in the molecular composition and properties of the activated endothelial cells, as described above (Fig. 2). Another fundamental difference to conventional liposome formulations is that binding and uptake of cationic liposomes occurs very fast. Therefore, the criteria for formulation development and the implications in vivo are fundamentally different from those for long circulation liposomes.

MOLECULAR SETUP OF EndoTAG®

EndoTAGR® is the technology platform of cationic nanoparticles for neovascular targeting, being developed at MediGene AG. The formulations are administered as IV infusion of the aqueous colloidal dispersion.

All products that are realized on this basis comprise a cationic nanoparticulate carrier and an active agent. For the cationic carrier, various types of colloidal particles comprising liposomes, micelles, emulsion droplets polymer particles, or any other type of nanoparticle can be chosen. Active agents may be small or large molecules, including polymeric compounds, proteins, peptides, or nucleic acids. Hydrophilic (water-soluble), lipophilic (water-insoluble), or amphiphilic compounds can be loaded to the carrier particle. Fundamental characteristics of the preparations are size, zeta-potential, and concentration of the colloidal particles, as well as composition and phase state of the particle-forming molecules.

In the following, only lipid-based (liposomal) formulations comprising small molecules as active agents are discussed. A selection of lipid components which are suitable for the assembly of liposomal EndoTAG® formulations is given in Fig. 3.

N-[1-(2,3-Dioleoyloxy)propyl]-*N*,*N*,*N*-trimethylammonium chloride (DOTAP) is a synthetic lipid, which comprises one positive charge at the headgroup. It is at room temperature in a fluid-like (liquid crystalline) state. *N*-[1-(2,3-Dioleoyloxy)propyl]-*N*,*N*,*N*-trimethylammonium chloride (DMTAP) is a CL where the headgroup is identical to that of DOTAP, but the alkyl chains are shorter and fully saturated. DMTAP displays a liquid crystalline–gel phase transition in the temperature range between room temperature and body temperature.

1,2-Dioleoyl-sn-glycero-3-phosphatidyl choline (DOPC) is a zwitterionic natural phospholipid, which, as DOTAP, comprises two oleic acid groups as hydrophobic part and is fluid like at room temperature. Cholines with different chain length and degree of saturation can be used to fine-tune the membrane dynamics and phase properties. With phosphatidylethanolamines (not shown), neutral phospholipids are available, where the headgroup cross section is smaller than that of cholines.

Cholesterol is a widely used constituent of liposome formulations. It is known to modify the properties of both liquid crystalline and gel phase of lipid membranes and to diminish the liquid crystalline/gel phase transition. In fact, the membrane fluidity and phase state is one of the crucial parameter for drug loading and targeting of the lipid complex.

DOTAP

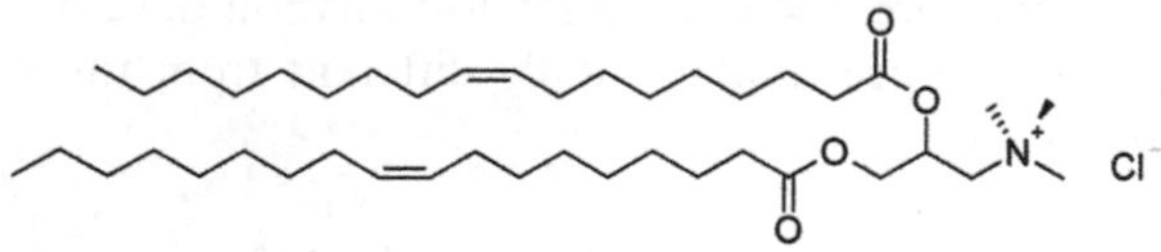

Molecular Weight = 663.11 + 35.45 = 698.56
Molecular Formula = $C_{42}H_{80}NO_4 Cl$

DOPC

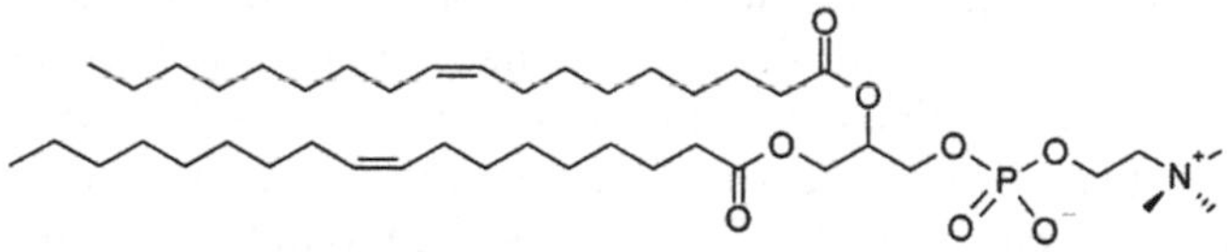

Molecular Weight = 786.14
Molecular Formula = $C_{44}H_{84}NO_8P$

Paclitaxel

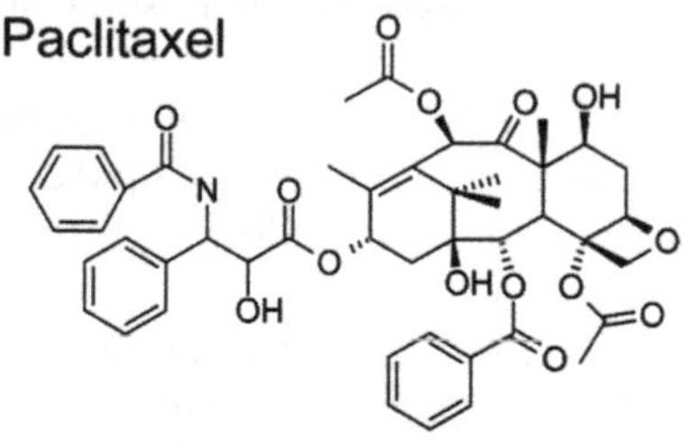

Molecular Weight = 853.93
Molecular Formula = $C_{47}H_{51}NO_{14}$

Na-camptothecin

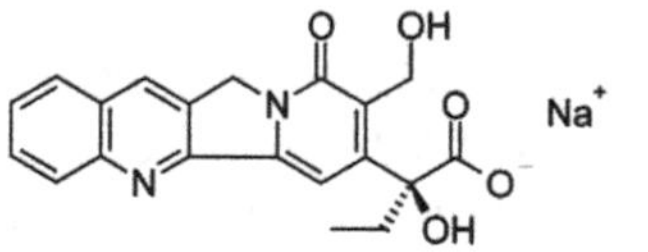

Molecular Weight = 365.37 + 22.99 = 388.36
Molecular Formula = $C_{20}H_{17}N_2O_5Na$

Figure 3 Important molecules for the formation of EndoTAG® formulations. DOTAP (cationic, high fluidity), DOPC (zwitterionic, high fluidity), and cholesterol (neutral, phase modulating) are lipid components. Paclitaxel and camptothecin are active components for therapeutic formulations. *Abbreviations*: DOTAP, *N*-[1-(2,3-Dioleoyloxy)propyl]-*N*,*N*,*N*-trimethylammonium chloride; DOPC, 1,2-Dioleoyl-sn-glycero-3-phosphatidyl choline.

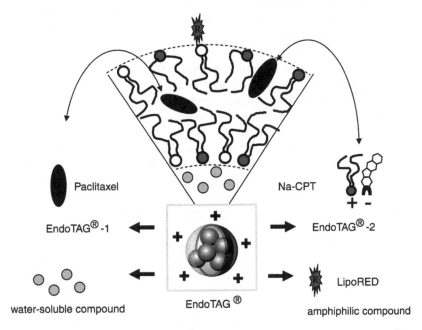

Paclitaxel

Na-CPT

EndoTAG®-1

EndoTAG®-2

EndoTAG®

water-soluble compound

LipoRED

amphiphilic compound

Figure 4 Methodologies for loading of active compounds to the EndoTAG® carrier. Classical approaches from liposome technology are to encapsulate water-soluble compounds in the aqueous compartment or to insert hydrophobic molecules in the lipid bilayer (*left side*). LipoRed comprises an amphiphilic fluorescent label, which forms an integral part of the bilayer. For the setup of EndoTAG®-2 advanced methods of loading camptothecin to the cationic carrier by favorable molecular interactions with a positively charged colipid are applied (*right side*).

The molecular composition and setup of these formulations is fine-tuned according to the properties of the active compound and the intended application. In Figure 4, various options for loading an active component to a liposomal EndoTAGR® carrier are depicted. Important factors in this context are if the molecule is hydrophilic, hydrophobic, or amphiphilic, and parameters like water-solubility, charge, size, or steric features. Such as with conventional liposomes, water-soluble compounds can be encapsulated into the aqueous compartment and hydrophobic molecules can be inserted in the lipid bilayers membrane (Fig. 4, *left side*). In addition, other means of loading have been established for the preparation of EndoTAG® formulations, which take into account favorable molecular interactions between the active component and the lipid matrix (Fig. 4, *right side*). Techniques for industrial manufacturing of the products in line with Good Manufacturing Practice (cGMP) requirements have been developed and in the subsequent text, as an example, the production scheme for EndoTAG®-1[b] will be outlined.

For targeted delivery of encapsulated water-soluble compounds, it is important to maintain the liposomal integrity in vivo, to keep the active component encapsulated in the aqueous compartment. With EndoTAG® liposomes, this is provided by adjusting membrane rigidity by selecting suitable lipid mixtures and by addition of moieties for steric hindrance of direct interactions with off-target molecules.

EndoTAG® formulations, where water-soluble contrast agents for magnetic resonance and X-ray imaging are encapsulated, have already been developed. With formulations comprising water-soluble gadolinium contrast agent, the distribution and tumor binding directly after injection can be determined in vivo (41) by K-edge imaging, a novel X-ray imaging technique (42). Measurements at the ID 17 medical beamline, European synchrotron radiation facility (ESRF), France enabled to determine the distribution of Gd with a temporal resolution of few seconds and with a spatial resolution in the order of 50 to 350 µm. These measurements are a tool for direct quantitative comparison and screening of different formulations with online control in an in vivo experiment.

Lipid-like or membrane-forming agents can be inserted as integral part into lipid bilayer membrane of the liposomes. For diagnostic applications, lipids that are functionalized with a marker for fluorescence or magnetic resonance imaging can be used, but also therapeutically active amphiphilic molecules can be inserted in the membrane. The vascular imaging agent LipoRed comprises a mixture of cationic and zwitterionic lipids and an amphiphilic fluorescent marker. LipoRed is a versatile tool for vascular targeting studies and screening. In contrast to the preparations comprising water-soluble compounds, here the lipid membrane is more fluid-like. Formulation development and screening experiments with LipoRed have been the basis for development of the subsequently described therapeutic product EndoTAG®-1. In preclinical studies, LipoRed and cationic liposomes showed strong association with the tumor blood vessel area without signs of significant extravasation. After IV administration, 20% to 50% of the administered cationic liposomes interacted with the endothelial compartment, whereby 2% to 5% of the total lipid dose accumulated in the tumor microvasculature (14), (MediGene AG. unpublished data, 2004.). This dramatic change in the intratumoral biodistribution can result in high therapeutically effective drug concentrations inside the tumor endothelial cells, because the tumor endothelial mass corresponds only to the estimated 0.1% to 1.0% of the total tumor tissue mass.

---

[b] EndoTAGR®-1 and EndoTAGR®-2 are the new names for MBT-0206 and MBT-0312

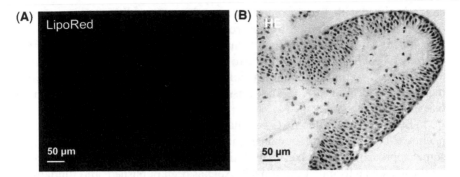

Figure 5 Accumulation of fluorescently labeled EndoTAG® (LipoRed) in stromal blood vessels in proximity to a papillary highly differentiated urothelial carcinoma (pT1, G3). LipoRed dose was 2 mg total lipid per kg bodyweight. (**A**) Fluorescence image of a cryo-section of tumor biopsy material. Strong distinct LipoRed fluorescence (light gray areas) is accumulated in stroma blood vessels surrounded by a papille of the carcinoma. (**B**) Hematoxylin/eosin staining (HE). After recording of the fluorescent image "A" the same section was directly subjected to HE staining for morphologic and histopathologic control using white light microscopy.

LipoRed was the first cationic liposome formulation for IV use in humans. Targeting to the human tumor neovasculature was demonstrated in clinical phase I trials in bladder cancer and head and neck squamous cell carcinoma. Histological analysis of bladder tumor biopsies revealed that liposome related fluorescence accumulated in capillaries and medium-sized tumor blood vessels. Interestingly, stromal capillaries in areas adjacent to the tumor were also labeled, indicating an association of targeting to neoangiogenesis (Fig. 5). In contrast, normal mucosa and deeper areas of the bladder wall with larger vessels were devoid of LipoRed fluorescence (43).

EndoTAG®-1

In EndoTAG®-1 the active component is hydrophobic. The drug is inserted into the lipid bilayer, and in particular its hydrophobic compartment (Fig. 4), which acts as a two-dimensional solvent for the compound. EndoTAG®-1 comprises the diterpenoid paclitaxel (44), a potent antimitotic agent widely used in cancer therapy (45). Paclitaxel has a very low solubility in water (in the order of 1 mg/L) and therefore its solubility must be increased for IV application. In Taxol® (Brystol-Myers Squibb), which is approved for treatment of advanced ovarian, breast, and non–small cell lung cancer in the United States and in Europe, paclitaxel is solubilized by a mixture of Cremophor® EL (BASF) and ethanol. However, Cremophor causes serious side effects such as hypersensitivity reactions and peripheral neuropathy (46,47). Prophylactic steroids and histamine receptor antagonists have to be coadministered with Taxol to reduce these effects. In addition, the maximum dose of paclitaxel is limited by neutropenia and neurotoxicity.

Great efforts to develop alternative formulations for paclitaxel are ongoing (48–50). Common goals are to provide sufficient solubility in aqueous environment, to improve pharmacokinetic and pharmacodynamic parameters, to reduce side effects and, possibly, to improve delivery to the target tissue of the drug.

Even though EndoTAG®-1 has another target than conventional paclitaxel-based products for cancer therapy, it can also be regarded as a liposomal approach for paclitaxel formulation (50). A significant number of studies on the liposomal formulation of paclitaxel has been published, and fundamental aspects of paclitaxel-membrane interactions have been investigated to detail by various methods (51–54). For development of EndoTAG®-1, an excessive screening of drug/lipid mixtures and formulation techniques has been carried out. Physicochemical characterization of drug-loaded model membranes was performed in order to get insight into general aspects of paclitaxel insertion into (cationic) lipid membranes. Inter alia, differential scanning calorimetry measurements, spectroscopic techniques, X-ray scattering, and Langmuir monolayer measurements (55,56) have been applied as tools to study paclitaxel membrane interactions and to define the formulation parameters.

Figure 6 shows the understanding of the molecular organization of paclitaxel in liposomal preparations which can be derived from such experiments (50). In addition to the paclitaxel, which is inserted into the liposomes,

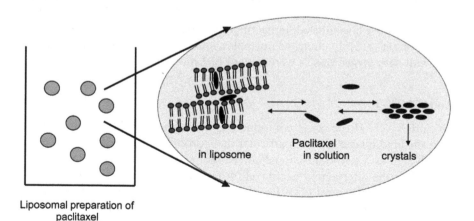

Liposomal preparation of
paclitaxel

Figure 6 Options for the partition of paclitaxel in liposome preparations. The drug is supposed to be inserted in the liposomal lipid bilayer. In addition, a fraction that is dissolved in water has to be taken into account. If the maximum solubility of paclitaxel is exceeded, formation of microcrystals as a colloidal dispersion or a precipitate can occur.

it may be present in the aqueous phase, or as precipitated or colloidally dispersed crystallites. In equilibrium, there is a constant ratio between the concentration of paclitaxel in the liposome and in the aqueous phase (53). If liposomes are loaded with an amount of paclitaxel, which is higher than the equilibrium value, paclitaxel release and subsequent crystallization of the drug may have to be taken into account. In a practical pharmaceutical preparation, the concentration of paclitaxel should be about two to three orders of magnitude higher than its maximum solubility in water.

EndoTAG®-1, which is currently tested in clinical studies, comprises about 3 mol% paclitaxel in a DOTAP/DOPC lipid matrix. For application to a patient it is present as a colloidal dispersion of particles of uniform size of about 200 nm, where the total lipid concentration is 10 mM. For storage, the formulations are lyophilized, and they are reconstituted with water for injection directly prior use. A robust industrial scale process for manufacturing and lyophilization of liposomal products was developed (57), and is summarized in Figure 7. Lipids and paclitaxel are dissolved in ethanol at the appropriate molar ratio, and this concentrated solution is injected into the aqueous phase under stirring. Thus, drug-loaded, polydisperse liposomes are formed spontaneously by a self-assembly process. The size distribution of the liposomes is adjusted by several consecutive extrusion cycles through membranes of defined pore size. After sterile filtration, the preparation is filled into moulded vials and freeze-dried. By lyophilization of the product, a shelf life of more than two years is provided. Regular cGMP production has been performed with a bulk size of about 70 L, resulting in a reproducible and consistent quality of the final product in more than 15 production batches.

In a broad preclinical program, the biological effects of EndoTAG®-1 were studied in comparison to different controls, including Taxol. EndoTAG®-1 showed superior antitumor activity in a variety of different species and

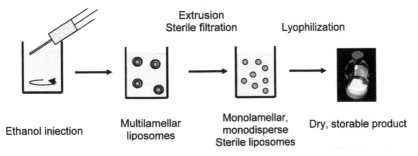

Figure 7 Production scheme of EndoTAG®-1. Multilamellar liposomes are formed by ethanol injection of the lipid and drug solution into the aqueous phase. By extrusion and sterile filtration, monolamellar, monodisperse, and sterile liposomes are formed. Subsequently, the preparation is freeze-dried for storage.

tumor models. It significantly inhibited tumor growth, delayed the onset of metastasis, inhibited infiltration of healthy tissue surrounding the tumor, and increased the survival of tumor-bearing animals (16,58,59). Importantly and indicative for a different mode of action, EndoTAG®-1 inhibited tumor growth also in Taxol-resistant animal tumor models, as for example, B16 melanoma and Sk-Mel 28 melanoma. EndoTAG®-1 demonstrated a strong antivascular effect on the preexisting tumor vasculature and affected several tumor microcirculatory parameters. It reduced the endothelial cell mitotic rate in the vicinity of the tumor (58), caused a dramatic lasting reduction of tumor perfusion (59) and tumor vessel damage (59,60). There is evidence that continuous EndoTAG®-1 treatment can lead to tumor vessel leakage and enhanced accessibility of the tumor tissue for low-molecular-weight substances (60,61). EndoTAG®-1 treatment of mice bearing an orthotopically grown human pancreatic carcinoma led to a total suppression of liver metastasis. Combination of EndoTAG®-1 with the conventional chemotherapeutic drugs cisplatin (60) and gemcitabine (62) further enhanced tumor growth inhibition.

EndoTAG®-1 has passed a phase I clinical program with more than 150 patients with advanced metastatic cancer for determination of the safety and tolerability, and for investigation of the pharmacokinetic parameters.

EndoTAG®-1 appears to be a safe drug with an overall response rate of 8% to 14%; a range which is typical for effective drugs in phase I cancer trials. A large phase II trial for EndoTAG®-1 in combination with gemcitabine in patients with advanced or metastatic adenocarcinoma of the pancreas has started in 2005.

EndoTAG®-2

Also EndoTAG®-2 is a preparation for tumor therapy, but the molecular target of the active compound and the mechanism of loading the drug to the liposome are essentially different to those in EndoTAG®-1. The active compound in EndoTAG®-2 is camptothecin (CPT), a quinoline-based alkaloid, which can be isolated from the Chinese tree Camptotheca acuminata. CPT is a topoisomerase inhibitor, i.e., binding to the topoisomerase I-DNA complex induces DNA breaks and cell death (64).

A fundamental molecular property of CPT is its pH-dependent equilibrium between the lactone and the carboxylate form (Fig. 8, *left side*). The lactone form is lipophilic, whereas the carboxylate, which predominates at physiological pH and above, is water-soluble. Both molecular forms are present as equilibrium and one form can be transformed into the other one, for example, by changing the pH. The carboxylate form is considered to be less active and responsible for severe side reactions such as neutropenia, thrombocytopenia, and hemorrhagic cystitis (63). Therefore, efforts in the development of CPT drugs concentrated on the stabilization of the lactone

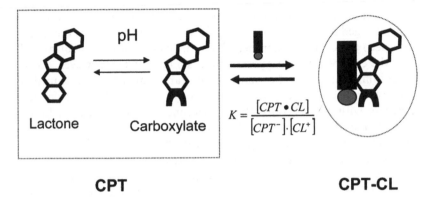

CPT   **CPT-CL**

Figure 8 The molecular properties of camptothecin (CPT) are determined by the pH-dependent equilibrium between the lactone and the carboxylate form. The lactone is poorly soluble in water and it is stable only at low pH. At physiological pH and higher, the carboxylate predominates. In EndoTAG®-2, CPT is used in the lipid-bound form, a new type of CPT drug with favorable molecular and pharmacological characteristics. *Abbreviations*: CPT, camptothecin; CL, cationic lipid.

form and on finding means for the administration of the compound in that form without difficulties (65). In a large number of studies, chemical modification of the parent drug and prodrug approaches have been pursued to obtain CPT lactones in an injectable form. This has been realized for the two components, which are on the market, both CPT derivatives, Topotecan (Hycamtin®, Glaxo Smith Kline) and Irinotecan (codeveloped and launched by Yakult, Japan; and Rhone-Poulenc Rorer, now Sanofi_Aventis, France, and Pharmacia, now Pfizer, U.S.A.). Also, liposomes have been used to stabilize CPT and derivatives thereof in the lactone form, either by encapsulation in acidic environment or by embedding the drug into lipid bilayers from certain phospholipids.

For the development of EndoTAG®-2, a strategy was followed that is opposite to the development within the last 30 years, because it bases on using the carboxylate form and not on the lactone form of a CPT drug. Instead of trying to control the equilibrium between the lactone and the carboxylate, such as in approaches for CPT drug development so far (Fig. 8, *left side*), a third, new CPT moiety is formed. The CPT-carboxylate binds by noncovalent interactions to a CL (Fig. 8, *right side*) and thus a new complex with altered properties is obtained. In contrary to the CPT-carboxylate, which is water-soluble, and the CPT-lactone, which is hydrophobic, the lipid-complexed CPT has amphiphilic properties and a high partition coefficient in the membrane. With this new compound, a key challenge in CPT drug development—to fulfill the conflicting requirements of transport to the target tissue and permeation across the cell membrane—has been solved. One reason for the low efficacy and the occurrence of side reaction with

CPT in its water-soluble (injectable) carboxylate form may be that, as a negatively charged molecule, it has a very low permeability across the (negatively charged) target cell membrane. CPT in its lactone form is neutral and, therefore, the energy barrier for permeation is much lower. However, due to its poor solubility in water, transport to the target membrane is impeded. Interestingly, the side effects with CPT-carboxylate were observed particularly in organs with low pH. Formation of lactone and subsequent facilitated uptake of CPT in these organs may have occurred.

With CL-complexed CPT, liposomes with high drug load can be produced while the manufacturing process can be greatly facilitated. On the basis of these molecular coherencies, a particular manufacturing schedule has been developed for EndoTAG®-2. Lipid composition, the procedure of liposome production, and the way to obtain long shelf life have been adopted.

Strong antitumor activity could be demonstrated for EndoTAG®-2. In several animal models, IV administration of EndoTAG®-2 caused remarkable tumor growth inhibition or even complete tumor regression. In contrast to treatment with the free-drug CPT, EndoTAG®-2 had strong and rapid effects on the tumor vasculature, causing tumor endothelial cell apoptosis even after single treatment (66) and strong reduction of tumor vessel density (67).

CONCLUSIONS

Cationic liposomes as VDAs open a new, promising approach for tumor therapy. In principle, all solid tumors should respond to EndoTAG® treatment and the risk of multidrug resistance development should be reduced. The potential of EndoTAG®-based formulations for treatment of chemotherapy-resistant tumors is of especially high importance. A further potential advantage of EndoTAG® therapeutics is that the tumor blood vessel targeting should lower organ toxicity, usually observed with conventional cytostatic drugs. For the future development of further products on the basis of the EndoTAG® technology platform, the control and understanding of the molecular interactions and the self-assembly processes will be a key asset.

ACKNOWLEDGMENT

The authors would like to thank Dr. Ulrich Delvos for critically reading the manuscript and for his comments.

REFERENCES

1. Siemann DW, Bibby MC, Dark GG, et al. Differentiation and definition of vascular-targeted therapies. Clin Cancer Res 2005; 11:416.
2. Tozer GM, Kanthou C, Baguley BC. Disrupting tumour blood vessels. Nat Rev Cancer 2005; 5:423.

3. Inai T, Mancuso M, Hashizume H, et al. Inhibition of vascular endothelial growth factor (VEGF) signaling in cancer causes loss of endothelial fenestrations, regression of tumor vessels, and appearance of basement membrane ghosts. Am J Pathol 2004; 165:35.

4. Willett CG, Boucher Y, diTomaso E, et al. Direct evidence that the VEGF-specific antibody bevacizumab has antivascular effects in human rectal cancer. Nat Med 2004; 10:145; erratum in: Nat Med 2004; 10:649.

5. Jain RK. Normalization of tumor vasculature: an emerging concept in anti-angiogenic therapy. Science 2005; 307:58.

6. Thorpe PE. Vascular targeting agents as cancer therapeutics. Clin Cancer Res 2004; 10:415.

7. Neri D, Bicknell R. Tumour vascular targeting. Nat Rev Cancer 2005; 5:436.

8. Dass CR. Improving anti-angiogenic therapy via selective delivery of cationic liposomes to tumour vasculature. Int J Pharm 2003; 267:1.

9. Eichhorn ME, Strieth S, Dellian M. Anti-vascular tumor therapy: recent advances, pitfalls and clinical perspectives. Drug Resist Updat 2004; 7:125.

10. Siemann DW, Chaplin DJ, Horsman MR. Vascular-targeting therapies for treatment of malignant disease. Cancer 2004; 100:2491.

11. Bergers G, Benjamin LE. Tumorigenesis and the angiogenic switch. Nat Rev Cancer 2003; 3:401.

12. Kerbel RS, Yu J, Tran J, et al. Possible mechanisms of acquired resistance to anti-angiogenic drugs: implications for the use of combination therapy approaches. Cancer Metastasis Rev 2001; 20:79.

13. Thurston G, McLaean JW, Rizen M, et al. Cationic liposomes target angiogenic endothelial cells in tumors and chronic inflammation in mice. J Clin Invest 1998; 101:1401.

14. Campbell RB, Fukumura D, Broen EB, et al. Cationic charge determines the distribution of liposomes between the vascular and extravascular compartments of tumors. Cancer Res 2002; 62:6831.

15. Krasnici S, Werner A, Eichhorn ME, et al. Effect of the surface charge of liposomes on their uptake by angiogenic tumor vessels. Int J Cancer 2003; 105:561.

16. Schmitt-Sody M, Strieth S, Krasnici S, et al. Neovascular targeting therapy: paclitaxel encapsulated in cationic liposomes improves antitumoral efficacy. Clin Cancer Res 2003; 9:2335.

17. Denekamp J. Review article: angiogenesis, neovascular proliferation and vascular pathophysiology as targets for cancer therapy. Br J Radiol 1993; 66:181.

18. Eberhard A, Kahlert S, Goede V, et al. Heterogeneity of angiogenesis and blood vessel maturation in human tumors: implications for antiangiogenic tumor therapies. Cancer Res 2000; 60:1388; erratum in Cancer Res 2000; 60:3668.

19. Ran S, Downes A, Thorpe PE. Increased exposure of anionic phospholipids on the surface of tumor blood vessels. Cancer Res 2002; 62:6132.

20. Ran S, Thorpe PE. Phosphatidylserine is a marker for tumor vasculature and a potential target for cancer imaging and therapy. Int J Rad Oncol Biol Phys 2002; 54:1479.

21. Ran S, He J, Huang X, et al. Antitumor effects of a monoclonal antibody that binds anionic phospholipids on the surface of tumor blood vessels in mice. Clin Cancer Res 2005; 11:1551.

22. Luster TA, Beck AW, Thorpe PE, et al. Inhibition of pancreatic tumor growth and metastasis in mice by targeting inside-out phospholipids on tumor vasculature. Proc Amer Assoc Cancer Res 2005; 46:3014.
23. Huang X, Bennett M, Thorpe PE. A monoclonal antibody that binds anionic phospholipids on tumor blood vessels enhances the antitumor effect of docetaxel on human breast tumors in mice. Cancer Res 2005; 65:4408.
24. Trowbridge JM, Gallo RL. Dermatan sulfate: new functions from an old glycosaminoglycan. Glycobiology 2002; 12:117.
25. Iozzo RV, San Antonio JD. Heparan sulfate proteoglycans: heavy hitters in the angiogenesis arena. J Clin Invest 2001; 108:349.
26. Schönherr E, Sunderkötter C, Schaeffer L, et al. Decorin deficiency leads to impaired angiogenesis in injured mouse cornea. J Vasc Res 2004; 41:499.
27. Nelimarkka L, Salminen H, Kuopio T, et al. Decorin is produced by capillary endothelial cells in inflammation-associated angiogenesis. Am J Pathol 2001; 158:345.
28. Qiao D, Meyer K, Mundhenke C, et al. Heparan sulfate proteoglycans as regulators of fibroblast growth factor-2 signaling in brain endothelial cells. Specific role for glypican-1 in glioma angiogenesis. J Biol Chem 2003; 278:16045.
29. Augustin HG, Kozian DH, Johnson RC. Differentiation of endothelial cells: analysis of the constitutive and activated endothelial cell phenotypes. Bioessays 1994; 16:901.
30. Doiron AL, Kirkpatrick AP, Rinker KD. TGF-beta and TNF-α affect cell surface proteoglycan and sialic acid expression on vascular endothelial cells. Biomed Sci Instrum 2004; 40:331.
31. Augustin HG, Braun K, Telemenakis I, et al. Ovarian angiogenesis. Phenotypic characterization of endothelial cells in a physiological model of blood vessel growth and regression. Am J Pathol 1995; 147:339.
32. St. Croix B, Rago C, Velculescu V, et al. Genes expressed in human tumor endothelium. Science 2000; 289:1197.
33. Christian S, Ahorn H, Koehler A, et al. Molecular cloning and characterization of EndoGlyx-1, an EMILIN-like multisubunit glycoprotein of vascular endothelium. J Biol Chem 2001; 276:48588.
34. Opavsky R, Haviernik P, Jurkovicova D, et al. Molecular characterization of the mouse Tem1/endosialin gene regulated by cell density in vitro and expressed in normal tissues in vivo. J Biol Chem 2001; 276:38795.
35. McDonald DM, Choyke PL. Imaging of angiogenesis: from microscope to clinic. Nat Med 2003; 9:713.
36. Tozer GM, Ameer-Beg SM, Baker J, et al. Intravital imaging of tumour vascular networks using multi-photon fluorescence microscopy. Adv Drug Deliv Rev 2005; 57:135.
37. Gregoriadis G. ed. Liposome Technology. Vol. 1–3. Boca Raton, U.S.A.: CRC, 1984.
38. Bangham AD, Standish MM, Watkins JC. Diffusion of univalent ions across the lamellae of swollen phospholipids. J Mol Biol 1965; 13:238.
39. Torchilin VP. Recent advances with liposomes as Pharmaceutical Carriers. Nat Rev 2005; 4:145.
40. Marucci F, Lefoulon F. Active targeting with particulate drug carriers in tumor therapy: Fundamental and recent progress. Drug Discovery Today 2004; 9:219.

41. Elleaume H et al. In vivo K-edge imaging with synchrotron radiation. Cell Mol Biol 2000; 46(6):1065.
42. Haas H, Werner A, Le Duc G. ESRF Experimental Report 26428_B, http:// ftp.esrf.fr/pub/UserReports/26428_B.pdf, 2003.
43. Knuechel R, Bartelheim K, Reich O, et al. Cationic liposomes for tumor imaging: first clinical results. Proc Amer Assoc Cancer Res 2003; 44:5397.
44. Wani MC, et al. Plant antitumor agents: VI the isolation and structure of taxol, a novel antileucemic and antitumor agent from *Taxus Bbrevifolia*. J Am Chem Soc 1971; 93:2325.
45. Rowinski EK, Donehower RC. Paclitaxel (Taxol). N Engl J Med 1995; 332:1004.
46. van Zuylen L, Verweij J, Sparreboom A. Role of formulation vehicles in taxane pharmacology. Invest New Drugs 2001; 19(2):125.
47. Tije AJ, Loos WJ, Sparreboom A. Pharmacological effects of formulation vehicles: implications for cancer chemotherapy. Clin Pharmacokinet 2003; 42(7): 665–685.
48. Nuijen B, Bouma M, Schellens JH, et al. Progress in the development of alternative pharmaceutical formulations of taxanes. Inv New Drugs 2001; 19:143.
49. Singla AK, Garg A, Deepika A. Paclitaxel and its formulations. Int J Pharm 2002; 235:179.
50. Haas H. Entwicklung neuer taxan-formulierungen. Pharm Unserer Zeit 2005; 2:115.
51. Balasubramanian SV, Straubinger RM. Taxol-lipid interactions: taxol-dependent effects on the physical properties of model membranes. Biochemistry 1994; 338941.
52. Bernsdorff C, Reszka R, Winter R. Interaction of the anticancer agent Taxol (paclitaxel) with phospholipid bilayers. J Biomed Mater Res 1999; 46(2):141.
53. Wenk MR, Fahr A, Reszka R, et al. Paclitaxel partitioning into lipid bilayers. J Pharm Sci 1995; 85(2):228.
54. Campbell RB, Balasubramanian SV, Straubinger RM, et al. Influence of cationic lipids on the stability and membrane properties of paclitaxel-containing liposomes. J Pharm Sci 2001; 90:1091–1105.
55. Gruber F. Dissertation, Ludwig-Maximilians-Universität München 2003. In: Gruber F, Haas H, Winter G, eds. Annual Meeting. Glasgow, UK: CRS, 2003.
56. Cavalcanti LP, Konovalov O, Torriani IL, et al. Drug loading to lipid-based cationic nanoparticles. Nucl Instr Methods Phys Res B 2005; 238(1–4):290–293.
57. Haas H, Drexler K, Welz C, et al. Industrial manufacturing of EndoTAG®-1, a novel liposomal product for tumor therapy. Miami, USA: Annual Meeting CRS, 2005.
58. Kunstfeld R, Wickenhauser G, Michaelis U, et al. Paclitaxel encapsulated in cationic liposomes diminishes tumor angiogenesis and melanoma growth in a "humanized" SCID mouse model. J Invest Dermatol 2003; 120:476.
59. Strieth S, Eichhorn ME, Sauer B, et al. Neovascular targeting chemotherapy: encapsulation of paclitaxel in cationic liposomes impairs functional tumor microvasculature. Int J Cancer 2004; 110:117.
60. Strieth S, Eichhorn ME, Sauer B, et al. Neovascular targeting paclitaxel encapsulated in cationic liposomes increases tumor microvessel permeability and enables effective combination therapy with cisplatin. Angiogenesis 2004; 7(suppl 1):35.
61. Eichhorn ME, Becker S, Strieth S, et al. Dynamic magnet resonance tomography for monitoring the efficacy of an anti-vascular tumor therapy by paclitaxel

encapsulating cationic lipid complexes (MBT-0206). Proc Amer Assoc Cancer Res 2004; 45:1310.

62. Papyan A, Werner A, Ischenko I, et al. Combination of standard chemotherapy with MBT-0206 enhances the anti-tumor efficacy in a highly metastatic human pancreatic cancer mouse model. Angiogenesis 2004; 7:22.

63. Wall ME, Wani MC. Camptothecin and taxol: from discovery to clinic. J Ethnopharmacol 1996; 51(1–3):239; discussion 253.

64. Hsiang YH, Liu LF. Identification of mammalian DNA topoisomerase I as an intracellular target of the anticancer drug camptothecin. Cancer Res 1988; 48(7): 1722.

65. Zunino F, Pratesi G. Camptothecins in clinical development. Expert Opin Investig Drugs 2004; 13(3):269.

66. Brill B, Werner A, Brunner C, et al. Neovascular targeting by EndoTAG-2 (MBT-0312) inhibits growth of subcutaneous mouse tumors. Angiogenesis 2004; 7(suppl 1):29.

67. Luedemann S, Eichhorn ME, Strieth S, et al. Vascular targeting chemotherapy: encapsulation of camptothecin in cationic lipid complexes (MBT-0312) effectively impairs tumor growth and tumor vasculature. Angiogenesis 2004; 7(suppl 1):19.

10

Folate Receptor-Targeted Liposomes

Xiaogang Pan and Robert J. Lee

Division of Pharmaceutics, College of Pharmacy, NCI Comprehensive Cancer Center, and NSF Nanoscale Science and Engineering Center, The Ohio State University, Columbus, Ohio, U.S.A.

Srikanth Shivakumar

Division of Pharmaceutics, College of Pharmacy, and NCI Comprehensive Cancer Center, The Ohio State University, Columbus, Ohio, U.S.A.

INTRODUCTION

Tumor cell selective targeting is an attractive approach to enhance the efficacy and therapeutic index of anticancer drugs. Targeted liposomal delivery is accomplished by attaching to the surface of drug-carrying liposomes a tumor cell–specific ligand via a lipophilic anchor. Folate receptor (FR) has been identified as a broad-spectrum tumor cell–surface marker. Lipophilic derivatives of folate, a high-affinity ligand for the FR, can be incorporated into liposomes for targeting tumor cells with amplified FR expression. This chapter will provide a brief overview of recent progress on FR-targeted liposomes, possible mechanisms for in vivo tumor targeting, and future directions.

FOLATE RECEPTOR AS A TUMOR MARKER

Folates [folic acid (Fig. 1) and its reduced derivatives] are essential coenzymes for DNA biosynthesis and one-carbon metabolism (1). Folates can be transported into cells via a low-affinity, high-capacity reduced-folate carrier (RFC) or a high-affinity FR. The RFC, which is ubiquitously expressed,

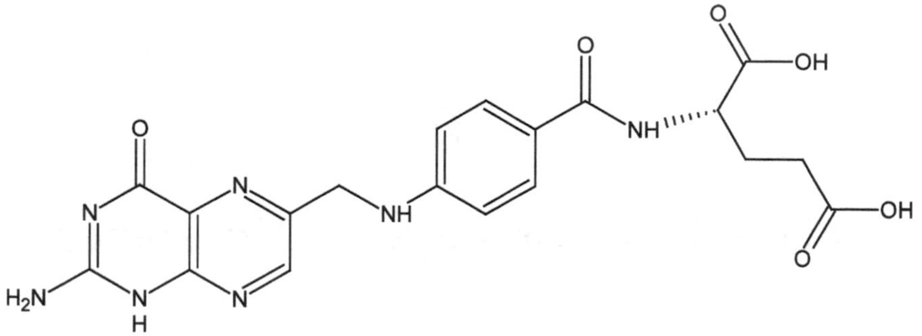

Figure 1 Structure of folic acid.

is an anion carrier that mediates the transmembrane transport of reduced folates, such as 5-methyltetrahydrofolate (5-MeTHF), or methotrexate, which is then followed by γ-polyglutamylation. With a K_t for folates in the 200–400 μM range, the RFC plays no role in targeted drug delivery.

In humans, FR is a 38 ~ 40 kDa N-glycosylated protein that has three isoforms α, β, and γ/γ′ (2–5). FR-α and FR-β are glycosylphosphatidyl-inositol-anchored membrane proteins, whereas FR-γ and FR-γ′ lack a membrane anchor and are constitutively secreted (6). The three FR isoforms share ~70% primary sequence homology (7,8) and exhibit high affinity for folic acid (K_d ~ 0.1 nM for FR-α, ~1 nM for FR-β, and ~0.4 nM for FR-γ). However, FR-α and FR-β show differential affinities to folate dia-stereomers (9,10). The α isoform of FR has a greater affinity for the physiological (6S)-5-MeTHF diastereomer, whereas the β isoform shows preference for the nonphysiological (6R)-5-MeTHF diastereomer.

Distribution of FRs in human tissues has been studied by various methods, including immunohistochemical staining (11), western blot, reverse transcription–polymerase chain reaction (12), and ^{3}H-folic acid binding (13). FR-α expression is found in a few normal tissues, including placenta, kidney (proximal tubules), fallopian tube, and choroids plexus. However, FR expression in these tissues is restricted to the luminal surface of certain epithelial cells, where it is not readily accessible to liposomes from systemic circulation (14). FR-α is frequently amplified in many malignant tissues, including ~90% of carcinomas of the ovary and the cervix, and to a lesser extent in other major cancers such as lung, breast, colon, brain, and renal cell (15,16). FR-α expression appears to be regulated by the estrogen and the glucocorticoid receptors and can be modulated by tamoxifen in estrogen receptor–positive cells (17) and by dexamethasone.

FR-β expression in normal tissue is restricted to placental and hema-topoietic cells. Mature neutrophils in peripheral blood express fivefold higher FR-β in an inactive form (nonfolate binding) than do myelomonocytic cells in the marrow (12). Therefore, FR-β is a differentiation maker in myelomonocytic lineage and during neutrophil maturation (12). In addition, FR-β in its active form is amplified in activated (not resting) monocytes and macrophages (18). Interestingly, functional FR-β is also expressed in about 70% acute myelogenous leukemia and in most chronic myelogenous leukemia (CML) blasts, which makes it a potential marker for targeting drug to myelogenous leukemias (12,19). FR-β expression is regulated via the retinoid receptors and can be modulated by *all-trans* retinoic acid (ATRA). Regulation of FRs in cancer and leukemia cells has recently been reviewed by Ratnam et al. (20)

The soluble FR-γ/γ′, mainly expressed at low levels in certain hematopoietic cells, has relatively insignificant role in the FR-targeted drug delivery. It, however, may potentially serve as serum markers for certain hematopoietic malignancies (6,21).

In summary, the tumor selectivity of FR lies in its absence in most normal tissues. In fact, physiological folate transport is primarily fulfilled through the RFC, which has much lower affinity for folates compared to the FR. The frequent overexpression among human tumors and highly restricted distribution among normal tissues suggest that both FR-α and FR-β can potentially be exploited as tumor-specific cell surface markers that can be used for targeted delivery of therapeutics in cancer and leukemia.

FOLATE AS A TUMOR-TARGETING LIGAND

Although early efforts on FR-targeting were focused on antibodies against the FR (22), folic acid conjugation has become the method of choice in more recent studies (23–27). Folic acid has two carboxyl groups (α and γ) in its glutamate moiety, which have pK_a's of 2.8 and 4.5, respectively. Derivatization via the γ-carboxyl group of folic acid preserves its affinity for the FR, thus providing a means to target this receptor (28–31). Derivatives of pteroic acid have also been shown to bind FR with high affinity. FR targeting has been evaluated as a strategy for enhancing tumor cell selective delivery of a wide variety of therapeutic agents. These include radiopharmaceuticals (32), chemotherapeutics (33), antisense oligodeoxyribonucleotides (ODNs) (34,35), prodrug-converting enzymes (36), anti-T cell receptor antibody (37), MRI and optical contrast agents (38,39), boronated neutron capture therapy agents (40), immunogenic hapten (41), gene transfer vectors (42), nanoparticles (43), and liposomal drug carriers (28). ^{111}In-DTPA-folate has been evaluated clinically as an imaging agent for detecting recurrent ovarian carcinomas. Preliminary results showed a sensitivity of 85% and a specificity of 82% for identifying malignant tumors (44). These findings demonstrate that FR-specific tumor uptake of a folate conjugate can occur despite the presence of physiological levels of folate and FR in the circulation and suggest that targeting of the FR in ovarian cancer is potentially feasible in the clinic.

Using folic acid as a ligand for targeted drug delivery has several advantages compared to polypeptide-based targeting ligands: (i) lack of immunogenicity; (ii) unlimited availability; (iii) functional stability; (iv) high affinity to FR ($K_d \sim 10^{-10}$ M); (v) low molecular weight; and (vi) defined conjugation chemistry. Moreover, it has been shown that folate conjugates can be efficiently and nondestructively internalized into cells via the FR-mediated endocytosis (28–31).

Notwithstanding these advantages, a potential pitfall for folate ligand targeting for low molecular weight conjugates is high renal uptake due to FR expression in the apical membrane of kidney proximal tubules (45). This site is, however, inaccessible to high molecular weight drug carriers, such as liposomes, that cannot pass through the glomerular membrane owing to their size (14).

FR-TARGETED LIPOSOMES

FR-targeted liposomes were first reported by Lee and Low in the early 1990s (28,46). Their studies showed that folate tethered liposomes have high affinity to FR+ cells and are efficiently internalized by receptor-mediated endocytosis. A lengthy spacer was found to be necessary between folic acid and its lipid anchor to enable FR binding, presumably to overcome steric hindrance encountered by the liposome when approaching the cellular surface. FR-targeted liposomes containing a long linker [e.g., poly(ethylene glycol) (PEG) 3350] appeared to be more effective in cellular uptake than those with a shorter linker (e.g., PEG 2000) (47). It was further shown that FR-targeted liposomes had much greater affinity to FR+ cells than did free folate, possibly due to multivalent binding between the liposomes and the cell (28,46). Binding of FR-targeted liposomes was saturable, although saturation appeared to be limited by the available cellular surface area rather than FR expression level. Since then, a wide variety of agents have been incorporated into FR-targeted liposomes and evaluated both in vitro and in vivo. These studies have been recently reviewed by Gabizon et al. (26).

Preparation of FR-Targeted Liposomes

To prepare FR-targeted liposomes, folate ligand is incorporated into the liposomal bilayer either during liposome preparation: by mixing a lipophilic folate ligand with other lipid components or by derivatizing distal termini of functionalized PEG-lipids, or by postinsertion of lipophilic folate ligand to preformed liposomes (28,48). The lipophilic anchor for the folate ligand can be either phospholipid or cholesterol. Two lipophilic folate derivatives, folate-PEG$_{3350}$-distearoylphosphatidylethanolamine (DSPE) and folate-PEG$_{3350}$-cholesterol (Fig. 2), have been synthesized as ready-to-use ligands for preparing FR-targeted liposomes (46,49).

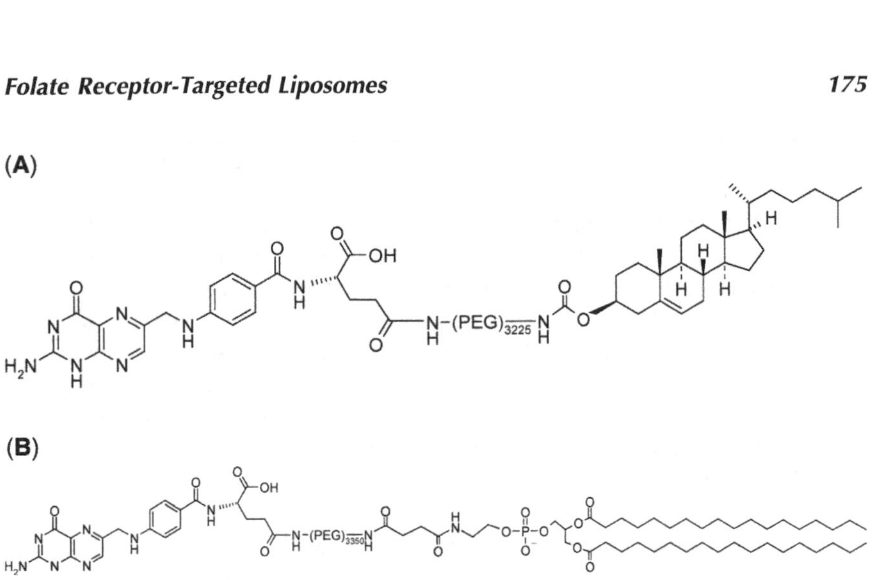

(A)

(B)

Figure 2 Structures of folate–PEG–cholesterol (**A**) and folate–PEG–DSPE (**B**). *Abbreviation*: PEG-DSPE, poly(ethyleneglycol)-distearoylphosphatidylethanolamine.

FR-Targeted Liposomes for Delivery of Chemotherapy and Photodynamic Agents In Vitro

Numerous in vitro studies have been reported evaluating FR-specific uptake of targeted liposomes by FR+ tumor cells. For example, when co-cultured FR+ HeLa human cervical cancer cells and FR− WI38 fibroblasts were treated with FR-targeted liposomes encapsulating calcein, a water-soluble fluorescence dye, only HeLa cells showed uptake of FR-targeted liposomes (46). In an FR-blocking study, uptake of FR-targeted liposomes was reduced by ∼70% by 1 mM free folic acid. In contrast, no reduction in cellular uptake of FR-targeted liposomes was observed in the presence of 5-MeTHF at the physiological concentration (20 nM) (46). To assess the potential application of FR-targeted liposomes in chemotherapy delivery, the liposomes were loaded with doxorubicin (DOX) by remote loading and evaluated in FR+ KB human oral carcinoma cells. The uptake of FR-targeted liposomal DOX was 45-fold higher than nontargeted liposomes and 1.6-fold higher than free DOX, and the cytotoxicity was 86- and 2.7-fold greater, respectively (46). In recent studies, FR-targeted paclitaxel formulation showed greater uptake and cytotoxicity in FR+ KB cells and M109 cells than nontargeted formulations (50,51).

Novel formulations of FR-targeted liposomes have also been developed to achieve improved intracellular drug delivery. FR-targeted pH-sensitive liposome, entrapping 200 mM anticancer agent araC, showed approximately 17-fold higher cytotoxicity in FR+ KB cells compared to araC delivered via FR-targeted non-pH–sensitive liposomes. This is because FR-mediated endocytosis leads to the trafficking of FR-targeted liposomes into an acidic compartment. pH-sensitive liposomes undergo acid-triggered destabilization in this compartment and increased endosomal drug release (52). In a separate report, FR-targeted liposomes composed mostly of DDPIsC, an acid-labile lipid, were evaluated for the delivery of chloroaluminum phthalocyanine tetrasulfonate ($AlPcS_4^{4-}$), a water-soluble photosensitizer. These liposomes showed substantially greater phototoxicity than free $AlPcS_4^{4-}$ and nontargeted liposomal $AlPcS_4^{4-}$ against FR+ KB cells (53). To increase the encapsulation rate, hematoporphyrin–stearylamine, a lipophilic derivative of hematoporphyrin, was prepared and incorporated into FR-targeted liposomes (54). The resulting liposomes exhibited fivefold greater phototoxicity than nontargeted formulation.

FR-Targeted Liposomal Delivery Bypasses MDR

Multidrug resistance (MDR) is a major clinical problem in chemotherapy treatment of cancer. It is frequently due to upregulation in tumor cells of plasma membrane pumps, such as P-glycoprotein (Pgp), that actively pumps out cytotoxic agents (55). Liposomal delivery of drugs has been shown to bypass MDR by possibly reducing exposure to the membrane efflux pump in vitro in leukemia cells (56,57) and in some solid tumor cells, including breast, ovarian, and small-cell lung carcinoma cells (58). FR-targeted liposomal delivery also shows the ability to overcome Pgp-mediated efflux in FR+ cancer cells (59,60). In a recent study by Gabizon et al., a MDR FR+ murine lung cell line (M109R-HiFR) was treated with either free DOX or FR-targeted liposomal DOX. Verapamil, a Pgp inhibitor, greatly enhanced the cellular uptake of free DOX, which otherwise would have been rapidly effluxed, while exhibiting no significant effect on the cellular accumulation of FR-targeted liposomes. Cellular fractionation analysis showed higher DOX concentration in the nuclear fraction of cells treated with FR-targeted liposomal DOX compared to the cells treated with free DOX. Furthermore, FR-targeted DOX showed greater tumor inhibitory activity than nontargeted liposomal DOX and free DOX in an in vivo adoptive study (59). In another study, thermosensitive FR-targeted DOX combined with hyperthermia was found to be 4.8 times more effective than free DOX on the MDR KB85 cells (60). These results suggest that FR-targeted liposomal delivery, alone or in combination with hyperthermia, is potentially more effective in circumventing the Pgp-mediated MDR than nontargeted liposomal delivery in FR+ tumors.

FR-Targeted Liposomes for Plasmid DNA and ODNs Delivery

Gene therapy is an emerging therapeutic modality for the treatment of cancers and genetic diseases. Efficient delivery of DNA is a limiting factor in the clinical adoption of gene therapy. Several FR-targeted vectors have been reported, including folate modified adenoviruses (42), cationic polymer/DNA complexes (polyplexes) (61), cationic lipid/DNA complexes (lipoplexes) (62,63), and polymer–lipid–DNA ternary complexes [lipopoly-plexes (LPD)] (42,64), that are linked to folate as a targeting ligand (reviewed in Refs. 25,44,65–67).

FR-Targeted Liposomes for Delivery of Plasmid DNA

FR-targeted lipoplexes have shown enhanced in vitro and in vivo transfection activity in FR+ tumor cells compared to nontargeted controls (68,69). Cationic liposomes consisting of RPR209120 (a lipopolyamine), dioleoyl phosphatidyl-ethanolamine (DOPE), and folate-PEG-Chol or folate-PEG-DSPE showed almost 1000-fold higher in vitro transfection than nontargeted lipoplexes in FR+ M109 cell line. In vivo study of this formulation also indicated greater gene delivery in tumor than in normal tissues, even though there was no signifi-cant increase in tumor uptake of the FR-targeted to nontargeted formulations.

LPDs are ternary complexes consisting of liposomes complexed with polycations [such as polylysine (PLL), polyethyleneimine, protamine, polya-midoamine dendrimers, etc.] condensed plasmid DNA. Condensed DNA complexed with cationic liposomes or anionic liposomes were defined as LPDI and LPDII, respectively (42,70). FR-targeted LPDI composed of cationic liposomes/protamine/DNA and folate-cys-PEG-PE showed greater transfection activity in FR+ M109 cell line. This formulation also showed 8- to 10-fold higher gene transfer activity in vivo compared to the nontargeted control. However, increasing the folate ligand density on liposome surface resulted in decreased gene transfer activity, presumably due to increased steric hindrance of PEG that prevented efficient endosomal release of the vector. The overall positively charged LPDI also might nonspecifically interact with negatively charged cell membranes and result in increased nonspecific gene transfer.

FR-targeted LPDII, first developed by Lee and Huang (42), has a net anionic character and showed reduced nonspecific binding to cell mem-branes, and therefore increased receptor dependent gene transfer. Prepara-tion of FR-targeted LPDII can be carried out by first condensing DNA with PLL at a ratio of 1:0.75 (w/w) to obtain an overall slightly positive charged DNA/PLL complex and then mixing it with anionic pH-sensitive lipo-somes composed of DOPE/cholesteryl hemisuccinate (CHMES)/folate-PEG-DOPE (6:4:0.01 mol/mol). The resulting FR-targeted LPDII spherical particles had a mean diameter of $\sim74 \pm 14$ nm under the electron micro-scope and exhibited superior transfection efficiency compared to nontar-geted LPDII. Similarly, another FR-targeted LPDII developed by Reddy et al. comprised of DNA/PLL complex and DOPE/Chol/C-DOPE (an acid-labile lipid for pH triggered endosomal release)/folate-PEG-DOPE exhibited efficient FR-mediated gene transfer (71).

FR-Targeted Liposomes for Delivery of Antisense ODNs

Based on electrostatic interaction, negatively charged antisense ODNs can also be encapsulated into FR-targeted liposomes for selective delivery to FR+ tumor cells in vitro and in vivo (35). FR-targeted cationic liposomes mediated the delivery of anti-HER-2 antisense oligonucleotide (AS HER-2 ODN), and inhibited cell growth and HER-2 expression (72). In vivo study also showed prolonged stability in blood and increased uptake in tumors (73). In a separate report, FR-targeted liposomes have been evaluated as a carrier of antisense oligodeoxynucleotides against the EGFR and shown to be much more efficient than a nontargeted liposomal formulation (34).

FR-Targeted Liposomal Delivery In Vivo

Animal Tumor Models

Available and stable tumor models are the prerequisite for in vivo targeting studies. Several stable FR-overexpressing animal tumor models have been developed, such as murine lung M109 carcinoma (47), J6456 lymphoma (74), L1210 leukemia (19), and human KB carcinoma xenograft (28). Because the high folate content found in normal rodent chow produces an artificially high plasma folate concentration, which may interfere with in vivo FR targeting, animals used in studies evaluating FR-targeted agents should be placed on a folate-free diet. However, profound folate deficiency is not seen in animals, as folate is produced by the intestinal microflora, unless antibiotics are also included.

FR-Targeted Liposomal Delivery to Leukemia

Leukemias are potentially suitable disease targets for folate-coated liposomes because the blast cells are readily accessible from systemic circulation. Approximately 70% of AMLs express functional FR-β that is absent in normal hematopoietic cells (12,19). FR-β can therefore serve as a maker for targeted delivery to AML (recently reviewed in Ref. 20). However, the heterogeneous and variable expression of FR-β poses a potential obstacle to FR-β-targeted therapeutics. The problem can be potentially overcome by selective induction of FR-β upregulation in the target cells via retinoid receptor ligands (75). For example, FR-β expression in KG-1 AML cells and primary AML blast cells (FAB-M2 and M4) can be upregulated by ATRA, and reach steady-state levels that are up to ~20-fold higher within a few days (75). ATRA-induced high FR-β differentiation does not cause terminal differentiation or growth inhibition in these cells. Furthermore, FR-β expression is restricted to the cell lines that are initially FR-β(+). FR-β(–) AML or other tumor cells, including FR-α(+) cells, cannot be induced by the ATRA to express FR-β (75).

A recent study showed that FR-targeted liposomal DOX was 25-fold more cytotoxic than nontargeted liposomal DOX to the FR-β(+) KG-1 cells and 63-fold more cytotoxic in ATRA pretreated KG-1 cells (19). In contrast, FR-β(−) cell lines did not show increased differential cytotoxicity when treated with ATRA (19). Furthermore, in vivo therapeutic activity of FR-targeted liposomal DOX was evaluated in two models, a DBA/2 mouse model containing syngeneic ascites tumor from L1210JF leukemia cells and a severe combined immunodeficient murine xenograft model with human KG-1 AML cells ascites tumor (19). In the latter model, FR-targeted liposomal DOX increased the median survival time from 35 days to more than 80 days. Moreover, the mice administered with ATRA and FR-targeted liposomal DOX showed further enhancement in antitumor efficacy with an increase in cure rate from 12.5% to 60% (19). As ATRA differentiation therapy is one of the standard treatments for APL subtype of AML and liposomal DOX has been approved for solid tumor treatment, the success of combined therapy using FR-targeted liposomal DOX and ATRA in this experiment suggests that further clinical studies may be warranted.

Targeting FR+ Solid Tumors—EPR Effect and Intratumoral Drug Distribution

Liposomes are known to preferentially accumulate in solid tumors due to the enhanced permeability and retention (EPR) effect. Because both FR-targeted and nontargeted liposomes are equally affected by this mechanism, these two types of liposomes often show similar levels of tumor uptake, at least within the first day following administration. Nonetheless, some preferential tumor uptake and improvement in therapeutic efficacy has been observed with the FR-targeted liposomes. For example, FR-targeted liposomal DOX showed greater antitumor efficacy in a FR+ KB cell BALB/c (nu/nu) murine xenograft model (76). Mice that received FR-targeted liposomal DOX exhibited greater tumor growth inhibition and longer lifespan than those that received nontargeted liposomal DOX (76). The improved efficacy might be due to the recognition and internalization of FR-targeted liposomes by the FR on the tumor cells following extravasation from the tumor vessels, which led to more efficient killing of the tumor cells. In a recent study using the J6456 ascitic tumor model, overall liposome deposition in tumors was shown to be similar for FR-targeted and nontargeted liposomes (77). However, FR-targeted liposomes appeared to have higher tumor cell association than nontargeted liposomes. These studies suggest that while FR targeting might not significantly alter the overall biodistribution of liposomes in solid tumors, greater therapeutic efficacy might be possible due to increased FR-dependent uptake by the targeted tumor cells.

Additional Potential Mechanisms for Targeted Liposomal
Drug Delivery to Solid Tumors

The relatively large size of liposomal drug carriers poses a significant barrier
for the targeting of cells in solid tumors. The breakdown of tumor vasculature
by perivascular accumulation of liposomes and local release of drug might provide an additional antitumor mechanism. The limitations of rates of diffusion
and convection pose a much greater barrier for the intratumoral distribution of
liposomes than low molecular weight agents. Intratumoral distribution of liposomes is, therefore, likely to be limited to the cell layers that are immediately
adjacent to the blood vessel, as indicated by studies using animal tumor models (78). The growth of tumor cells depends on nutrition and oxygen supplied
by the blood vessels. In turn, initiation and maintenance of blood vessels rely
on growth factors provided by tumor or host cells (79–82). Endothelial cells in
the tumor are more susceptible to cytotoxic agents and apoptotic effects than
quiescent endothelial cells in the normal tissue, due to their increased proliferation rates and lack of p53 mutation (83). It is possible that FR-targeted
liposomes preferentially accumulate around the blood vessels within the
tumor and effectively eliminate adjacent tumor cells via active FR-mediated
uptake, as well as destroy the endothelial cells by direct cytotoxicity and
bystander effect (Figs. 3 and 4) (84). Therefore, FR-targeted liposomes might
exhibit antitumor activity by facilitating tumor cell death via FR-mediated

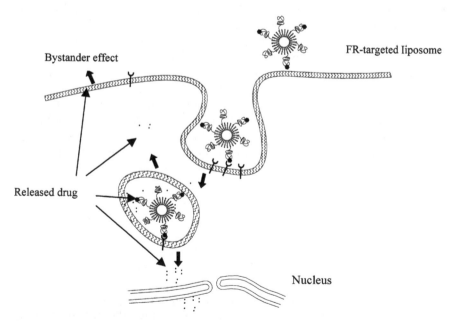

Figure 3 A schematic diagram of the folate receptor (FR)–mediated endocytosis
pathway.

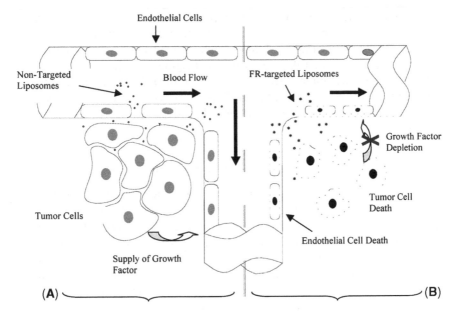

Figure 4 Mechanisms for direct and indirect inhibition of tumor cells and endothelial cells. (**A**) Nontargeted liposomes extravasate into tumor through the leaky blood vessels. However, most of them remain in the extracellular interstitial space. (**B**) (i) Folate receptor (FR)-targeted liposomes exert *direct* cytotoxicity towards the tumor cells that are close to the blood vessel and endothelial cells. (ii) *Indirectly*, distal tumor cells were killed by antiangiogenic effect, whereas the endothelial cells were killed by bystander effect of released free drug and growth factor depletion due to tumor cell death.

liposomal uptake, which shuts down the supply of vascular growth factors to adjacent endothelial cells. The killing of endothelial cells, in turn, serves to suffocate the tumor by diminishing oxygen and nutrition supply. The synergistic interplay of cytotoxicity and antiangiogenic activity may be jointly responsible for the observed antitumor efficacy of liposomal chemotherapy in vivo. Furthermore, FR-mediated liposomal internalization may also promote drug release from the liposome and facilitate secondary intratumoral distribution by diffusion following cell death. In addition, FR-targeted liposomes should be able to effectively target tumor cells that are in the circulation, such as leukemia cells, or micrometastases that are directly accessible from circulation by means of diffusion.

Another population of cells that express functional FR is activated macrophages (18). In fact, a recent report on an ascites tumor model showed that FR-targeted liposomes were taken up most efficiently by peritoneal macrophages rather than ascitic tumor cells, which were FR+. It is likely that tumor-infiltrating macrophages express FR and can take up folate-tethered liposomes and facilitate its penetration by tumor infiltration and its release upon liposome breakdown following phagocytosis. This brings out the possibility that folate-derivatized imaging and therapeutic agents, including liposomes, might target tumors that are FR−, due to FR expression in the infiltrating macrophages. This, therefore, potentially broadens the therapeutic potential of FR-targeted drug delivery.

Pharmacokinetic Considerations for FR-Targeted Liposomes

In assessing targeted liposomal delivery, pharmacokinetic considerations must be taken into account. For example, FR-targeted liposomal DOX showed similar cytotoxicity in KB cells compared to free DOX but superior to nontargeted liposome DOX. This might be partially attributed to the use of equal incubation time during the in vitro assay. In vivo, liposomal DOX has a much longer mean residence time (MRT). Comparing cytotoxicity in vitro using drug exposure based on MRT in vivo might produce a very different relative IC_{50} value for the FR-targeted liposomal DOX and free DOX.

In vivo, DOX in FR-targeted liposomes must be bioavailable to the tumor cells to be therapeutically active. There is a significant barrier to the intratumoral distribution of the bulky liposomal drug carriers. This might tip the therapeutic efficacy towards free DOX, although this is partially offset by the potential cardiac toxicity of free DOX. FR targeting might have a significant impact on the rate of DOX release due to increased phagocytosis by activated tumor infiltrating macrophages that express the FR. In addition, it is important to note that the pharmacokinetic behavior of folate-tethered liposomes is heavily influenced by serum opsonins that might display moderate affinity for folate, e.g., albumin and low-density lipoprotein (LDL). This might result in significantly altered clearance rate and tropism of these liposomes following plasma exposure after intravenous administration.

As with any targeted drug delivery system, it is also important to assess the suitability of the payload carried by FR-targeted liposomes. Assuming that therapeutic effect is correlated with intracellular drug concentration, it is possible to predict the effectiveness of a targeting strategy by comparing the capacity of receptor-mediated uptake ($B_{\text{total}} = \frac{B_{\max} * C}{C + Kd} + k * C$) versus nonspecific uptake ($B_{nsp} = k * C$). Figure 5 is a simulation plotted on a double log scale. In this plot, an assumption was made that a certain level of cellular uptake was required for effective cytotoxicity, i.e., an effective uptake level (B_{eff}). It is apparent that differential cytotoxicity due to targeted delivery ($\Delta \log C_{\text{eff}}$) is only significant if the $B_{\max}$ is well above the effective uptake level. Therefore, only drugs with low micromolar to nanomolar

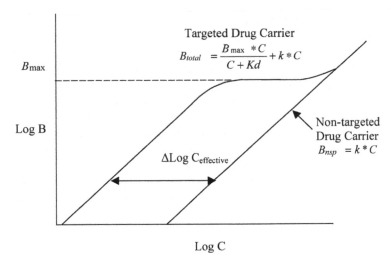

$$B_{total} = \frac{B_{max} * C}{C + Kd} + k * C$$

Targeted Drug Carrier

Non-targeted Drug Carrier

$$B_{nsp} = k * C$$

$\Delta Log\ C_{effective}$

B_{max}

Log B

Log C

Figure 5 Drug uptake via receptor-targeted versus nontargeted delivery strategy.

IC_{50} values, including anthracyclines and paclitaxel, are good candidates for targeted liposomal delivery.

SUMMARY

Many FR-targeted therapies have been evaluated both in vitro and in vivo and have consistently shown excellent tumor-cell targeting properties. Radionuclide conjugates of folic acid for whole-body imaging of ovarian and endometrial cancer have been evaluated clinically with promising initial results (44).

There have not yet been clinical studies evaluating therapeutic potential of FR-targeted liposomes. Recent promising data on FR-targeted liposomes in animal tumor models and similar findings in studies on anti-HER2-immunoliposomes suggest that targeted liposomes are therapeutically superior to nontargeted liposomes, even though the effect of targeting on overall tumor accumulation is usually moderate due to the overriding EPR effect (85). These data, combined with the many theoretical advantages of FR-targeting such as nonimmunogenicity, Pgp avoidance, and possibility for in vivo upregulation, suggest that FR-targeted liposomes may have potential for future clinical applications.

FUTURE DIRECTIONS

Although FR-targeted liposomal delivery has consistently produced satisfying results in vitro, relatively few in vivo studies have been reported to date.

For solid tumors, these studies seem to suggest that, rather than increasing overall tumor localization, elevated cellular internalization, bypassing of Pgp, and antiangiogenic effects might play critical roles for the superior efficacy of targeted liposomes. FR-expression in tumor negative infiltrating macrophages and opsonization of folate-coated liposomes by LDL might extend tumor-targeting properties of these liposomes to tumors that are FR−. In addition, there is compelling rationale for exploring targeted liposomal delivery to leukemias due to the relative accessibility of the target cells and the long circulating properties of liposomal drug carriers. The clinical potential of this promising delivery strategy has yet to be explored. Because FR expression in cancer and leukemia patients is likely to vary, effective targeting might benefit from co-administration of ATRA (for FR-β upregulation in AMLs via the retinoid receptors) and tamoxifen or dexamethasone (for FR-α upregulation in solid tumors via the estrogen and gluocorticoid receptors), possibly along with histone deacetylase inhibitors (e.g., valproic acid), which further increases the effects of these agents. In addition, prescreening of patients using serum FR assay and/or imaging using FR-targeted radiopharmaceuticals might be necessary to determine the potential benefit for targeted therapy. In addition to cancer and leukemia, FR is also overexpressed among activated macrophages in rheumatoid arthritis, which constitutes another potential disease target for FR-targeted liposomes. Further preclinical and clinical studies are clearly warranted to assess the potential role of FR-targeted liposomes in the management of cancer, leukemia, and rheumatoid arthritis.

ACKNOWLEDGMENT

This work was supported in part by NSF grant EEC-02425626 and NIH grant CA095673 to R.J. Lee.

REFERENCES

1. Voet D. Biochemistry. In: Donald V, Judith GV, eds. Biochemistry Vol. 1. 3rd ed. Wiley, New York: Chichester, 2003 (Chapters 1–19).
2. Elwood PC. Molecular cloning and characterization of the human folate-binding protein cDNA from placenta and malignant tissue culture (KB) cells. J Biol Chem 1989; 264(25):14893–14901.
3. Lacey SW, Sanders JM, Rothberg KG, Anderson RG, Kamen BA. Complementary DNA for the folate binding protein correctly predicts anchoring to the membrane by glycosyl-phosphatidylinositol. J Clin Invest 1989; 84(2): 715–720.
4. Ratnam M, Marquardt H, Duhring JL, Freisheim JH. Homologous membrane folate binding proteins in human placenta: cloning and sequence of a cDNA. Biochemistry 1989; 28(20):8249–8254.
5. Antony AC. Folate receptors. Annu Rev Nutr 1996; 16:501–521.

6. Shen F, Wu M, Ross JF, Miller D, Ratnam M. Folate receptor type gamma is primarily a secretory protein due to lack of a efficient signal for glycosylphosphatidylinositol modification: protein characterization and cell type specificity. Biochemistry 1995; 34(16):5660–5665.

7. Shen F, Ross JF, Wang X, Ratnam M. Identification of a novel folate receptor, a truncated receptor, and receptor type beta in hematopoietic cells: cDNA cloning, expression, immunoreactivity, and tissue specificity. Biochemistry 1994; 33(5): 1209–1215.

8. Maziarz KM, Monaco HL, Shen F, Ratnam M. Complete mapping of divergent amino acids responsible for differential ligand binding of folate receptors alpha and beta. J Biol Chem 1999; 274(16):11086–11091.

9. Wang X, Shen F, Freisheim JH, Gentry LE, Ratnam M. Differential stereospecificities and affinities of folate receptor isoforms for folate compounds and antifolates. Biochem Pharmacol 1992; 44(9):1898–1901.

10. Shen F, Zheng X, Wang J, Ratnam M. Identification of amino acid residues that determine the differential ligand specificities of folate receptors alpha and beta. Biochemistry 1997; 36(20):6157–6163.

11. Garin-Chesa P, Campbell I, Saigo PE, et al. Trophoblast and ovarian cancer antigen LK26. Sensitivity and specificity in immunopathology and molecular identification as a folate-binding protein. Am J Pathol 1993; 142(2):557–567.

12. Ross JF, Wang H, Behm FG, et al. Folate receptor type beta is a neutrophilic lineage marker and is differentially expressed in myeloid leukemia. Cancer 1999; 85(2):348–357.

13. Ross JF, Chaudhuri PK, Ratnam M. Differential regulation of folate receptor isoforms in normal and malignant tissues in vivo and in established cell lines. Physiologic and clinical implications. Cancer 1994; 73(9):2432–2443.

14. Weitman SD, Weinberg AG, Coney LR, et al. Cellular localization of the folate receptor: potential role in drug toxicity and folate homeostasis. Cancer Res 1992; 52(23):6708–6711.

15. Wu M, Gunning W, Ratnam M. Expression of folate receptor type alpha in relation to cell type, malignancy, and differentiation in ovary, uterus, and cervix. Cancer Epidemiol Biomarkers Prev 1999; 8(9):775–782.

16. Weitman SD, Lark RH, Coney LR, et al. Distribution of the folate receptor GP38 in normal and malignant cell lines and tissues. Cancer Res 1992; 52(12): 3396–3401.

17. Kelley KM, Rowan BG, Ratnam M. Modulation of the folate receptor alpha gene by the estrogen receptor: mechanism and implications in tumor targeting. Cancer Res 2003; 63(11):2820–2888.

18. Nakashima-Matsushita N, Homma T, Yu S, et al. Selective expression of folate receptor beta and its possible role in methotrexate transport in synovial macrophages from patients with rheumatoid arthritis. Arthritis Rheumat 1999; 42(8):1609–1616.

19. Pan XQ, Zheng X, Shi G, et al. Strategy for the treatment of acute myelogenous leukemia based on folate receptor beta-targeted liposomal doxorubicin combined with receptor induction using all-trans retinoic acid. Blood 2002; 100(2):594–602.

20. Ratnam M, Hao H, Zheng X, et al. Receptor induction and targeted drug delivery: a new antileukaemia strategy. Expert Opin Biol Ther 2003; 3(4):563–574.

21. Corrocher R, Bambara LM, Pachor ML, et al. Serum folate binding capacity in leukemias, liver diseases and pregnancy. Acta Haematol 1979; 61(4):203–208.
22. Coney LR, Tomassetti A, Carayannopoulos L, et al. Cloning of a tumor-associated antigen: MOv18 and MOv19 antibodies recognize a folate-binding protein. Cancer Res 1991; 51(22):6125–6132.
23. Sudimack J, Lee RJ. Targeted drug delivery via the folate receptor. Adv Drug Deliv Rev 2000; 41(2):147–162.
24. Lu Y, Low PS. Folate-mediated delivery of macromolecular anticancer therapeutic agents. Adv Drug Deliv Rev 2002; 54(5):675–693.
25. Gosselin MA, Lee RJ. Folate receptor-targeted liposomes as vectors for therapeutic agents. Biotechnol Annu Rev 2002; 8:103–131.
26. Gabizon A, Shmeeda H, Horowitz AT, Zalipsky S. Tumor cell targeting of liposome-entrapped drugs with phospholipid-anchored folic acid-PEG conjugates. Adv Drug Deliv Rev 2004; 56(8):1177–1192.
27. Leamon CP, Reddy JA. Folate-targeted chemotherapy. Adv Drug Deliv Rev 2004; 56(8):1127–1141.
28. Lee RJ, Low PS. Delivery of liposomes into cultured KB cells via folate receptor-mediated endocytosis. J Biol Chem 1994; 269(5):3198–3204.
29. Leamon CP, Low PS. Delivery of macromolecules into living cells: a method that exploits folate receptor endocytosis. Proc Natl Acad Sci USA 1991; 88(13): 5572–5576.
30. Lee RJ, Wang S, Low PS. Measurement of endosome pH following folate receptor-mediated endocytosis. Biochim Biophys Acta 1996; 1312(3):237–242.
31. Rijnboutt S, Jansen G, Posthuma G, et al. Endocytosis of GPI-linked membrane folate receptor-alpha. J Cell Biol 1996; 132(1–2):35–47.
32. Guo W, Hinkle GH, Lee RJ. 99mTc-HYNIC-folate: a novel receptor-based targeted radiopharmaceutical for tumor imaging. J Nucl Med 1999; 40(9): 1563–1569.
33. Leamon CP, Pastan I, Low PS. Cytotoxicity of folate-Pseudomonas exotoxin conjugates toward tumor cells. Contribution of translocation domain. J Biol Chem 1993; 268(33):24847–24854.
34. Wang S, Lee RJ, Cauchon G, Gorenstein DG, Low PS. Delivery of antisense oligodeoxyribonucleotides against the human epidermal growth factor receptor into cultured KB cells with liposomes conjugated to folate via polyethylene glycol. Proc Natl Acad Sci USA 1995; 92(8):3318–3322.
35. Leamon CP, Cooper SR, Hardee GE. Folate-liposome-mediated antisense oligodeoxynucleotide targeting to cancer cells: evaluation in vitro and in vivo. Bioconjug Chem 2003; 14(4):738–747.
36. Lu JY, Lowe DA, Kennedy MD, Low PS. Folate-targeted enzyme prodrug cancer therapy utilizing penicillin-V amidase and a doxorubicin prodrug. J Drug Target 1999; 7(1):43–53.
37. Kranz DM, Patrick TA, Brigle KE, Spinella MJ, Roy EJ. Conjugates of folate and anti-T-cell-receptor antibodies specifically target folate-receptor-positive tumor cells for lysis. Proc Natl Acad Sci USA 1995; 92(20):9057–9061.
38. Konda SD, Aref M, Wang S, Brechbiel M, Wiener EC. Specific targeting of folate-dendrimer MRI contrast agents to the high affinity folate receptor expressed in ovarian tumor xenografts. Magma 2001; 12(2–3):104–113.

39. Moon WK, Lin Y, O'Loughlin T, et al. Enhanced tumor detection using a folate receptor-targeted near-infrared fluorochrome conjugate. Bioconjug Chem 2003; 14(3):539–545.
40. Shukla S, Wu G, Chatterjee M, et al. Synthesis and biological evaluation of folate receptor-targeted boronated PAMAM dendrimers as potential agents for neutron capture therapy. Bioconjug Chem 2003; 14(1):158–167.
41. Lu Y, Low PS. Folate targeting of haptens to cancer cell surfaces mediates immunotherapy of syngeneic murine tumors. Cancer Immunol Immunother 2002; 51(3):153–162.
42. Lee RJ, Huang L. Folate-targeted, anionic liposome-entrapped polylysine-condensed DNA for tumor cell-specific gene transfer. J Biol Chem 1996; 271(14): 8481–8487.
43. Oyewumi MO, Mumper RJ. Influence of formulation parameters on gadolinium entrapment and tumor cell uptake using folate-coated nanoparticles. Int J Pharm 2003; 251(1–2):85–97.
44. Leamon CP, Low PS. Folate-mediated targeting: from diagnostics to drug and gene delivery. Drug Discov Today 2001; 6(1):44–51.
45. Christensen EI, Birn H, Verroust P, Moestrup SK. Membrane receptors for endocytosis in the renal proximal tubule. Int Rev Cytol 1998; 180:237–284.
46. Lee RJ, Low PS. Folate-mediated tumor cell targeting of liposome-entrapped doxorubicin in vitro. Biochim Biophys Acta 1995; 1233(2):134–144.
47. Gabizon A, Horowitz AT, Goren D, et al. Targeting folate receptor with folate linked to extremities of poly(ethylene glycol)-grafted liposomes: in vitro studies. Bioconjug Chem 1999; 10(2):289–298.
48. Saul JM, Annapragada A, Natarajan JV, Bellamkonda RV. Controlled targeting of liposomal doxorubicin via the folate receptor in vitro. J Control Release 2003; 92(1–2):49–67.
49. Guo WJ, Lee T, Sudimack J, Lee RJ. Receptor-specific delivery of liposomes via folate-PEG-Chol. J Liposome Res 2000; 10(2–3):179–195.
50. Stevens PJ, Lee RJ. A folate receptor-targeted emulsion formulation for paclitaxel. Anticancer Res 2003; 23(6C):4927–4931.
51. Stevens PJ, Sekido M, Lee RJ. A folate receptor-targeted lipid nanoparticle formulation for a lipophilic paclitaxel prodrug. Pharm Res 2004; 21(12): 2153–2157.
52. Sudimack JJ, Guo W, Tjarks W, Lee RJ. A novel pH-sensitive liposome formulation containing oleyl alcohol. Biochim Biophys Acta 2002; 1564(1):31–37.
53. Qualls MM, Thompson DH. Chloroaluminum phthalocyanine tetrasulfonate delivered via acid-labile diplasmenylcholine-folate liposomes: intracellular localization and synergistic phototoxicity. Int J Cancer 2001; 93(3):384–392.
54. Stevens PJ, Sekido M, Lee RJ. Synthesis and evaluation of a hematoporphyrin derivative in a folate receptor-targeted solid-lipid nanoparticle formulation. Anticancer Res 2004; 24(1):161–165.
55. Gottesman MM, Fojo T, Bates SE. Multidrug resistance in cancer: role of ATP-dependent transporters. Nat Rev Cancer 2002; 2(1):48–58.
56. Michieli M, Damiani D, Ermacora A, et al. Liposome-encapsulated daunorubicin for PGP-related multidrug resistance. Br J Haematol 1999; 106(1): 92–99.

57. Rahman A, Husain SR, Siddiqui J, et al. Liposome-mediated modulation of multidrug resistance in human Hl-60 leukemia-cells. J Natl Cancer Inst 1992; 84(24):1909–1915.
58. Sadava D, Coleman A, Kane SE. Liposomal daunorubicin overcomes drug resistance in human breast, ovarian and lung carcinoma cells. J Liposome Res 2002; 12(4):301–309.
59. Goren D, Horowitz AT, Tzemach D, Tarshish M, Zalipsky S, Gabizon A. Nuclear delivery of doxorubicin via folate-targeted liposomes with bypass of multidrug-resistance efflux pump. Clin Cancer Res 2000; 6(5):1949–1957.
60. Gaber MH. Modulation of doxorubicin resistance in multidrug-resistance cells by targeted liposomes combined with hyperthermia. J Biochem Mol Biol Biophys 2002; 6(5):309–314.
61. Felgner PL, Gadek TR, Holm M, et al. Lipofection: a highly efficient, lipid-mediated DNA-transfection procedure. Proc Natl Acad Sci USA 1987; 84(21): 7413–7417.
62. Gao X, Huang L. Potentiation of cationic liposome-mediated gene delivery by polycations. Biochemistry 1996; 35(3):1027–1036.
63. Gao X, Huang L. Cationic liposome-mediated gene transfer. Gene Ther 1995; 2(10):710–722.
64. Lee RJ, Huang L. Lipidic vector systems for gene transfer. Crit Rev Ther Drug Carrier Syst 1997; 14(2):173–206.
65. Chiu SJ, Ni J, Lee RJ. Targeted gene delivery via the folate receptor, in polymeric gene delivery. Principles and Applications (edited by M.M. Amiji), CRC Press, New York, 2005, 523–535.
66. Wang S, Low PS. Folate-mediated targeting of antineoplastic drugs, imaging agents, and nucleic acids to cancer cells. J Control Release 1998; 53(1–3):39–48.
67. Reddy JA, Low PS. Folate-mediated targeting of therapeutic and imaging agents to cancers. Crit Rev Ther Drug Carrier Syst 1998; 15(6):587–627.
68. Hofland HE, Masson C, Iginla S, et al. Folate-targeted gene transfer in vivo. Mol Ther 2002; 5(6):739–744.
69. Xu L, Pirollo KF, Chang EH. Tumor-targeted p53-gene therapy enhances the efficacy of conventional chemo/radiotherapy. J Control Release 2001; 74(1–3): 115–128.
70. Reddy JA, Abburi C, Hofland H, et al. Folate-targeted, cationic liposome-mediated gene transfer into disseminated peritoneal tumors. Gene Ther 2002; 9(22):1542–1550.
71. Reddy JA, Low PS. Enhanced folate receptor mediated gene therapy using a novel pH-sensitive lipid formulation. J Control Release 2000; 64(1–3):27–37.
72. Rait AS, Pirollo KF, Ulick D, Cullen K, Chang EH. HER-2-targeted antisense oligonucleotide results in sensitization of head and neck cancer cells to chemotherapeutic agents. Ann NY Acad Sci 2003; 1002:78–89.
73. Rait AS, Pirollo KF, Xiang L, Ulick D, Chang EH. Tumor-targeting, systemically delivered antisense HER-2 chemosensitizes human breast cancer xenografts irrespective of HER-2 levels. Mol Med 2002; 8(8):475–486.
74. Gabizon A, Goren D, Cohen R, Barenholz Y. Development of liposomal anthracyclines: from basics to clinical applications. J Control Release 1998; 53(1–3): 275–279.

75. Wang H, Zheng X, Behm FG, Ratnam M. Differentiation-independent retinoid induction of folate receptor type beta, a potential tumor target in myeloid leukemia. Blood 2000; 96(10):3529–3536.
76. Pan XQ, Wang H, Lee RJ. Antitumor activity of folate receptor-targeted liposomal doxorubicin in a KB oral carcinoma murine xenograft model. Pharm Res 2003; 20(3):417–422.
77. Gabizon A, Horowitz AT, Goren D, et al. In vivo fate of folate-targeted polyethylene-glycol liposomes in tumor-bearing mice. Clin Cancer Res 2003; 9(17):6551–6559.
78. Jain RK. Delivery of molecular medicine to solid tumors: lessons from in vivo imaging of gene expression and function. J Control Release 2001; 74(1–3):7–25.
79. Carmeliet P, Jain RK. Angiogenesis in cancer and other diseases. Nature 2000; 407(6801):249–257.
80. Helmlinger G, Endo M, Ferrara N, Hlatky L, Jain RK. Formation of endothelial cell networks. Nature 2000; 405(6783):139–141.
81. Hanahan D, Folkman J. Patterns and emerging mechanisms of the angiogenic switch during tumorigenesis. Cell 1996; 86(3):353–364.
82. Holash J, Maisonpierre PC, Compton D, et al. Vessel cooption, regression, and growth in tumors mediated by angiopoietins and VEGF. Science 1999; 284(5422):1994–1998.
83. Folkman J. Angiogenesis and apoptosis. Semin Cancer Biol 2003; 13(2): 159–167.
84. Pan X, Lee RJ. Tumour-selective drug delivery via folate receptor-targeted liposomes. Expert Opin Drug Deliv 2004; 1(1):7–17.
85. Park JW, Hong K, Kirpotin DB, Papahadjopoulos D, Benz CC. Immunoliposomes for cancer treatment. Adv Pharmacol 1997; 40:399–435.

11

Methods for Tracking Radiolabeled Liposomes After Injection in the Body

Beth A. Goins and William T. Phillips

Department of Radiology, The University of Texas Health Science Center at San Antonio, San Antonio, Texas, U.S.A.

INTRODUCTION

The blood clearance kinetics and tissue biodistribution of liposome-based agents after administration into the body can be determined noninvasively using scintigraphic imaging. With this imaging modality, liposomes labeled with gamma (photon)-emitting radionuclides can be monitored in vivo. More information concerning techniques for radiolabeling liposomes with gamma-emitting radionuclides can be found in Volume II, Chapter 9 of this book series.

Scintigraphic imaging is proving to be a valuable tool for liposome researchers, especially in the following areas:

1. Tracking the distribution of liposomes in the blood and other organs in the body during both preclinical liposome product development and clinical testing stages.
2. Localizing the site of radiolabeled liposome uptake for disease diagnosis.
3. Monitoring the distribution and therapeutic response of liposome-encapsulated pharmaceuticals during and after treatment.
4. Investigating the physiological responses associated with liposome administration.

This chapter describes the methodology for conducting scintigraphic imaging studies with radiolabeled liposome-based agents including a review of the instrumentation used for these studies as well as acquisition and analysis methods. For comparison, a section is also included describing evaluation of tissue distribution by measuring the radioactivity in ex vivo tissue samples. Some examples describing the use of scintigraphic imaging methods in the development of liposome-based agents will be presented. A comprehensive review of the uses of scintigraphic imaging in liposome research is outside of the scope of this chapter, but a number of reviews discussing these applications have been published (1–8).

ADVANTAGES OF SCINTIGRAPHIC IMAGING COMPARED WITH OTHER IMAGING MODALITIES

There are several advantages of scintigraphic imaging compared with other noninvasive imaging modalities. First, scintigraphic imaging provides the ability to noninvasively track and quantitate the distribution of liposomes in the body using a gamma-emitting radionuclide label. While invasive studies of liposomes labeled with beta-emitting radionuclides such as carbon-14 may be suitable for animal studies, human studies with noninvasive photon imaging is a significantly more efficient approach for liposome-based drug development. Second, scintigraphy requires only a small amount of actual matter (usually in the nanogram range), that does not interfere with either the biodistribution of the labeled liposome or the physiological processes involved in its distribution (9). Other imaging modalities, such as magnetic resonance imaging (MRI) and computed tomographic (CT) imaging, provide higher resolution images than scintigraphy, but require the administration of a significantly higher amount of matter to achieve image contrast (milligrams for MRI and grams for CT) (9,10). The greater amount of contrast material required with these other imaging modalities can alter the normal biodistribution of the agent being tracked as well as increase the risk for an adverse reaction induced by the contrast agent. Third, scintigraphic imaging has been used to depict a wide variety of physiological processes, ranging from changes in glucose, protein, and fatty acid metabolism to the demonstration of gene expression and detection of changes in the concentration of cell signaling receptors (9,11–17) whereas clinicians have generally utilized MRI and CT contrast agents to demonstrate changes in vascular permeability and blood flow.

Fourth, compared with other imaging modalities, scintigraphic imaging has the ability to image the total organism in a single whole body scan. Fifth, with scintigraphic imaging, a time course of the movement of the radiolabeled liposomes can be easily obtained by acquiring images at varying time points after administration. Sixth, scintigraphic imaging methods are inherently quantitative because each scintillation is recorded and scanned into an image matrix for assembly of the visual image. Seventh, multiple agents can be tracked simultaneously in the same organism using scintigraphic imaging (18–20). The radionuclides typically used in clinical scintigraphic imaging have energies ranging from 70 kiloelectron volts (keV) to 511 keV, although the use of iodine-125 with 30 keV energy is increasing being used in preclinical drug development in mice (20). Thus, with careful selection of the radionuclides so that there is an adequate separation of photon energies, several radiolabeled agents can be tracked simultaneously in the same organism. For example, the liposome vehicle labeled with one radionuclide can be tracked simultaneously with the encapsulated drug labeled with a different radionuclide.

OVERVIEW OF SCINTIGRAPHIC IMAGING INSTRUMENTATION AND EXAMPLES

Different instruments have been designed to noninvasively monitor the photons emitted from the body after injection depending on the characteristics of the radionuclides used to label the liposomes (21). This section will review the instruments available for tracking radiolabeled liposomes in vivo and present some examples.

Single Photon Instrumentation

The instrument used for the scintigraphic imaging of single photon radionuclides is known as a gamma camera (22). In the past decade, several advances in gamma camera design have provided the capability for more sophisticated preclinical and clinical testing of liposome agents (21,23–26). One advance is that three-dimensional single photon images can more easily be obtained by rotating the gamma camera around the human or animal while acquiring a set of images at each angle (27,28). This set of images is then processed in order to reconstruct a tomographic image slice. The single photon emission computed tomography (SPECT) images provide improved localization of the source of radioactivity in the body. A schematic diagram of the acquisition of images using a SPECT camera is shown in Figure 1B. As with traditional gamma cameras, SPECT cameras use collimators made of lead to determine the position of the emitted photon within the body (28). After passing through a lead collimator, the photon strikes a sodium iodide crystal producing scintillation at a particular location on the crystal. This small scintillation is amplified into an electronic signal by photomultiplier tubes, converted from an analogue signal to a digital signal, and localized in an image matrix that is stored in a computer data bank. The total size of the image matrix can be varied from 64×64 to 512×512. Image acquisition over a fixed time period allows for the summation of counts in the different matrix boxes or pixels, which produces an image of radionuclide activity

(A)

(B)

(C)

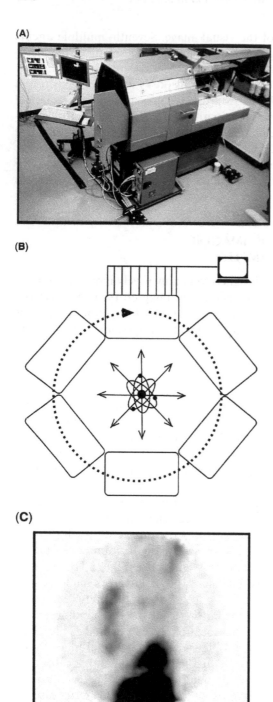

Figure 1 (*Caption on facing page*)

within the body. A typical 3-D projection image of a rat with head and neck xenograft located on the nape of the neck acquired 24 hours after intravenous injection of neutral liposomes labeled with technetium-99m (^{99m}Tc) is shown in Figure 1C. Using image-processing software, transaxial, sagittal, and coronal views of the 3-D volume can be displayed. Figure 2 shows the different image views for the same rat as Figure 1C.

SPECT imaging has been used to monitor the distribution of Caelyx® radiolabeled with ^{99m}Tc-diethylenetriamine pentaacetic acid in thirty patients with either non-small cell lung cancer or head and neck cancer (29). The ^{99m}Tc-Caelyx® was given concurrently with conventional fractionated radiotherapy. The authors of this study conclude that scintigraphic imaging for assessment of the amount and localization of Caelyx to the tumors could help identify the patients that would better respond to chemoradiotherapy with liposome-based agents (29). Harrington et al. have also incorporated whole body gamma camera as well as SPECT imaging into their studies to determine the targeting of pegylated liposomes radiolabeled with indium-111 to various locally advanced solid tumors in 17 patients (30). From their study they observed the heterogencity of uptake of liposomes between different tumor types and different patients with the same tumor type, and suggest these differences seen in the imaging studies could help explain some of the treatment response rates seen in clinical trials (30).

The time necessary for acquisition of an image using a gamma camera ranges from 2 seconds to 20 minutes, depending on the process to be imaged and the amount of radionuclide activity administered. The time required for SPECT imaging is generally more than for planar imaging because the camera must be rotated around the subject.

Another advance has been the miniaturization of SPECT cameras for acquiring high resolution images of small animals and the commercial availability of these microSPECT units for researchers interested in preclinical drug development (9,27,31,32). These microSPECT systems have image resolutions that are in the 1 mm range, which is 10 times smaller than standard clinical systems which have 1 cm resolution (9,31). Because many genetic and disease models are only available in mice, the resolution of these miniaturized instruments can provide the opportunity to study these novel mouse models that would not be possible with standard clinical imaging systems

---

Figure 1 (*Figure on facing page*) Overview of single photon emission computed tomography (SPECT) instrumentation. (**A**) Photograph of commercial microSPECT/CT unit dedicated to small animal imaging studies. (**B**) Schematic diagram outlining the components of a SPECT camera. (**C**) Typical projection image of rat obtained with commercial microSPECT following injection of neutral ^{99m}Tc-liposomes in a rat xenograft model of head and neck cancer. This image is one still frame of a dynamic rotating 3-D projection image. *Abbreviation*: CT, computed tomography.

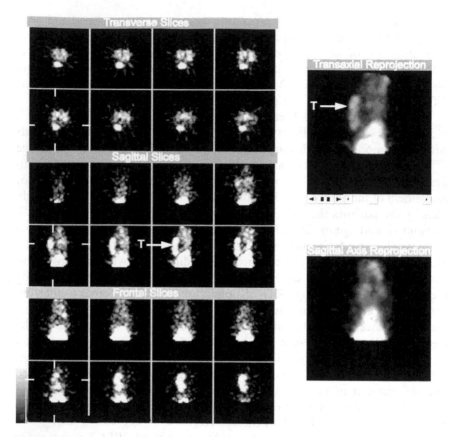

Figure 2 Display of the transaxial, sagittal, and coronal (frontal) images as well as reprojection images of a rat with head and neck xenograft acquired 24 hours after intravenous injection of neutral ^{99m}Tc-liposomes. *Abbreviation*: T, tumor.

(31–35). Figure 1A shows a photograph of the XSPECT^{TM} commercial instrument developed by Gamma Medica (Northridge, California, U.S.A.).

A third advance in instrumentation is the ability to house both a SPECT gamma camera and CT X-ray unit in the same gantry for both clinical and dedicated small animal scanners (9,23,25,36–38). This allows coregistration of the SPECT images, which is normally tracking physiological events using a radiotracer, to be fused with anatomical CT images. Figure 3 shows the individual microCT and microSPECT images in the transaxial, sagittal, and coronal views as well as an overlay of microSPECT and microCT images for a rat with head and neck xenograft acquired 24 hours after intravenous injection of neutral ^{99m}Tc-liposomes. The ability to verify the location and pattern of distribution of ^{99m}Tc-liposomes in the tumor in vivo is very useful for comparing different liposome formulation

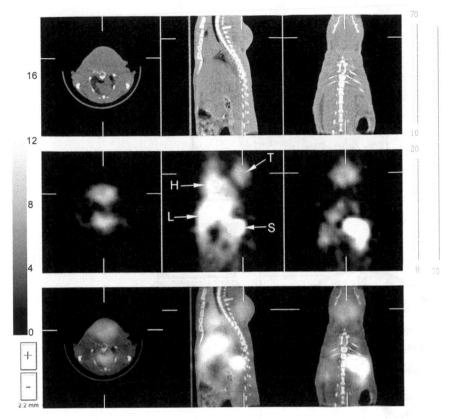

Figure 3 Display of the transaxial, sagittal, and coronal views for individual micro computed tomography (microCT) (*upper panel*), micro single photon emission computed tomography (SPECT) (*middle panel*), and fused microCT and microSPECT images (*lower panel*) for a rat with head and neck xenograft implanted on nape of neck. Images were acquired 24 hours after intravenous injection of neutral ^{99m}Tc-liposomes. *Abbreviations*: T, tumor; H, heart; L, liver; S, spleen.

parameters for liposome-based drug delivery. As expected for neutral liposomes of 110 nm diameter at 24 hours postinjection, there is prominent accumulation of the ^{99m}Tc-liposomes in the reticuloendothelial system organs of liver and spleen. A portion of the ^{99m}Tc-liposomes is still circulating in the blood stream as observed by the activity remaining in the blood pool of the heart.

Positron Emission Tomography Instrumentation

Positron-emitting radionuclides produce imaging photons by emitting a positron which travels a short distance and then collides with an electron in the tissue (39,40). This collision results in the annihilation of the positron

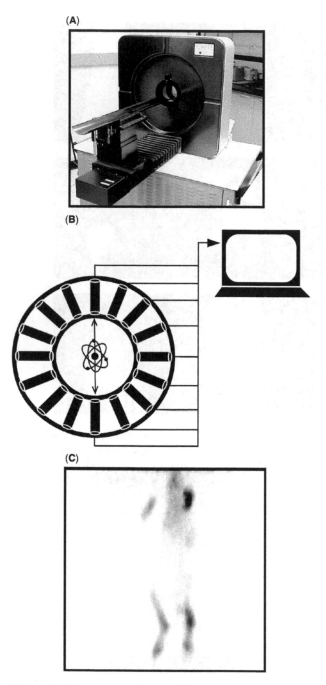

Figure 4 (*Caption on facing page*)

and the electron, and the subsequent release of two annihilation photons that are emitted at 180° angles from each other in a straight line. Positron emission tomography (PET) cameras, as shown in Figure 4B, take advantage of the two 180° photon emissions by determining a line of activity when photons are detected simultaneously on two different detectors. This simultaneous detection known as coincidence detection allows localization of the radiotracer without the need of lead collimators. The coincidence detection process is much more efficient for localization of the radionuclide activity than single photon imaging so that many more photons are detected per a given amount of radioactivity than with single photon imaging, resulting in a higher quality image compared to a single photon image. This higher efficiency is due to the fact that a filtering lead collimator is not required. Detection of multiple lines of activity at different angles from the same source permits the construction of a three-dimensional PET image (Fig. 4C). As with the SPECT projection images, PET images can be displayed in transaxial, sagittal, and coronal views.

One of the principal advantages of PET imaging is the ability to use biologically important radionuclides such as fluorine-18 (^{18}F), carbon-11, nitrogen-13, and oxygen-15 (13,21). Although the isotopes are fairly short-lived at 2 hours, 20 minutes, 10 minutes and 2 minutes, respectively, they can be used to monitor therapeutic responses to liposomally delivered therapeutic agents. For instance, ^{18}F-fluorodeoxyglucose (^{18}F-FDG) is frequently used to diagnosis cancer in patients because tumors have a very elevated rate of glucose metabolism (40–42). Studies have shown that a rapid decrease in ^{18}F-FDG uptake indicates a good response to tumor therapy (43). A study describing the use of ^{18}F-FDG tomoscintigraphy to monitor tumor growth delay in a rat osteosarcoma model following intravenous administration of endostatin cDNA/cationic liposome complexes has recently been reported (44).

The recent development of commercially available dedicated small animal PET cameras promises to accelerate liposome research. For example, microPET has been used to monitor reporter gene expression of mutant herpes simplex virus thymidine kinase using a cationic liposome delivery system (45). Figure 4A shows a photograph of the Concorde microPET R4 unit (Knoxville, Tennessee, U.S.) located at our University. Three-dimensional images provided by the dedicated small animal PET cameras have much higher resolution than is possible with large bore PET cameras currently

---

Figure 4 (*Figure on facing page*) Overview of positron emission tomography (PET) instrumentation. (**A**) Photograph of commercial microPET camera dedicated to small animal imaging studies. (**B**) Schematic diagram outlining the components of a PET camera. (**C**) Typical 3-D projection image of mouse obtained with commercial microPET following injection of ^{64}Cu-labeled polyclonal antibody to glucose-6-phosphate isomerase.

in use for human imaging. As with SPECT instrumentation, PET can also benefit from anatomical coregistration with CT. PET/CT scanners are commercially available for both clinical and small animal imaging (36,46–48).

Several studies have reported the use of liposome encapsulated ^{18}F-FDG for monitoring the distribution of liposomes with PET imaging in animal models (49–51). A disadvantage of this isotope is that it has a fairly short half-life of two hours, making it difficult to track liposome distribution for more than eight hours. Two longer lived positron emitting isotopes, copper-64 (^{64}Cu) and iodine-124 ([124]I), with half-lives of 12 hours and 4.1 days, respectively, are now becoming widely available. Both of these isotopes have been used to label engineered antibody fragments because their longer physical half-lives match with the longer residence times of the antibody fragments in the body (48,52). An image acquired with the microPET of a mouse intravenously injected with ^{64}Cu-labeled polyclonal antibody to glucose-6-phosphate isomerase is shown in Figure 4C (53). In addition to antibody labeling, both ^{64}Cu and [124]I would be promising radioisotopes for tracking the distribution of liposomes with PET since their half-lives better match the pharmacokinetic properties of most liposome formulations. To date, no methods of efficiently labeling liposomes with ^{64}Cu or [124]I have been described.

Our group has used the short-lived isotope of oxygen-15 to study the delivery of oxygen to the brain with liposome encapsulated hemoglobin (LEH) (54,55). The short two-minute half-life of oxygen-15 requires that the oxygen be loaded directly onto the LEH from a nearby cyclotron and then rapidly injected into a rat or mouse for imaging with the microPET scanner. The studies clearly demonstrate the transport of oxygen from the LEH, and release and metabolism in the brain where the oxygen is converted to water. No other imaging technique can so clearly demonstrate the effective function of LEH.

SCINTIGRAPHIC IMAGING METHODOLOGY

The general technique for performing a noninvasive biodistribution study in a human subject or animal injected with radiolabeled liposomes is described in Figure 5. Although the intravenous route of administration is the most common for approved liposome drugs, these imaging methods are very versatile and can be used to study other liposome administration routes such as intracavitary, intramuscular, subcutaneous, and local administration.

Image Acquisition Procedures

A diagram outlining the steps involved in setting up an imaging study is shown in Figure 5. Since the photomultiplier tubes and electronics of the cameras can drift over time, it is important to verify that the camera is calibrated prior to beginning the imaging study. In most cases it is helpful

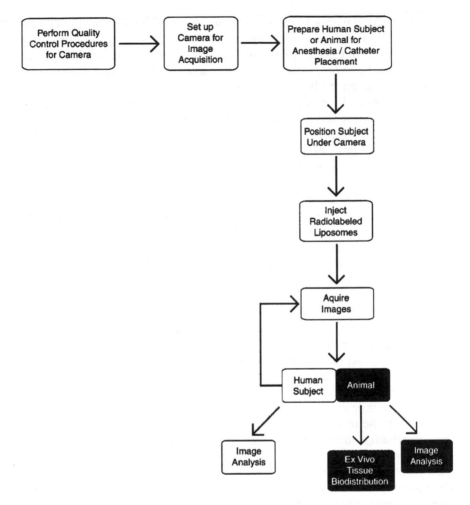

Figure 5 Diagram outlining the procedure for performing an imaging study for either human or animal testing using a single photon emission computed tomography or positron emission tomography camera.

to perform the necessary quality control procedures the day before the imaging study is to begin.

On the day of the imaging study, the subject is prepared to receive the injection of radiolabeled liposomes. For preclinical animal studies, the animals are typically anesthetized prior to catheter placement for intravenous infusion. It is very difficult to perform imaging studies without immobilization of the animal using appropriate anesthesia. The anesthesia chosen will depend on the length of each image acquisition session. Gas delivery systems using isoflurane are often the most convenient. Some small animal imaging scanners are sold with a built-in system for gas anesthesia delivery. Typical locations for catheter placement are in the tail vein of mice and rats or ear vein of rabbits. In certain situations, it may be useful to surgically place an indwelling catheter in a large vessel of the animal several days prior to the day of the imaging study.

The subject is then positioned under the camera so that images of the area of interest can be acquired. For most clinical imaging studies the camera head is positioned over the human subject lying in the supine position on an imaging table. With the newer preclinical microSPECT and microPET cameras as well as when scanning larger animals with a clinical unit, the animals are also places in the supine position on an imaging table. When using a clinical planar gamma camera for scanning small animals such as rats and mice, the camera head can be facing up and the animal laid in the prone position directly on top of the camera head.

Next, the computer workstation is set up for acquisition and the study is given a unique file name so images can be retrieved after acquisition. The radiolabeled liposomes are then injected and the computer is activated to begin image acquisition. The rate of infusion may be varied depending on the quantity to be infused. Small quantities relative to the weight of the animal are generally infused manually through a syringe while larger quantities may be infused over a slower time period using a syringe pump.

Typically, images are acquired as often and as long as desired until the activity of the injected agent is no longer sufficient to produce an image. Images can be acquired either in a static, dynamic or tomographic mode. A common dynamic image acquisition method is to acquire one-minute images for the first one to two hours after liposome administration. This time can vary depending on the physical half-life of the radionuclide used to label the liposomes and the rate of clearance of the liposomes from the blood. This time period can be determined by visual image assessment to estimate the best time of terminating the study. The camera matrix is set in a matrix size appropriate to the count rate, typically 64×64 or 128×128. Images can also be zoomed.

After completion of the acquisition, the animal is allowed to regain consciousness. If images are to be acquired the next day, the anesthesia will need to be readministered to the animal and image acquisition performed at all time points desired. After the final images have been acquired, a tissue biodistribution study can be performed if desired as described in section V below.

Image Analysis Procedures

One of the most powerful attributes of scintigraphic imaging is that it is inherently quantitative. In addition, scintigraphy is a computer-based system so software has been developed to perform a variety of analysis options. The most common analysis tool is the region-of-interest (ROI) analysis that determines the number of counts in a designated area in an image after being highlighted by the investigator. This method consists of drawing a region around the entire organ of interest, and permits a global assessment of total organ deposition. ROI can generally be obtained for large organs that can be clearly distinguished on the images such as the liver, spleen, and heart, and around a specifically targeted region such as tumor or infection. Figure 6 depicts a planar image of a rat after ROI analysis has been performed. Clearance of the liposome-based agent from the blood can also be estimated by placing a fixed region over the heart as shown in Figure 6, because the normal heart does not have a significant deposition of radiolabeled liposomes. A second method is to place fixed size boxes on the organ

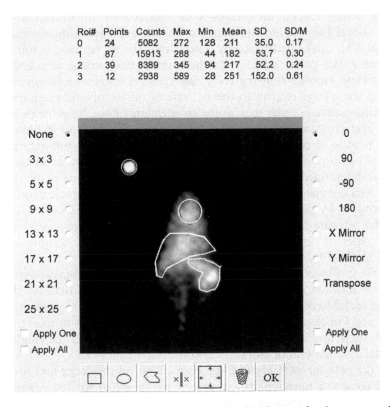

Roi#	Points	Counts	Max	Min	Mean	SD	SD/M
0	24	5082	272	128	211	35.0	0.17
1	87	15913	288	44	182	53.7	0.30
2	39	8389	345	94	217	52.2	0.24
3	12	2938	589	28	251	152.0	0.61

Figure 6 Planar static image of rat acquired 24 hours after intravenous injection of neutral ^{99m}Tc-liposomes demonstrating the methodology used for quantitation of scintigraphic images. Region-of-interest (ROI) boxes are placed on various organs and the total counts in a region displayed. An ROI box is also drawn around a known standard amount of ^{99m}Tc-liposomes placed outside of the animal field of view. Images acquired at various times are analyzed using this methodology to determine the percentage of injected dose that accumulates in the various organs.

of interest to observe the changes in counts in the box of that organ over time. The fixed box is less subjective but it only indicates the pattern of uptake of the liposomes over time, whereas the drawing of regions around the whole organ provides a better estimate of total organ uptake at each time point. It is sometimes useful to place a box around the whole animal on the first image acquired after liposome infusion is completed to estimate the total counts injected into the animal.

After all regions have been drawn, the counts in all ROIs are recorded. Data can be transferred to a spreadsheet. Several correction factors including decay, background, and blood pool can be applied depending on the imaging study. If the counts measured in the same region in images acquired at different times are to be compared, it is necessary to correct the images for decay to the time of the initial image based on the physical half-life of the radionuclide. If the counts in the images to be analyzed are low in comparison to the natural background radiation, a background correction should be performed. This can be performed by determining the background counts per pixel from a ROI placed adjacent to the animal's body and subtracting these counts from each pixel in the ROI. Background correction becomes important on late images relative to the half-life of the isotope that require a long acquisition time in order to acquire an adequate image. Some organs such as the liver contain a substantial pool of blood. The activity in these organs may need to be corrected for this blood pool activity by subtracting the estimated blood pool activity of each organ region. The activity in the blood can be determined by counting blood samples at each time point in the gamma well counter. The percentage of the blood pool in each organ can be determined by performing a study with labeled red blood cells and recording the percentage of labeled red blood cell activity in each organ or by using reported organ blood pool contribution data. This technique is described in detail by Rudolph et al. (56).

The data can be analyzed by plotting as a percentage of total counts after liposome infusion in a targeted site (30,57,58). In many situations it may be more useful to compare the liposome accumulation in a targeted site to a control site. For example, target-to-background analysis may be used to compare liposome uptake in a tumor or other lesion located in the thigh to the normal contralateral thigh (19,57,59,60).

When the data for each organ have been corrected for decay and for background counts, a time-activity curve can be generated for the organ of interest. Most image analysis systems will automatically generate this curve if the images have been acquired dynamically. This curve provides a fairly accurate estimation of the amount of radioactivity in the organ over time. From this data, an estimation of the percentage of the total injected dose (%ID) in each organ can be made. In human studies, the appropriate volume of the organ or targeted region can be estimated by computed tomography or MRI so that a %ID/kg of tissue can be estimated (61).

The standard by which all the regions can be compared is obtained by drawing a large box, if possible, over the whole body image of the animal on the first image acquired immediately after the radiolabeled liposomes have been infused. This provides an estimation of the total detectable counts in the injected dose.

TISSUE BIODISTRIBUTION METHODOLOGY

Traditionally, tissue biodistribution studies have used beta-emitting radionuclides, which cannot be detected by scintigraphic imaging. These traditional studies require the use of many more animals because animals must be sacrificed at each desired time point. A better approach is to use tissue biodistribution to validate the imaging results by performing the distributions only after images have been acquired at the final time point. All other biodistribution information can be determined from the noninvasive imaging study.

In a typical biodistribution study, enough tubes are weighed in advance so that each tissue sample can be placed in a separate tube and the weights recorded. After acquiring images at the last desired time point or at the designated time point if performing no imaging, the animal is euthanized. The tissues of interest are collected, rinsed with saline and each tissue sample is placed in a separate weighed tube. The entire organ is also removed and weighed. After tissue collection, each of the tubes containing the tissue samples is weighed and the weights are recorded. The radioactivity of each tissue sample and reserved standard sample are measured in a gamma well counter, and the results recorded. A standard of a known percentage of the injected dose should be saved for counting so that total counts injected can be calculated. This permits the determination of the percentage of the injected dose in each organ. The standard should be counted in the same approximate volume as the volume of the tissue samples.

It is important that the samples be allowed to decay until the count rates are below the saturation count rate of the gamma well counter. Samples should also be distributed as homogeneously as possible and be distributed so that they do not exceed the geometric size limit of the gamma well counter. These factors are best determined experimentally by the investigator prior to performing tissue biodistribution studies. For tissues that are not discrete organs such as skin, muscle, and blood, an estimation of the total weight of these tissues in the body can be estimated from the body weight of the animal using previously determined correction factors (62).

The weight and radioactive data can be entered in a spreadsheet and the percentage of injected dose in the organ (%ID/organ) calculated using Eq. (1).

$$\%ID/organ = (((\text{tissue cpm} \div \text{tissue wt}) \times \text{organ wt}) \div$$
$$((\text{std cpm} \div \text{std vol}) \times (\text{sample vol injected}))) \times 100$$

$$(1)$$

The percentage of injected dose per gram of tissue (%ID/g tissue) can also be calculated using Eq. (2)

$$\%ID/g \text{ tissue} = ((\text{tissue cpm} \div \text{tissue wt}) \div$$
$$((\text{std cpm} \div \text{std vol}) \times (\text{sample vol injected}))) \times 100$$

$$(2)$$

A spreadsheet of the %ID/organ and %ID/g tissue can be generated and the values averaged and differences compared statistically.

CONCLUSIONS

Scintigraphic imaging can be very useful for monitoring the distribution of liposomes radiolabeled with gamma-emitting photon radionuclides. This imaging modality can be used in both the preclinical and human testing stages of new liposome-based agent development because of its noninvasive and inherently quantitative nature. Incorporation of scintigraphic imaging into liposome drug development protocols should become more widespread with the increasing availability and improvement of miniaturized versions of clinical SPECT and PET cameras, allowing faster translation of preclinical animal results to clinical testing phases.

REFERENCES

1. Boerman OC, Laverman P, Oyen WJ, et al. Radiolabeled liposomes for scintigraphic imaging. Prog Lipid Res 2000; 39(5):461–475.
2. Boerman OC, Rennen H, Oyen WJ, et al. Radiopharmaceuticals to image infection and inflammation. Semin Nucl Med 2001; 31(4):286–295.
3. Goins B, Phillips WT. Radiolabelled liposomes for imaging and bio distribution studies. In: Torchilin VP, Weissig V, eds. Liposomes A Practical Approach. 2nd ed. Oxford, U.K.: Oxford University Press, 2003:319–336.
4. Goins BA, Phillips WT. The use of scintigraphic imaging as a tool in the development of liposome formulations. Prog Lipid Res 2001; 40(1–2):95–123.
5. Laverman P, Boerman OC, Storm G. Radiolabeling of liposomes for scintigraphic imaging. Meth Enzymol 2003; 373:234–248.
6. Phillips WT, Goins B. Assessment of liposome delivery using scintigraphic imaging. J Liposome Res 2002; 12(1–2):71–80.
7. Tilcock C. Delivery of contrast agents for magnetic resonance imaging, computed tomography, nuclear medicine and ultrasound. Adv Drug Deliv Rev 1999; 37(1–3):33–51.
8. Torchilin VP. Polymeric contrast agents for medical imaging. Curr Pharm Biotechnol 2000; 1(2):183–215.
9. Blankenberg FG, Strauss HW. Nuclear medicine applications in molecular imaging. J Magn Reson Imaging 2002; 16(4):352–361.

10. Wolf GL. Targeted delivery of imaging agents: an overview. In: Torchilin VP, ed. Handbook of Targeted Delivery of Imaging Agents. Boca Raton, FL: CRC Press, 1995:3–22.
11. Blasberg R. Imaging gene expression and endogenous molecular processes: molecular imaging. J Cereb Blood Flow Metab 2002; 22(10):1157–1164.
12. Hargreaves R. Imaging substance P receptors (NK1) in the living human brain using positron emission tomography. J Clin Psychiatry 2002; 63(suppl 11):18–24.
13. Phelps ME. PET: the merging of biology and imaging into molecular imaging. J Nucl Med 2000; 41(4):661–681.
14. Ray P, Bauer E, Iyer M, et al. Monitoring gene therapy with reporter gene imaging. Semin Nucl Med 2001; 31:312–320.
15. Sharma V, Luker GD, Piwnica-Worms D. Molecular imaging of gene expression and protein function in vivo with PET and SPECT. J Magn Reson Imaging 2002; 16(4):336–351.
16. Van de Wiele C, Lahorte C, Oyen W, et al. Nuclear medicine imaging to predict response to radiotherapy: a review. Int J Radiat Oncol Biol Phys 2003; 55(1):5–15.
17. Winnard P Jr., Raman V. Real time non-invasive imaging of receptor-ligand interactions in vivo. J Cell Biochem 2003; 90(3):454–463.
18. Awasthi V, Goins B, Klipper R, et al. Imaging experimental osteomyelitis using radiolabeled liposomes. J Nucl Med 1998; 39(6):1089–1094.
19. Awasthi VD, Goins B, Klipper R, et al. Dual radiolabeled liposomes: biodistribution studies and localization of focal sites of infection in rats. Nucl Med Biol 1998; 25(2):155–160.
20. Marsee DK, Shen DH, MacDonald LR, et al. Imaging of metastatic pulmonary tumors following NIS gene transfer using single photon emission computed tomography. Cancer Gene Ther 2004; 11(2):121–127.
21. Perkins AC, Frier M. Radionuclide imaging in drug development. Curr Pharm Des 2004; 10(24):2907–2921.
22. Graham LS, Muehllehner G. Anger Scintillation Camera. In: Sandler MP, ed. Diagnostic Nuclear Medicine. Baltimore: Williams & Wilkins, 1996:81–92.
23. Bocher M, Balan A, Krausz Y, et al. Gamma camera-mounted anatomical X-ray tomography: technology, system characteristics and first images. Eur J Nucl Med 2000; 27(6):619–627.
24. Humm JL, Rosenfeld A, Del Guerra A. From PET detectors to PET scanners. Eur J Nucl Med Mol Imaging 2003; 30(11):1574–1597.
25. Keidar Z, Israel O, Krausz Y. SPECT/CT in tumor imaging: technical aspects and clinical applications. Semin Nucl Med 2003; 33(3):205–218.
26. Phelps ME, Cherry SR. The changing design of positron imaging systems. Clin Positron Imaging 1998; 1(1):31–45.
27. Acton PD, Kung HF. Small animal imaging with high resolution single photon emission tomography. Nucl Med Biol 2003; 30(8):889–895.
28. Galt JR, Faber T. Principles of single photon emission computed tomography (SPECT) imaging. In: Christian PE, Bernier DR, Langan JK, eds. Nuclear Medicine and PET: Technology and Techniques. 5th ed. St. Louis, MO: Mosby, 2004:242–284.
29. Koukourakis MI, Koukouraki S, Giatromanolaki A, et al. Liposomal doxorubicin and conventionally fractionated radiotherapy in the treatment of locally

advanced non-small-cell lung cancer and head and neck cancer. J Clin Oncol 1999; 17(11):3512–3521.

30. Harrington KJ, Mohammadtaghi S, Uster PS, et al. Effective targeting of solid tumors in patients with locally advanced cancers by radiolabeled pegylated liposomes. Clin Cancer Res 2001; 7(2):243–254.

31. Green MV, Seidel J, Vaquero JJ, et al. High resolution PET, SPECT and projection imaging in small animals. Comput Med Imaging Graph 2001; 25(2):79–86.

32. Lewis JS, Achilefu S, Garbow JR, et al. Small animal imaging: current technology and perspectives for oncological imaging. Eur J Cancer 2002; 38(16): 2173–2188.

33. Herschman HR. Molecular imaging: looking at problems, seeing solutions. Science 2003; 302(5645):605–608.

34. Hildebrandt IJ, Gambhir SS. Molecular imaging applications for immunology. Clin Immunol 2004; 111(2):210–224.

35. Lyons SK. Advances in imaging mouse tumour models in vivo. J Pathol 2005; 205(2):194–205.

36. Del Guerra A, Belcari N. Advances in animal PET scanners. Q J Nucl Med 2002; 46(1):35–47.

37. Ritman EL. Molecular imaging in small animals—roles for micro-CT. J Cell Biochem Suppl 2002; 39:116–124.

38. Ritman EL. Micro-computed tomography-current status and developments. Annu Rev Biomed Eng 2004; 6:185–208.

39. Cherry SR, Phelps ME. Positron emission tomography: methods and instrumentation. In: Sandler MP, ed. Diagnostic Nuclear Medicine. Baltimore: Williams & Wilkins, 1996:139–159.

40. Rohren EM, Turkington TG, Coleman RE. Clinical applications of PET in oncology. Radiology 2004; 231(2):305–332.

41. Couturier O, Luxen A, Chatal JF, et al. Fluorinated tracers for imaging cancer with positron emission tomography. Eur J Nucl Med Mol Imaging 2004; 31(8):1182–1206.

42. Delbeke D, Martin WH. Metabolic imaging with FDG: a primer. Cancer J 2004; 10(4):201–213.

43. Avril NE, Weber WA. Monitoring response to treatment in patients utilizing PET. Radiol Clin North Am 2005; 43(1):189–204.

44. Dutour A, Monteil J, Paraf F, et al. Endostatin cDNA/cationic liposome complexes as a promising therapy to prevent lung metastases in osteosarcoma: study in a human-like rat orthotopic tumor. Mol Ther 2005; 11(2):311–319.

45. Iyer M, Berenji M, Templeton NS, et al. Noninvasive imaging of cationic lipid-mediated delivery of optical and PET reporter genes in living mice. Mol Ther 2002; 6(4):555–562.

46. Beyer T, Townsend DW, Blodgett TM. Dual-modality PET/CT tomography for clinical oncology. Q J Nucl Med 2002; 46(1):24–34.

47. Fahey FH. Instrumentation in positron emission tomography. Neuroimaging Clin N Am 2003; 13(4):659–669.

48. Robinson MK, Doss M, Shaller C, et al. Quantitative immuno-positron emission tomography imaging of HER2-positive tumor xenografts with an iodine-124 labeled anti-HER2 diabody. Cancer Res 2005; 65(4):1471–1478.

49. Kondo M, Asai T, Katanasaka Y, et al. Anti-neovascular therapy by liposomal drug targeted to membrane type-1 matrix metalloproteinase. Int J Cancer 2004; 108(2):301–306.
50. Oku N. Delivery of contrast agents for positron emission tomography imaging by liposomes. Adv Drug Deliv Rev 1999; 37(1–3):53–61.
51. Oku N, Namba Y. Glucuronate-modified, long-circulating liposomes for the delivery of anticancer agents. Meth Enzymol 2005; 391:145–162.
52. Wu AM, Yazaki PJ, Tsai S, et al. High-resolution microPET imaging of carcinoembryonic antigen-positive xenografts by using a copper-64-labeled engineered antibody fragment. Proc Natl Acad Sci USA 2000; 97(15):8495–8500.
53. Wipke BT, Wang Z, Kim J, et al. Dynamic visualization of a joint-specific autoimmune response through positron emission tomography. Nat Immunol 2002; 3(4):366–372.
54. Goins B, Klipper R, Martin C, et al. Use of oxygen-15-labeled molecular oxygen for oxygen delivery studies of blood and blood substitutes. Adv Exp Med Biol 1998; 454:643–652.
55. Phillips WT, Lemen L, Goins B, et al. Use of oxygen-15 to measure oxygen-carrying capacity of blood substitutes in vivo. Am J Physiol 1997; 272(5 Pt 2): H2492–H2499.
56. Rudolph AS, Klipper RW, Goins B, et al. In vivo biodistribution of a radiolabeled blood substitute: 99mTc-labeled liposome-encapsulated hemoglobin in an anesthetized rabbit. Proc Natl Acad Sci USA 1991; 88(23):10976–10980.
57. Goins B, Klipper R, Rudolph AS, et al. Use of technetium-99m-liposomes in tumor imaging. J Nucl Med 1994; 35(9):1491–1498.
58. Matteucci ML, Anyarambhatla G, Rosner G, et al. Hyperthermia increases accumulation of technetium-99m-labeled liposomes in feline sarcomas. Clin Cancer Res 2000; 6(9):3748–3755.
59. Boerman OC, Oyen WJ, Storm G, et al. Technetium-99m labelled liposomes to image experimental arthritis. Ann Rheum Dis 1997; 56(6):369–373.
60. Goins B, Klipper R, Rudolph AS, et al. Biodistribution and imaging studies of technetium-99m-labeled liposomes in rats with focal infection. J Nucl Med 1993; 34(12):2160–2168.
61. Harrington KJ. Liposomal cancer chemotherapy: current clinical applications and future prospects. Expert Opin Investig Drugs 2001; 10(6):1045–1061.
62. Frank DW. Physiological data of laboratory animals. In: Melby ECJ, ed. Handbook of Laboratory Animal Science. Boca Raton, FL: CRC Press, 1976:23–64.

12

Cytoplasmic Tracking of Liposomes Containing Protein Antigens[†]

Mangala Rao, Kristina K. Peachman, and Carl R. Alving

*Division of Retrovirology, Department of Vaccine Production and Delivery,
U.S. Military HIV Research Program, Walter Reed Army Institute of Research,
Silver Spring, Maryland, U.S.A.*

INTRODUCTION

Recombinant protein vaccines and synthetic peptides generally do not induce robust immune responses when administered in the absence of an adjuvant. By delivering these vaccines with appropriate adjuvants and delivery systems, not only can the immune response be significantly improved but also the antigen can be channeled into the major histocompatibility complex (MHC) class I or MHC class II pathways to induce a Th1 or Th2 type of immune response (1,2). In 1974, liposomes were proposed as carriers of antigens to augment antibody responses in vivo (3,4). The use of liposomes as potential carriers of antigens for vaccines in combination with a variety of adjuvants including lipid A, muramyl dipeptide and its derivatives, and several cytokines has been explored (1,2,5). In fact, the first liposomal hepatitis A vaccine has been licensed in Europe (6,7).

Antigens can be reconstituted within the lipid bilayers of the liposomes, encapsulated within the internal aqueous spaces, or covalently attached to the outer surface. One of the attractive features and rationale for using

---

[†] *Disclaimer*: The information contained herein reflects the views of the authors and should not be construed to represent those of the Department of the Army or the Department of Defense.

liposomes as vehicles for vaccines has been the rapid uptake of liposomes by macrophages and immature dendritic cell (8–10). Liposome-formulated vaccines have the added potential to simultaneously gain entry into both the conventional MHC class I and MHC class II pathways, thus inducing both antibody and cellular immune responses (2,11).

Antigens are processed and presented by the highly polymorphic MHC class I and class II membrane proteins (12,13). MHC class I and class II molecules bind and transport peptide fragments of intact proteins to the surface of antigen-presenting cells and provide a continuously updated display of an array of peptide fragments derived from endogenous and exogenous proteins, respectively, for interaction with T cells. The peptide-MHC complex interacts with either CD8[+] or CD4[+] T lymphocytes to induce the appropriate immune response (12,14).

In most cells, exogenous antigens cannot be presented by MHC class I molecules because of the inability of antigens to gain access to the cytosol, unless the antigen is channeled into the cytoplasm by artificial means such as osmotic loading (15), conjugation with lipid carriers (16–18), latex beads, biodegradable microspheres (19–21), or encapsulated in liposomes (1,2,22–24). Among these methods, liposomes have proven to be an efficient delivery system for entry of exogenous protein antigens into the MHC class I pathway due to their particulate nature (25). In this chapter, we will address the intracellular fate of liposomal protein antigens in antigen-presenting cells utilizing fluorophore-labeled proteins encapsulated in liposomes.

COUPLING OF FLUORESCENT DYES TO PROTEIN ANTIGENS

Proteins can be labeled with many different fluorochromes. The availability of Alexa Fluor® dyes and quantum dots has revolutionized the field of fluorescence microscopy. One of the advantages of Alexa Fluor® dyes is that the dye is water soluble, and when conjugated to the protein, the fluorescence is very bright and stable over a wide range of pH. In addition, Alexa Fluor® dyes ranging from visible to infrared spectrum are available providing for multicolor analysis. Proteins can also be labeled with quantum dots. Again this is also available in multiple colors and is being extensively used in fluorescence microscopy. We have successfully labeled several different proteins, peptides, and viruses with Texas Red and Alexa Fluor® dyes.

Labeling of Protein with Texas Red and Separation of Labeled Protein

Proteins can be labeled with Texas Red by covalent coupling of the amino groups on proteins that occurs when the protein and the Texas Red reagent are mixed together (26). The efficiency of labeling of the protein with Texas Red is dependent on several factors. The most critical factors being the pH of the buffer solution and the temperature at which the coupling reaction is carried out. Maximal conjugation is obtained at pH 9.0. However, this dye may not be suitable if the protein is sensitive to alkaline pH. Texas Red hydrolyses rapidly in aqueous solution. Therefore, Texas Red should be kept dry prior to addition to the protein solution. The detailed procedure that we use for coupling proteins with Texas Red for our in vitro studies has been described earlier (27). A brief outline for coupling, using ovalbumin as a model antigen, is presented below.

One milliliter of sterile ovalbumin solution (10 mg/mL in 0.2 M carbonate-bicarbonate buffer, pH = 9.0), is placed in a 12 × 75 cm sterile plastic tube containing a small stir bar and placed on ice and, while stirring, 1 mg of Texas Red sulfonyl chloride (Molecular Probes, Eugene, Oregon, U.S.A.) is added. After three hours, the protein solution is loaded onto a Sephadex G-25 column and 0.5 mL fractions are collected until all the purple color is eluted. The purple dye is Texas Red conjugated to protein. The unbound Texas Red is red in color and is retarded on the column. The absorbance at 280 nm (protein) and at 596 nm (Texas Red) is measured in a spectrophotometer and the amount of Texas Red conjugated to the protein is calculated. The Texas Red-labeled protein is now ready to be encapsulated in liposomes.

Labeling of Protein with Alexa Fluor® and Separation of Labeled Protein

As mentioned earlier, there is a whole gamut of Alexa Fluor® dyes. We have successfully labeled several different peptides, proteins (including dengue envelope proteins), and viruses (dengue, yellow fever, and irradiated Ebola virus). Protein (900 µL of a 15 mg/mL solution) or whole viruses (1×10^6–1×10^8 pfu) are transferred to an Eppendorf tube and 100 µL of a 1 M sodium bicarbonate solution pH 8.5 is added to the antigen solution. The antigen solution is then transferred to the vial containing Alexa Fluor® dye. The vial is capped and inverted several times to fully dissolve the dye. Depending upon the stability of the protein, the reaction mixture is stirred (flea stirrer or orbital shaker) for a minimum of one hour at room temperature or for better results overnight at 4°C (this is preferable when labeling viruses to retain infectivity). The free unbound dye is removed by overnight dialysis at 4°C using a Slide-A-Lyzer® dialysis cassette (3500 molecular cut off) (Peirce Biotechnology, Inc., Rockford, Illinois, U.S.A.) followed by column purification as described above.

PREPARATION OF LIPOSOMES

A liposome formulation, termed as the Walter Reed Army Institute of Research (WRAIR) liposomes, was developed in our laboratory. This formulation has been used in human clinical trials and contains dimyristoyl phosphatidylcholine, dimyristoyl phosphatidylglycerol, and cholesterol (Avanti Polar Lipids, Alabaster, Alabama, U.S.A.) in molar ratios of

1.8:0.2:1.5, and 1.6 mg/mL of lipid A and an encapsulated protein antigen (28,29). This formulation of liposomes has also been shown to be an effective vehicle for delivery of proteins or peptides to antigen-presenting cells for presentation via the MHC class I pathway in mice (30,31). Detailed procedures for the preparation of liposomes have been previously described (28) and are also described by Matyas and Alving in Chapter 19.

Encapsulation of Protein in Liposomes

Multilamellar liposomes are prepared by dispersion of lyophilized mixtures of lipids at a phospholipid concentration of 100 mM in Dulbecco's phosphate-buffered saline (PBS) containing either unlabeled antigen, or Texas Red-labeled antigen or Alexa Fluor® labeled antigen (32). To obtain fluorescent liposomes, a fluorescent phospholipid, N-nitrobenzoxadiazole (NBD)-phosphoethanolamine (Molecular Probes, Eugene, Oregon, U.S.A.; 2 mol% with respect to the phospholipid concentration) can be used (33). To encapsulate the proteins in liposomes, the vial containing lyophilized lipids stored at $-70°C$ is allowed to come to room temperature and the required amount of protein is added into the vial, mixed, and kept in the refrigerator. After 48 hours, the contents of the tube are transferred into a sterile capped tube and centrifuged at 7500 g for 30 minutes at RT in a Sorvall RC-5B refrigerated superspeed centrifuge using SA600 rotor. At the end of the run, the supernatant is carefully decanted or removed with a sterile pipette and discarded. The pellet of liposomes is washed twice in sterile saline and then resuspended in saline or buffer of choice to give a final phospholipid concentration of 30 mM, and stored at 4°C until used.

Determination of Encapsulation Efficiency

In order to determine that the protein is not degraded as a result of conjugation with the fluorescent dye, the labeled proteins can be analyzed for the degree of degradation by sodium dodeeyl sulfate (SDS) polyacrylamide gel electrophoresis. The protein is first released from liposomes by chloroform treatment (34) and then separated by electrophoresis on 4% to 20% precast polyacrylamide gels using a Tris (25 mM)-glycine (192 mM)-SDS (20%) electrode buffer. At the end of the run, the fluorescent protein band is viewed under UV light. The amount of antigen encapsulated in liposomes is determined by a modified Lowry procedure (28) using an aliquot of liposomes (10–50 µL) and protein standard (generally the same protein used for encapsulation, 0–80 µg). The protein is released from the liposomes by dissolving the lipids in chloroform (0.5 mL), drying the samples and then dissolving the dried pellet in 200 µ;L deoxychlolate (15% w/v in deionized water). The tubes are vortexed and the amount of protein present is determined by the standard Lowry procedure. The percent encapsulation is calculated as follows: encapsulation = amount of protein measured by the

Lowry assay/amount of protein initially added to the lipids $\times$ 100. The encapsulation efficiency varies from protein to protein. With ovalbumin and conalbumin the percent encapsulation is between 60% and 70%. With HIV proteins p24 and ogp140, the encapsulation efficiencies are 34% and 56%, respectively.

PREPARATION OF ANTIGEN-PRESENTING CELLS

We have developed an in vitro antigen presentation system, consisting of bone marrow–derived macrophages or dendritic cells as the antigen-presenting cells. Our system is well suited for studying intracellular trafficking because we begin with precursor cells that can be differentiated into either dendritic cells or macrophages. This system has the advantage of eliminating fixation and permeabilization steps, and living cells can thus be observed. B cells can also be used as antigen-presenting cells to study intracellular trafficking of antigens. In our studies, we have mainly used bone marrow–derived macrophages and immature dendritic cells because of their inherent phagocytic properties. An added advantage of using macrophages is their ability to adhere to plastic dishes. This also facilitates all the washing steps involved and permits easy microscopic observation of living cells. These are not synchronized cultures and there is variability in the phagocytic and antigen processing rates from cell to cell. Therefore, it is critical to observe and count cells from multiple fields to obtain representative results. The antigen processing efficiency of macrophages decreases with the age of the mice. Consistent results are obtained with bone marrow–derived macrophages from mice that are not older than three months.

Preparation of Bone Marrow–Derived Mouse Macrophages

Typically, the femurs from three mice are processed for each macrophage harvesting. The femurs are removed under sterile conditions and placed in a 60-mm Petri dish containing 2 mL Dulbecco's PBS without Ca^{2+} and Mg^{2+} (PBS, BioWhittaker Inc., Walkersville, Maryland, U.S.A.). It is important to remove all the fat or cartilage attached to the femurs to prevent fibroblasts from growing and taking over the culture. The ends of the bones are snipped off and the marrows flushed into a sterile 50-mL conical tube with 10 mL of PBS using a syringe and a 22-gauge needle. A single cell suspension is made by drawing up the cells with the syringe and passing it three times through the needle. The cell suspension is transferred into a fresh sterile 50-mL conical tube and spun at 800 g for 10 minutes at 4°C. The supernatant is discarded and the cell pellet is resuspended in 10 mL bone marrow macrophage growth media (RPMI-1640 containing 10% heat inactivated fetal bovine serum, 10% L-cell conditioned media, 100 U/mL penicillin and 100 µg/mL streptomycin [P/S] and 8 mM glutamine). To prepare L-cell conditioned media, L-929 cells (obtained from American Type Culture Collection) are grown in a tissue culture flask for a week in RPMI-1640 containing 10% FBS and 8 mM glutamine. The supernatant is clarified by centrifugation in a Sorvall RT6000 refrigerated centrifuge at 800 g for 10 minutes at 4°C and the supernatant stored at −70°C. All the media components are readily available from several sources like Invitrogen-Gibco, BioWhittaker, or Cellgro.

Cell Counting

The viability of the cells is determined by staining an aliquot of the cells with trypan blue (2% solution made in PBS). The cells are counted in a hemocytometer. The number of live cells (those that did not take up the stain) is multiplied with the cell dilution factor and a constant factor (10^4) to obtain cells/mL. The cell concentration is adjusted to achieve 2×10^6 cells/mL and 100 μL cells is seeded on acid-washed coverslips that are placed in 35-mm Petri dishes. The cells are then spread over the area of the coverslip with a sterile pipette tip. After 20 minutes of adherence at RT, 2 mL bone marrow macrophage media/dish is carefully added and the dishes are placed in a humidified CO_2 incubator at 37°C. On alternate days, until day 9, 1 mL media is removed and replaced with 1 mL fresh media. On day 9, macrophage cultures are supplemented with 10 U/mL murine IFN-γ (R&D Systems, Inc.) and used for trafficking experiments the next day (10,29).

Preparation of Bone Marrow–Derived Mouse Dendritic Cells

Marrows are isolated, processed, and single-cell suspensions made from the femurs of mice as described above for the preparation of macrophages except that the red blood cells are lysed using ACK lysis buffer (Gibco) for 10 minutes at RT followed by washing the cells thoroughly with media. Then 3 mL of cells (1×10^6 cells/mL) are cultured in six-well dishes (Corning-Costar) in the presence of 0.1 μg Granulocyte macrophage-colony stimulating factor (GM-CSF) and 0.04 μg IL-4 for six days. On day five, 1.5 mL of media containing GM-CSF and IL-4 is added to each well. On day 6 to 7, the aggregates are dislodged (gently pipette RPMI-1640), pooled, and centrifuged at RT for 10 minutes at 1500 rpm. The cell pellet is resuspended in media and an aliquot of cells counted as described above. Typically the cell yield is 1.3×10^7 and >60% of the cells have surface markers characteristic of dendritic cells. For higher purity, dendritic cells can be positively selected on Miltenyi CD11c$^+$ beads (Miltenyi Biotec Inc., Auburn, California, U.S.A.).

Generation of Human Macrophages and Dendritic Cells

CD14$^+$ monocytes are isolated from peripheral blood mononuclear cells of healthy volunteers by positive selection using CD14 microbeads and a magnetic cell separator (Miltenyi Biotec Inc., Auburn, California, U.S.A.). The selection, carried out according to the manufacturer's recommendations, results in $\geq 95\%$ purity. For generation of dendritic cells, enriched CD14$^+$ monocytes are cultured in complete RPMI media (RPMI 1640, 1% L-glutamine, 1% penicillin/streptomycin, 1% sodium pyruvate, 1% essential amino acids, 50 mM 2-mercaptoethanol and 10% heat inactivated FBS) in six-well plates (1×10^6/mL) for six days at 37°C and 5% CO_2 in the presence of Leukine (rhGM-CSF) (100 ng/mL, Amgen, Thousand Oaks, California, U.S.A.) and rhIL-4 (50 ng/mL, Pharmingen, San Diego, California, U.S.A.). In contrast, for the generation of macrophages, cells are cultured from the beginning of culture on coverslips placed in six well plates containing rh-M-CSF (50 ng/mL). On day 3, half of the medium is replaced with fresh media. By day 7, macrophages and dendritic cells are ready to be used for in vitro studies.

Dendritic Cell and Macrophage Characterization

For evaluation of dendritic cell and macrophage phenotypes, cells are re-suspended at a cell count of 2×10^5 in PBS containing 0.5% bovine serum albumin and stained for expression of surface markers for 30 minutes at 4°C. Cells are then washed twice with PBS and incubated with monoclonal antibodies directed against the respective murine or human surface antigens like costimulatory molecules, MHC class I molecules, dendritic cell and macrophage-specific surface markers. The corresponding isotypes are used

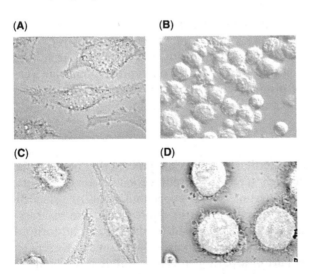

Figure 1 Macrophages and dendritic cells. Panels (**A**) and (**B**) are bright field images of purified bone marrow–derived murine macrophages and dendritic cells, respectively. Panels (**C**) and (**D**) are bright field images of purified CD14$^+$ blood derived human macrophages and dendritic cells, respectively.

as control antibodies. The samples are analyzed on a flow cytometer (Becton Dickinson, Mountain View, California, U.S.A.) to determine the purity of the cell preparations. Figure 1 shows the bright field images of murine and human macrophages and dendritic cells purified as described above. Morphologically, macrophages and dendritic cells are very easy to distinguish. Macrophages are typically flat, elongated cells containing phagocytic vacuoles. Dendritic cells are round or elliptical with dendritic processes.

INTRACELLULAR VISUALIZATION OF LIPOSOMAL ANTIGEN

For a long time, macrophages were thought to be the predominant antigen-presenting cells responsible for processing and presentation of exogenous antigens, including particulate antigens such as liposomal antigens (8–11, 20,29). Several lines of evidence support the conclusion that macrophages serve as the predominant antigen presenting cells for processing and presentation of liposomal antigens in vivo: (i) one of the earliest and most well-known observations in the field of liposome research is that parentally injected liposomes are rapidly ingested by macrophages, particularly in the liver and spleen, where they are gradually degraded in lysosomal vacuoles (35,36); (ii) in vivo depletion of macrophages caused suppression of the immune response to liposomal antigens (8); and (iii) injection of liposomes containing dichloromethylene diphosphate, a substance that is toxic to macrophages, caused the suppression of the immune response to liposome-encapsulated albumin (8).

Numerous studies have demonstrated that immature dendritic cells also efficiently uptake and process soluble and particulate antigens (37,38), although the phagocytic activity is significantly less than that seen with macrophages. In fact, dendritic cells are the only professional antigen-presenting cells capable of priming naive T cells in vivo (39) and are also the ideal cells for mediating crosspresentation (38).

Routes of Entry

Antigen processing and presentation is a dynamic process that involves the movement of antigens through several different vesicular compartments. Dendritic cells and macrophages internalize exogenous antigens primarily through four routes: phagocytosis, endocytosis, pinocytosis, and macropinocytosis. The route of entry is largely dictated by the nature and form of the antigen, i.e., whether it is a soluble or a particulate antigen, present as an immune complex or as part of a pathogen. These different mechanisms can be distinguished by the size of the ingested particle and by the presence or absence of receptor-mediated mechanisms. Macrophages (Fig. 2A) and immature dendritic cells (Fig. 2B) are extremely efficient at phagocytosis, a process that is used by cells to internalize large particles such as debris,

(A)   (B)

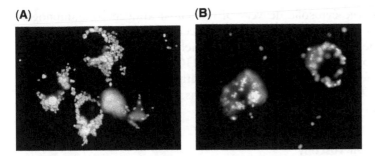

Figure 2 Phagocytosis. Murine macrophages (**A**) and human dendritic cells (**B**) were incubated with FITC-labeled *Escherichia coli* K-12 bio particles for 40 minutes at 37°C. Cells were washed and examined by a Leitz Orthoplan fluorescence microscope.

apoptotic cells and pathogens greater than 0.2 μm. In addition, phagocytosis may be associated with the presence of specific receptors such as complement receptors or Fc receptors (40). In Figure 2, murine macrophages (Fig. 2A) and human dendritic cells (Fig. 2B) were incubated with fluorescein isothiocyanate (FITC)-labeled-*Escherichia coli* K-12 bioparticles for 40 minutes at 37°C. The phagocytic cells engulf the particulate antigen in a clathrin-independent process that involves the plasma membrane to form a phagosome around the particulate material that is engulfed. In each cell type, multiple particles have been phagocytosed. Antigens can be taken up by endocytosis through the formation of coated pits formed by clustering specific cell-surface receptors. Clathrin-mediated endocytosis usually involves the binding of a surface receptor before the invagination of the plasma membrane. Pinocytosis can be either clathrin dependent or independent and is restricted to soluble proteins and small particles less than 0.2 μm. Dendritic cells are highly efficient in macropinocytosis and this route plays a major role in processing and presentation of exogenous antigens through the MHC class I pathway. Macropinocytosis is a nonphagocytic process for the ingestion of soluble antigens or particles greater that 0.2 μm and is not associated with any receptor activity (41). It is a form of regulated endocytosis that involves the formation of large endocytic vesicles (200–500 nm in diameter) after closure of the cell-surface membrane ruffles.

The internalized antigen follows different scenarios in the antigen-presenting cell depending upon the mechanism of uptake. These different routes of uptake could result in either a partial or complete degradation of the antigen and consequently the antigen has the potential to enter different processing compartments such as the endosomes, lysosomes, or the cytoplasmic compartment (42). For many years, it was believed that phagocytosed liposomes did not reach the cytoplasmic compartment but were degraded in the endosomes and lysosomes of macrophages. Others and we have demonstrated the cytosolic delivery of liposomal antigen by immunogold electron microscopy (10,43) and by fluorescence microscopy (29–33). The delivery of the antigen in the cytosol ultimately leads to the processing and presentation of the antigen through the MHC class I pathway (30,32). In this chapter, we will focus on fluorescent methods to track the intracellular trafficking of liposome-encapsulated antigens and its entry into the MHC class I pathway.

Fluorescence Microscopy

Fluorescence-labeled antigens can be visualized microscopically by epifluorescence or by a confocal microscope. Fluorescence microscopes are available through several companies including Bio-Rad, Olympus, Zeiss, and Leitz. Both epi and confocal microscopes come in upright and inverted forms enabling the users to view samples in tissue culture wells, on slides or on coverslips. The image obtained by epifluorescence is the composite of the fluorescence found in the entire cell. This method has the advantage of quick image gathering times. It has the disadvantage of faster photo bleaching and the inability to see individual planes within a single cell. In confocal microscopy, the laser beam is directed through a specific intracellular plane as instructed by the user. By directing the laser beam through a more specific plane, it is possible to use less laser power per scan and thus decrease the likelihood of photo bleaching other cells in close proximity to the cell being viewed. In a confocal microscope, the images of the cell can be viewed and collected at 0.5-μm thick sections. Therefore, more detailed data on internal and cell surface localization of the antigen can be obtained by this method when compared to epifluorescence. This method becomes important for colocalization studies involving two variables such as liposome-encapsulated antigens and a specific organelle, such as the Golgi complex. Additionally, the software also allows for 3-D visualization of the cell. The disadvantages to confocal microscopy are the length of time for data collection and the requirement for fixation for highly motile samples. However, the time for data collection can be decreased by either decreasing the number of sections obtained through the cell or by increasing the laser speed, which in turn can decrease pixel quality. Depending on the exact hypothesis being tested in the experiment, these modifications can be adjusted accordingly.

In our studies, we have used both epifluorescence and confocal microscopy. When using epifluorescence, cells are examined with a Leitz Orthoplan (Leica, Deerfield, Illinois, U.S.A.) microscope equipped with differential interference contrast objectives and a Leitz 63× oil immersion lens designed for fluorescence microscopy. Fluorescence signals are generated using fluorescence filters from Omega Optical that are optimized for Texas Red (excitation wavelength: 595; emission wavelength: 615), fluorescein (excitation wavelength: 494; emission wavelength: 518), Alexa Fluor 594 (excitation wavelength: 594; emission wavelength: 617), or Alexa Fluor 488 (excitation wavelength: 488; emission wavelength: 519) fluorochromes. Images are collected with a color digital camera (Model DEI-470, Optronics Engineering, Goleta, California, U.S.A.) coupled to an Apple Macintosh computer and are stored as Adobe Photoshop files (Adobe Systems, Inc., San Jose, California, U.S.A.). For confocal microscopy, we use a BioRad Radiance 2100 confocal microscope equipped with a 60× oil immersion objective. Images are collected with the manufacturer's software for the confocal images. Brightfield images are contrast-enhanced electronically by the video camera to permit viewing of living cells.

Visualization of Liposomal Protein in Macrophages and Dendritic Cells

Procedure

Coverslips containing macrophages are washed twice in Hanks balanced salt solution (HBSS) without phenol red, pH 7.4, and incubated in a total volume of 1 mL HBSS containing 30 µg of liposome-encapsulated Texas Red-labeled antigen. Macrophages are adherent and the coverslips can be easily washed and mounted on depression slides to view the cells. Because dendritic cells are nonadherent, they need to be collected and spun down. All the washings also require centrifugation. Dendritic cells are collected from six-well dishes, spun down, resuspended in a small volume of PBS and incubated with liposome-encapsulated labeled antigen (approximately 50 µg/mL) in 6-mL polypropylene tubes. After incubation at 37°C in a CO_2 incubator for various time periods, the coverslips containing macrophages are washed and mounted cell-side down on a depression slide containing a small quantity of buffer. In our studies, we have incubated the cells with liposomal antigen for 90 minutes, washed and incubated the cells in buffer for an additional 90 minutes (chase period). The tubes containing dendritic cells and antigen are washed by centrifugation, and then the cells are gently centrifuged onto coverslips. The coverslips are mounted on depression glass slides. The cells are viable under these conditions for at least two hours and can be put back in culture, if needed. In our studies (27), using macrophages and liposomal ovalbumin and conalbumin as the antigens, we have observed the uptake of liposomal antigen as early as five minutes. Areas with diffuse fluorescence can be seen within 15 minutes, suggesting the presence of protein in the cytoplasm of the macrophages. Internalization of the liposomal antigen continues, and by 45 minutes the liposomal antigen begins to concentrate in the perinuclear/Golgi area of the cells. After 90 minutes, the protein is mainly localized to a perinuclear region with some diffuse staining. This localization is distinctly visualized by washing the cells in HBSS and incubating for a further time period (90 minutes chase at 37°C) in HBSS. An example of liposomal antigen localizing to the

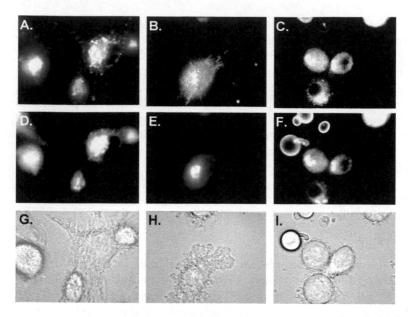

Figure 3 Uptake and localization of liposome-encapsulated antigen to the trans-Golgi complex. Murine and human macrophages and dendritic cells were incubated with liposome-encapsulated Texas Red-labeled synthetic peptide [L(TR-50 AA-peptide)] consisting of 50 amino acids for 90 minutes followed by a 90 minutes chase at 37°C. Cells were washed and stained for trans-Golgi with N-(ε-NBD-aminohexanoyl)-D-erythro-sphingosine (C_6NBD-ceramide). [L(TR-50 AA-peptide)] concentrates in the perinuclear region of mouse macrophages (**A**), human macrophages (**B**), and human dendritic cells (**C**). The liposomal antigen fluorescence showed a similar pattern as the C_6 NBD-ceramide fluorescence staining for the trans-Golgi (panels **D**, **E**, and **F**, respectively). The corresponding brightfield images are shown in panels (**G**), (**H**), and (**I**), respectively.

perinuclear region of murine macrophages (Fig. 3A), human macrophages (Fig. 3B), and human dendritic cells (Fig. 3C) can be seen in Figure 3.

Labeling of Cellular Organelles

The unique observation that liposomal antigens can spill from endosomal vesicles into the cytoplasm (10,43) raises the question of the ultimate fate of the intracytoplasmic liposomal antigen. The cytoplasmic liposomal antigens can thus gain access to the endoplasmic reticulum (ER) or to the Golgi apparatus, major cellular organelles that contain MHC class I molecules.

Organelle specific fluorescent markers are readily available from Molecular Probes, Inc. Alternatively, antibodies to a marker enzyme or a protein specific for a particular organelle can be used followed by a fluorescence-labeled secondary antibody. In the former case, live cells can be used, whereas, in the latter case, the cells have to be fixed and permeabilized for staining with the appropriate antibody.

Colocalization of the antigen with the Golgi can be demonstrated in several ways (30). At the end of the chase period, trans-Golgi is visualized by staining the cells with a green fluorescent analog of ceramide [N-(ε-NBD-aminohexanoyl)-D-erythro-sphingosine, abbreviated C_6NBD-ceramide] (44). C_6NBD-cer is a vital stain for the Golgi apparatus that is known to specifically stain the trans-Golgi of a number of different cell types. Coverslips containing macrophages are incubated on ice with 2 nmol/mL of C_6NBD-ceramide for 30 minutes, then washed twice with HBSS and transferred to 37°C for 15 minutes. After washing twice with HBSS, cells are mounted and viewed as described previously (29). The liposomal antigen fluorescence can be superimposed on the Golgi fluorescence to demonstrate colocalization. Our studies using several different protein antigens and macrophages as the antigen-presenting cell have demonstrated that soluble antigens are excluded from the trans-Golgi area and that localization or exclusion of the proteins from the trans-Golgi is determined by the particulate nature of the antigens (29). An example of trans-Golgi labeling can be seen in murine macrophages (Fig. 3D), human macrophages (Fig. 3E), and human dendritic cells (Fig. 3F).

Macrophages and immature dendritic cells avidly phagocytose liposomal antigens in a process that involves cell membrane engulfment and cytoskeletal rearrangement (45). The importance of cytoskeletal elements in the intracellular trafficking of liposomal antigens can be evaluated by staining F-actin with fluorescein phalloidin as described below.

Macrophages or dendritic cells are incubated with liposomal antigen for 90 minutes as described above (no chase in this case). At the end of the incubation period, cells are washed with PBS before fixation with formaldehyde (3.7% in PBS) for 10 minutes at RT. Cells are washed, placed in acetone at −20°C for five minutes, washed, and stained with fluorescein phalloidin (165 nM, Molecular Probes, Eugene, Oregon, U.S.A.) for 20 minutes at RT. After washing, cells are viewed under a fluorescence microscope as described above. Our studies have demonstrated the colocalization of liposome-encapsulated ovalbumin and F-actin at the point of uptake of the antigen in the cell membrane and that a rearrangement of cytoskeleton occurs to facilitate the uptake of liposomal antigens (45).

To determine the importance of microtubules in liposome antigen trafficking, microtubule destabilizers such as colchicine or microtubule stabilizers such as paclitaxel can be used. Macrophages grown on coverslips are pretreated with 10 µ;g/mL of colchicine or 10 µ;g/mL paclitaxel, at 37°C for 30 minutes. With dendritic cells, all incubations are done in polypropylene tubes. After incubation, cells are washed three times with PBS. Localization of liposomal antigen is determined by incubating the cells with liposome-encapsulated antigen (50 µg/mL) in a total volume of 1 mL for 90 minutes at 37°C followed by a 90-minute chase. Cells can also be stained for trans-Golgi as described above. After washing, the dendritic cells are centrifuged onto coverslips. The coverslips containing macrophages or dendritic cells are mounted on depression slides and are viewed as described previously. In cells treated with paclitaxel or colchicine, internalized liposomal antigen should not be concentrated if a functional microtubule-dependent translocation system is important for antigen processing. Our studies indicate that functional microtubules are required for the intracellular trafficking of liposomal antigens (45).

Involvement of Proteasome Complex in Antigen Processing

Exogenous antigens that reach the cytoplasmic compartment are subjected to proteolytic degradation by the proteasome complex in preparation for presentation of the peptides thus generated through the MHC class I pathway. Processing of the antigens through the conventional MHC class I pathway requires the proteasome complex and the transporter associated with antigen processing (TAP) proteins. The role of proteasomes can be studied by the use of reversible or irreversible proteasome inhibitors. Macrophages to be used in the proteasome inhibitor studies are incubated with the irreversible proteasome inhibitor, lactacystin (10 µM) (BIOMOL Research Laboratories, Inc., Plymouth Meeting, Pennsylvania, U.S.A.), for 30 minutes before incubation with liposome-encapsulated labeled antigen. Following the chase period, the cells are stained with the Golgi–specific stain, C_6NBD ceramide. The cells are then washed in HBSS, mounted, and viewed. In untreated macrophages, the antigen should localize in the Golgi area. In contrast, the lactacystin-treated cells should not show localization but exhibit a diffuse, granular pattern. The lactacystin treated cell should also be stained for the Golgi complex to document that treatment with lactacystin does not affect the integrity of the Golgi itself (30).

Requirement of TAP Proteins

Once the exogenous antigens undergo proteolytic degradation by the proteasome complex, in order for the peptides to bind to the MHC class I molecules, they need to be translocated into the ER. This is achieved by the heterodimeric transporter associated with antigen-processing proteins, which is composed of TAP1, TAP2, and tapasin (46,47). Both TAP1 and TAP2 proteins are required for the transport of peptides into the ER.

The easiest and the best way to determine whether peptides derived from the liposomal proteins utilize TAP proteins for their transport is to prepare macrophages from TAP1 $(-/-)$ knockout mice (obtained from Jackson Laboratories) as well as from the corresponding TAP1 $(+/+)$ wild-type mice as the positive control. In our studies, we have used TAP1 knockout mice on a C57BL/6 background, and the antigen of choice for trafficking experiments has been ovalbumin (30). Bone marrow–derived macrophages are incubated with fluorescent-labeled antigen encapsulated in liposomes and the experiment carried out as described above. The same cells are also stained with the Golgi–specific marker. In macrophages derived from TAP1 knock-out mice, if TAP proteins play a central role in transporting proteasome processed ovalbumin protein, then the antigen should not be localized but exhibit a punctate distribution (antigen is probably in endosomes and lysosomes) and be excluded from the area of the trans-Golgi. In macrophages derived from the wild-type mice, ovalbumin peptides should exclusively localize in the trans-Golgi area. A punctate distribution of liposomal antigen and exclusion of liposomal antigen from the trans-Golgi demonstrates the requirement of TAP proteins for the transport of processed peptides into the ER-Golgi complex.

Cell Surface Presentation of MHC-Peptide Class I Complexes

The expression of MHC-peptide complexes on the cell surface generated as a result of in vivo processing of the liposomal antigen can be visualized using fluorescence microscopy, or can be quantitatively measured using flow cytometry. An antibody that specifically recognizes only the MHC-peptide complexes is required. In our studies, we have used mouse monoclonal antibody 25-D1.16, which specifically binds to MHC class I-SIINFEKL complexes (SIINFEKL peptide processed from ovalbumin) generated by Drs. Porgador and Germain at NIAID, NIH, Bethesda, Maryland, U.S.A. (48). The positive control for these experiments is macrophages incubated with the ovalbumin peptide, SIINFEKL (500 µg) for 2.5 hours at 37°C. The negative controls are macrophages incubated with media alone or a liposomal antigen that is not recognized by this antibody.

Bone marrow–derived macrophages grown on coverslips are incubated with ovalbumin encapsulated in liposomes for 90 minutes, followed by a 90-minute-chase. This time frame allows liposomal ovalbumin to undergo proteolysis through the proteasome complex to generate the SINFEKL peptide, translocation into the ER and Golgi complex by the TAP proteins, binding of the peptide to MHC class I molecules, and transport to the cell surface. Before staining with the antibody, it is important to block the Fc receptors by incubating with normal goat serum (1/100 dilution in 100 µL PBS) for 30 minutes on ice followed by adding the 25-D1.16 antibody (1 mL culture supernatant) directly to the cells without washing. Cells are incubated overnight at 4°C, washed three times with PBS, and then incubated with FITC-goat-antimouse IgG (5 µg/mL diluted in PBS containing 1/100 normal goat serum) for one hour on ice. Cells are washed three times in PBS and the coverslips are mounted on depression slides as described above. Cells are then observed under a fluorescence microscope. For performing flow cytometry, the procedure for detecting cell surface expression of MHC class I-peptide complexes is similar to that described above except that macrophages are not grown on coverslips but in 35-mm Petri dishes. After the final wash with PBS, the cells are gently scraped from the Petri dishes with a rubber policeman and cells are collected by

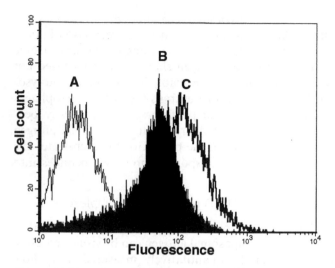

Figure 4 Detection of MHC class I-peptide complexes on the surface of murine macrophages. Murine macrophages were incubated with liposome-encapsulated ovalbumin or with the ovalbumin peptide SIINFEKL for 90 minutes followed by a 90-minute chase. Cells were processed for labeling with 25D.1.16 antibody followed by FITC conjugated antimouse antibody (Boehringer Manheim, Indianapolis, Indiana, U.S.A.). The monoclonal antibody, 25 D.1.16 recognizes ovalbumin only in its processed form (SIINFEKL) in the context of MHC class I. Cells were analyzed by flow cytometry. The surface staining increased by 1 to 2 log after incubation with liposomal ovalbumin or SIINFEKL peptide. Peak A represents unstained cells (buffer control); Peak B represents incubation with liposomal ovalbumin and peak C represents incubation with SIINFEKL peptide (positive control).

centrifugation. The cell-associated fluorescence is measured using a flow cytometer and the results are analyzed with the manufacturer's software. An example of a flow cytometer histogram can be seen in Fig. 4. Peak A represents macrophages not incubated with liposomal antigen or the antibody 25-D1.16 (buffer control). Peak B represents macrophages incubated with the liposomal ovalbumin followed by binding of the antibody 25-D1.16 to the peptide (generated as a result of intracellular processing of liposomal ovalbumin)-MHC class I complex transported to the cell surface. Peak C represents macrophages incubated with ovalbumin peptide SIINFEKL followed by binding of the antibody 25-D1.16 to the peptide-MHC complex (positive control). Because flow cytometry requires more cells than fluorescence microscopy, if macrophage cell lines are available that behave just like primary macrophages then they can be used for these studies. We have used the macrophage hybridoma cell line, C2.3 (generated in Dr. K.L. Rock's laboratory, Harvard School of Medicine, Boston, Massachusetts, U.S.A.) for flow cytometric analysis (30).

IN VIVO PROCESSING OF LIPOSOMAL ANTIGENS

To determine if the trafficking of liposomal antigen into the Golgi observed in vitro with bone marrow–derived macrophages also occurs in vivo, one can inject mice intravenously with the liposomal antigen (150–200 μg in a total volume of 0.5 mL). At various time periods, euthanize the mice, remove the spleens, and prepare a single cell suspension by mashing the spleen with a syringe plunger. Collect the cells by centrifugation (800 g, 4°C, 10 minutes). Plate 2×10^6 cells on a glass coverslip and allow cells to adhere for one hour at 37°C. Wash away the nonadherent cells, mount the coverslip on a glass slide, and examine by fluorescence microscopy.

We have demonstrated that one hour after intravenous injection of B10.BR mice with liposome-encapsulated Texas Red-labeled conalbumin, the splenic macrophages collected on the coverslip had avidly phagocytosed liposomal conalbumin and that the fluorescence was localized to a perinuclear area consistent with the Golgi localization seen in in vitro experiments described above (27,30). The fluorescence seen in macrophages was specific because neither neutrophils nor lymphocytes phagocytosed or concentrated the fluorescence-labeled protein.

CONCLUSIONS

Liposomes have been widely used as carriers of protein and peptide antigens. Here we have presented details that would allow one to study antigen trafficking in vitro using live cells. We have described an in vitro antigen presentation system that can be utilized to study intracellular trafficking patterns of liposomal antigens in living cells. The system utilizes either bone marrow–derived macrophages or dendritic cells in the case of the murine system or peripheral blood derived macrophages and dendritic cells in humans. Our system is well suited for studying intracellular trafficking because we begin with precursor cells that can be differentiated into either dendritic cells or macrophages. This system has the added advantage of eliminating fixation and permeabilization steps, and living cells can thus be observed. The system is very amenable to studies utilizing both fluorescence microscopy and flow cytometry. The system can be used to dissect out the various components necessary for the processing and presentation of liposomal antigens by using specific inhibitors that block intracellular trafficking or processing. We have used this system to explore the role of cholesterol in liposomal antigen trafficking in macrophages using specific inhibitors of cholesterol biosynthesis. In cholesterol-depleted cells, the liposomal antigen fails to localize to the Golgi complex probably due to a defect in the transport of liposomal proteins from the endocytic vesicles to the cytoplasm. Using this system, we have also demonstrated that functional microtubules are essential for antigen transport to the Golgi complex in both macrophages and dendritic cells. Liposome-encapsulated antigens have the unique property of gaining access to both the MHC class I and class II pathways and therefore have the potential to stimulate both arms of the immune response simultaneously. Understanding the trafficking patterns and the role of the various membrane bound organelles is essential for developing effective liposomal vaccines.

REFERENCES

1. Alving CR. Liposomes as carriers of antigens and adjuvants. J Immunol Meth 1991; 140:1–13.
2. Alving CR, Koulchin V, Glenn GM, Rao M. Liposomes as carriers of peptide antigens: induction of antibodies and cytotoxic T lymphocytes to conjugated and unconjugated peptides. Immunol Rev 1995; 145:5–31.
3. Allison AG, Gregoriadis G. Liposomes as immunological adjuvants. Nature 1974; 252:252.
4. Uemura K, Nicolotti RA, Six HR, Kinsky SC. Antibody formation in response to liposomal model membranes sensitized with N-substituted phosphatidyletha-nolamine derivatives. Biochemistry 1974; 13:1572–1578.
5. Gregoriadis G. Immunological adjuvants: a role for liposomes. Immunol Today 1990; 11:89–97.
6. Loutan L, Bovier P, Althaus B, Gluck R. Inactivated virosome hepatitis A vaccine. Lancet 1994; 343:322–324.
7. Gluck R. Adjuvant activity of immunopotentiating reconstituted influenza virosomes (IRIVs). Vaccine 1999; 17:1782–1787.
8. Su D, Van Rooijen N. The role of macrophages in the immunoadjuvant action of liposomes: effects of elimination of splenic macrophages on the immune response against intravenously injected liposome-associated albumin antigen. Immunology 1989; 66:466–470.
9. Verma JN, Rao M, Amselem S, et al. Adjuvant effects of liposomes containing lipid A: enhancement of liposomal antigen presentation and recruitment of macrophages. Infect Immun 1992; 60:2438–2444.
10. Verma JN, Wassef NM, Wirtz RA, et al. Phagocytosis of liposomes by macro-phages: intracellular fate of liposomal malaria antigen. Biochim Biophys Acta 1991; 1066:229–238.
11. Rao M, Alving CR. Delivery of lipids and liposomal proteins to the cytoplasm and Golgi of antigen-presenting cells. Adv Drug Deliv Rev 2000; 41:171–188.
12. Braciale TJ, Morrison LA, Sweetser MT, et al. Antigen presentation pathways to class I and class II MHC-restricted T lymphocytes. Immunol Rev 1987; 98:95–114.
13. Germain RN, Margoulies DH. The biochemistry and cell biology of antigen processing and presentation. Ann Rev Immunol 1993; 11:403–450.
14. Townsend A, Bodmer H. Antigen recognition by class I-restricted T lympho-cytes. Ann Rev Immunol 1989; 7:601–624.
15. Moore MW, Carbone FR, Bevan MJ. Introduction of soluble protein into the class I pathway of antigen processing and presentation. Cell 1988; 54:777–785.
16. Deres K, Schild H, Wiesmufler KH, Jung G, Rammensee HG. In vivo priming of virus-specific cytotoxic T lymphocytes with synthetic lipopeptide vaccine. Nature 1989; 342:561–564.

17. Schild H, Norda M, Deres K, et al. Fine specificity of cytotoxic T lymphocytes primed in vivo either with virus or synthetic lipopeptide vaccine or primed in vitro with peptide. J Exp Med 1991; 174:1665–1668.

18. Martinon F, Gras-Masse H, Boutillon C, et al. Immunization of mice with lipopeptides bypasses the prerequisite for adjuvant. Immune response of BALB/c mice to human immunodeficiency virus envelope glycoprotein. J Immunol 1992; 149:3416–3422.

19. Harding CV, Song R. Phagocytic processing of exogenous particulate antigens by macrophages for presentation by class I MHC molecules. J Immunol 1994; 153:4925–4933.

20. Kovacsovics-Bankowski M, Rock KL. A phagosome-to-cytosol pathway for exogenous antigens presented on MHC class I molecules. Science 1995; 267:243–246.

21. Men Y, Tamber H, Audran R, Gsnder B, Corradin G. Induction of a cytotoxic T lymphocyte response by immunization with a malaria specific CTL peptide entrapped in biodegradable polymer microspheres. Vaccine 1997; 15:1405–1412.

22. Lopes LM, Chain BM. Liposome-mediated delivery stimulates a class I restricted cytotoxic T cell response to soluble antigen. Eur J Immunol 1992; 22:287–290.

23. Reddy R, Zhou F, Nair S, Huang L, Rouse BT. In vivo cytotoxic T lymphocyte induction with soluble proteins administered in liposomes. J Immunol 1992; 148:1585–1589.

24. White K, Krzych U, Gordon TD, et al. Induction of cytolytic and antibody responses using *Plasmodium falciparum* repeatless circumsporozoite protein encapsulated in liposomes. Vaccine 1993; 11:1341–1346.

25. Alving CR, Wassef NM. Cytotoxic T lymphocytes induced by liposomal antigens: Mechanisms of immunological presentation. AIDS Res Human Retrovir 1994; 10:S91–S94.

26. Titus JA, Haugland R, Sharrow SO, Segal DM. Texas Red, a hydrophilic red-emitting fluorophore for use with fluorescein in dual parameter flow microfluorometric and fluorescence microscopic studies. J Immunol Meth 1982; 50:193–204.

27. Rao M, Rothwell SW, Alving CR. Trafficking of liposomal antigens to the trans-Golgi complex in macrophages. Meth Enzymol 2004; 373:16–33.

28. Alving CR, Shichijo S, Mattsby-Baltzer I, Richards RL, Wassef NM. In: Gregoriadis G, ed. Liposome Technology. Vol. 3. 2nd ed. Boca Raton, FL: CRC Press, 1993:317–343.

29. Rao M, Rothwell SW, Wassef NM, Pagano RE, Alving CR. Visualization of peptides derived from liposome-encapsulated proteins in the trans-Golgi area of macrophages. Immun Lett 1997; 59:99–105.

30. Rothwell SW, Wassef NM, Alving CR, Rao M. Proteasome inhibitors block the entry of liposome-encapsulated antigens into the classical MHC class I pathway. Immunol Lett 2000; 74:141–152.

31. Rao M, Wassef NM, Alving CR, Krzych U. Intracellular processing of liposome encapsulated antigens by macrophages depends upon the antigen. Infect Immun 1995; 63:2396–2402.

32. Peachman KK, Rao M, Alving CR, Palmer DR, Sun W, Rothwell SW. Human dendritic cells and macrophages exhibit different intracellular processing

pathways for soluble and liposome-encapsulated antigens. Immunobiology. 2005; 210:321–333.

33. Rao M, Rothwell SW, Wassef NM, Koolwal AB, Alving CR. Trafficking of liposomal antigen to the trans-Golgi of murine macrophages requires both liposomal lipid and liposomal protein. Exp Cell Res 1999; 246:203–211.

34. Wessel D, Flügge UI. A method for the quantitative recovery of protein in dilute solution in the presence of detergents and lipids. Anal Biochem 1984; 138:141–143.

35. Segal AW, Willis EJ, Richmond JE, et al. Morphological observations on the cellular and subcellular destination of intravenously administered liposomes. Br J Exp Pathol 1974; 55:320–327.

36. Woodle MC, Lasic DD. Sterically stabilized liposomes. Biochim Biophys Acta 1992; 1113:171–199.

37. Nair S, Zhou F, Reddy R, Huang L, Rouse BT. Soluble proteins delivered to dendritic cells via pH-sensitive liposomes induce primary cytotoxic T lymphocyte responses in vitro. J Exp Med 1992; 175:609–612.

38. Brode S, Macary PA. Cross-presentation: dendritic cells and macrophages bite off more than they can chew. Immunology 2004; 112:345–351.

39. Guermonprez P, Valladeau J, Zitvogel L, Thery C, Amigorena S. Antigen presentation and T cell stimulation by dendritic cells. Annu Rev Immunol 2002; 20:621–667.

40. Allen LA, Aderem A. Mechanism of phagocytosis. Curr Opin Immunol 1996; 8:36–40.

41. Swanson JA, Watts C. Macropinocytosis. Trends Cell Biol 1995; 5:424.

42. Rock KL. A new foreign policy; MHC class I molecules monitor the outside world. Immunol Today 1996; 17:131–137.

43. Zhou F, Watkins SC, Huang L. Characterization and kinetics of MHC class I-restricted presentation of a soluble antigen delivered by liposomes. Immunobiology 1994; 190:35–52.

44. Pagano RE, Sepanski MA, Martin OC. Molecular trapping of a fluorescent ceramide analogue at the Golgi apparatus of fixed cells: interaction with endogenous lipids provides a trans-Golgi marker for both light and electron microscopy. J Cell Biol 1989; 109:2067–2079.

45. Peachman KK, Rao M, Alving CR, et al. Functional microtubules are required for antigen processing by macrophages and dendritic cells. Immunol Lett 2004; 95:13–24.

46. Androlewicz MJ, Anderson KS, Cresswell P. Evidence that transporters associated with antigen processing translocate a major histocompatibility complex class I-binding peptide into the endoplasmic reticulum in an ATP-dependent manner. Proc Natl Acad Sci USA 1992; 90:9130–9134.

47. Li S, Paulsson K, Sjogren H, Wang P. Peptide-bound major histocompatibility complex class I molecules associate with tapasin before dissociation from transporter associated with antigen processing. J Biol Chem 1999; 274:8649–8654.

48. Porgador A, Yewdell JW, Deng Y, Bennink JR, Germain RN. Localization, quantitation, and in situ detection of specific peptide-MHC class I complexes using a monoclonal antibody. Immunity 1997; 6:715–726.

13

Targeting of Liposomes to Lymph Nodes

William T. Phillips and Beth A. Goins
Department of Radiology, The University of Texas Health Science Center at San Antonio, San Antonio, Texas, U.S.A.

Luis A. Medina
Institute of Physics, Universidad Nacional Autonoma de Mexico, Mexico City, Mexico

INTRODUCTION

Recently, liposomes have received much attention as lymph-node drug-delivery agents. This interest in the development of new methods of lymph-node drug-delivery stems from the increasing awareness of the importance of lymph nodes in cancer prognosis, their importance for vaccine immune stimulation and the realization that the lymph nodes harbor human immunodeficiency virus (HIV) as well as other infectious disease (1–4). New methods of delivering drugs and antigens to lymph nodes are currently under investigation.

The lymphatic system consists of a network of lymphatic vessels and lymph nodes that serve as a secondary vascular system to return fluid that leaks from the blood vessels in the extremities and other organs back to the vasculature (5). The lymphatic system also moves substantial volumes of fluid from the peritoneal cavity and pleural cavity back into the blood circulation.

Lymph fluid originating from the interstitial spaces between tissue cells and from within the body's cavities moves into lymphatic capillaries through lymph nodes and back into the blood circulation. Lymph fluid of different organs and the body's extremities in addition to body cavities is collected by large lymphatic trunks that feed into one of two lymphatic ducts: the thoracic duct and right lymphatic duct. From these ducts, the lymph fluid then returns to the bloodstream through veins in the neck region (internal jugular and subclavian veins) (6,7).

The lymphatic vessels also serve as a major transport route for antigens, microorganisms, immune cells, and disseminating tumor cells along with interstitial macromolecules that have gained entry to the interstitial space (8). The lymphatic vessels are traversed by immune cells such as dendritic cells, macrophages, and, as their name reveals, lymphocytes. As a part of this system that recycles fluid from the interstitial spaces and the body's cavity back to the arteriovenous vascular system, the lymph nodes are ideally positioned to serve as surveillance organs to monitor microbial invasion and to defend the body against these invading microorganisms.

Liposomes are ideal structures for delivering therapeutic agents to the lymph nodes. Their ideal features are based on their size, which prevents their direct absorption into the blood, the large amount of drugs and other therapeutic agents that liposomes can carry, and their biocompatibility.

Although liposomes are too large to be directly absorbed into the bloodstream, they are small enough to enter the lymphatic vessels and lymph nodes following subcutaneous (SC) injection, intradermal injection, intramuscular injection, injection directly into organs or tumors, and injection into the body's cavities. Following SC injection or other injection directly into tissue, it appears that a certain portion of liposomes are taken up locally and retained for a prolonged time, whereas another portion of the liposomes are cleared from this local site and move into the lymphatic vessels where they can be trapped in lymph nodes or else move completely through the lymphatic system and return to the blood at the thoracic duct.

Lymphatic fluid enters the lymph node through the afferent lymphatic vessels and it leaves the lymph node through an efferent lymphatic vessel as shown in Figure 1. There are estimated to be 400 to 600 lymph nodes in the human body. One of the major functions of the lymph nodes is to help defend the body against diseases by filtering bacteria and viruses from the lymph fluid, and to support the activities of the lymphocytes, which furnish resistance to specific disease causing agents. However, in abnormal conditions, as in the case of cancer and some infections, it is well known that lymph nodes can, act as holding reservoirs from where tumor cells, bacteria either or viruses can spread into other organs and regions of the body (5,7). For example, in the case of cancer, disseminating tumor cells can take root in lymph nodes and form residual metastatic tumors that are difficult to detect and treat.

Considering the importance of the lymphatics in relationship to many disease processes, the number of studies investigating drug delivery or targeting of other therapeutic agents to the lymphatics has been relatively modest (9). This chapter will focus on a review of the literature relevant to the delivery of liposomes to lymph nodes following SC injection.

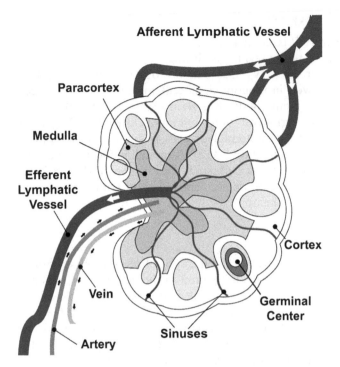

Figure 1 This diagram illustrates the structure of the lymph node. Efferent lymphatic vessels deliver lymph fluid to the lymph node and afferent lymphatic vessels take the lymphatic fluid from the lymph node. Each lymph node is supplied the fluid by an artery and a vein. Lymphatic fluid is filtered through the sinuses of the lymph nodes that are lined with macrophages to phagocytize foreign particulate agents. Lymph nodes also contain cortical, paracortical, and medullary regions that contain different immune cells.

Lymph-node delivery of liposomes following injection into body cavities will be addressed in Chapter 15 of Volume III in this book.

IMPORTANT LYMPH-NODE TARGETS

Cancer

The majority of solid cancers spread primarily by lymph-node dissemination (10). The status of the lymph node in regards to cancer metastasis is a major determinant of the patient's prognosis. Accurate lymph-node staging is the most important factor that determines the appropriate care of the patient (11). Therapeutic interventions that treat metastatic cancer in lymph nodes with either surgery or local radiation therapy have been shown to improve patient survival (12).

Human Immunodeficiency Virus

Primary infection with HIV is characterized by an early viremia followed by a specific HIV immune response and a dramatic decline of virus in the plasma (13). Long after the HIV virus can be found in the blood, HIV can be found in high levels in mononuclear cells located in lymph nodes. Viral replication in these lymph nodes has been reported to be 10- to 100-fold higher than in the peripheral blood mononuclear cells (14). Drug delivery to these lymph node mononuclear cells is difficult with standard oral or intravenous drug administration (15). Although highly active antiretroviral therapy (HAART) reduces plasma viral loads in HIV infected patients by 90%, active virus can still be isolated from lymph nodes even after 30 months of HAART therapy.

Filaria

Lymph nodes are an important part of the life cycle of several parasite organisms, including filaria. Adult worms are found in the lymphatic vessels and lymph nodes of infected patients. These adult filaria are responsible for the obstruction of lymphatic drainage that causes swelling of extremities that are distal to the infected lymph node. These very swollen limbs that are found in patients with filarial disease have been termed elephantiasis. Frequently, eradication of adult worms in lymph nodes is not possible and to be successful, it commonly takes a very extended course of medical therapy (16). Liposome drug delivery has potential for drug delivery in filarial disease, particularly in the case before the lymphatics have not become totally obstructed.

Anthrax

New methods of treating anthrax have become of urgent interest following the recent outbreak of anthrax infections and deaths in the United States as a result of terrorism. In anthrax infection, endospores from Bacillus anthracis that gain entrance into the body are phagocytosed by macrophages and carried to regional lymph nodes where the endospores germinate inside the macrophages and become vegetative bacteria (17). Computed tomography of the chest was performed on eight recent patients infected with inhalational anthrax. Mediastinal lymphadenopathy was present in seven of the eight patients (18). In a recent case report of one patient, the anthrax bacillus was shown to be rapidly sterilized within the blood stream after initiation of antibiotic therapy. However, viable anthrax bacteria were still present in postmortem mediastinal lymph node specimens (19). This case demonstrates the difficulty that drugs have in penetrating the mediastinal lymph nodes. A potential use of liposomes could be for delivery of antianthrax drugs to the mediastinal lymph nodes for therapy or prevention of anthrax extension to the lymph nodes.

Tuberculosis

The tuberculosis infection is caused by mycobacteria that invade and grow chiefly in phagocytic cells. Tuberculosis is frequently found to spread from the lungs to lymph nodes so that lymph node tuberculosis is the most common form of extrapulmonary tuberculosis. In one study, 71% of the tuberculosis lymph node involvement was located in the intrathoracic lymph nodes while 26% of the cervical lymph nodes and 3% of the axillary lymph nodes were involved with tuberculosis (20). The development of methods to target antituberculosis drugs to these lymph nodes could potentially decrease the amount of time that drug therapy is required. Currently, patients with tuberculosis are required to take medicine for more than six months. One likely reason for this lengthy treatment is the difficulty in delivering drugs into these tubercular lesions. Liposomes could be used to carry high levels of drugs to lymph nodes containing tuberculosis. Liposomes encapsulating antituberculosis drugs have already been developed as potential intravenous therapeutic agents for treatment of tuberculosis (21).

LYMPH-NODE ANTIGEN DELIVERY FOR DEVELOPMENT OF AN IMMUNE RESPONSE

The importance of the lymph nodes in the development of an immune reaction induced by vaccines is gradually becoming recognized. Experimental evidence suggests that induction of immune reactivity depends upon the antigen reaching and being available in lymphoid organs in a dose- and time-dependent manner (22). This concept has been termed the geographical concept of immune reactivity (22–25). The delivery of antigen to a lymph node in a manner that resembles an actual microbial invasion may be one of the most important functions of a vaccine adjuvant. The adjuvants are considered effective if they either enhance or prolong expression of antigen components to reactive T cells in lymph nodes (23). Antigen-presenting cells are thought to be of critical importance in transporting antigen from the periphery to local organized lymphoid tissue. However, delivery of antigen to the lymph node by any means may be more important. Several studies have investigated the immune response following direct injection of antigen into lymph nodes. Instead of injecting peptide-based vaccines subcutaneously or intradermally, researchers injected these agents directly into the lymph nodes (24). This intralymphatic injection enhanced immunogenicity by as much as 10^6 times when compared to SC and intradermal vaccination. Intralymphatic administration induced CD8 T cell responses with strong cytotoxic activity and interferon (IFN)-gamma production that conferred long-term protection against viral infections and tumors. This greatly increased response based on direct delivery to the lymph node has also been reported with naked DNA vaccines. Naked DNA vaccines are usually administered either intramuscularly or intradermally. When naked DNA was injected directly into a peripheral lymph node, immunogenicity was enhanced by 100- to 1000-fold, inducing strong and biologically relevant CD8(+) cytotoxic T lymphocyte responses (26).

Liposomes can be used to greatly increase the delivery of an antigen to the lymph node (24). For instance, animal experiments have shown that immunization by the intramuscular or the SC route with liposome-entrapped plasmid DNA encoding the hepatitis B surface antigen leads to much greater humoral [immunoglobulin (IgG) subclasses] and cell mediated (splenic IFN-gamma) immune responses than with naked DNA. In other experiments with a liposome-encapsulated plasmid DNA encoding a model antigen (ovalbumin), a cytotoxic T lymphocyte response was also observed. These results could be explained by the ability of liposomes to protect their DNA content from local nucleases and direct it to antigen presenting cells in the lymph nodes draining the injected site (25).

USE OF IMAGING TO TRACK SUBCUTANEOUSLY INJECTED LIPOSOMES

Scintigraphic imaging of the distribution of liposomes labeled with technetium-99m (^{99m}Tc) following SC injection was first performed by Osborne, et al. (27). The liposomes were labeled using a method in which the ^{99m}Tc was reduced and associated with the outer surface of the liposome. Liposome distributions were determined in rats following injection in the ^{99m}Tc liposomes in the rat hind footpads. In these studies, 1% to 2% of the injected dose of neutral and cationic liposomes was found to localize in the draining lymph nodes. Negatively charged liposomes did not show good accumulation in the lymph nodes (27). Soon after these studies were performed, the reliability of these previous studies for representing the actual distribution of liposomes was questioned and it was suggested that much of the ^{99m}Tc activity localized in the lymph nodes was not associated with liposomes due to instability of the ^{99m}Tc label (28). This article recommended that new methods of labeling liposomes with ^{99m}Tc be developed. Follow up studies demonstrated that the type of labeling used in the prior studies, in which ^{99m}Tc was labeled to the outer surface of liposomes following reduction of the ^{99m}Tc with stannous chloride, was not stable (29). These studies demonstrate the importance of label stability in the tracking of liposomes for quantitation of targeting to the lymph nodes.

Since those early studies, more stable methods of labeling liposomes with ^{99m}Tc have been developed and applied to lymph node imaging. A method developed by our group uses hexamethylpropyleneamine oxime (HMPAO), a clinically approved and commercially available chelator of ^{99m}Tc used for brain imaging (30). In this method, ^{99m}Tc-pertechnetate, which is readily available from a generator, is incubated for five minutes with HMPAO, which chelates the ^{99m}Tc into lipophilic ^{99m}Tc-HMPAO (31). The ^{99m}Tc-HMPAO is then added to previously manufactured liposomes that encapsulate glutathione. It is generally believed that lipophilic HMPAO carries the ^{99m}Tc into the liposomes, where it interacts with the encapsulated glutathione, resulting in its conversion to hydrophilic ^{99m}Tc-HMPAO. The hydrophilic ^{99m}Tc-HMPAO is irreversibly trapped in the aqueous phase of the liposome because it is unable to cross the lipid membrane. A similar mechanism has been proposed to explain the process whereby ^{99m}Tc-HMPAO becomes trapped in brain cells for use as a brain-imaging agent. This liposome label is very stable with minimal dissociation of the ^{99m}Tc from the liposomes. It has been used to study the distribution of intravenously administered liposome (32) as well as subcutaneously injected liposomes (33,34).

QUANTITATION OF LYMPH-NODE DELIVERY

Methods of Reporting Lymph-Node Delivery

The research literature reporting liposome uptake in lymph nodes can be very confusing. This is due to the tendency of many investigators to only report the percent uptake in the lymph node as a percent of the injected dose per gram of tissue. Although reporting uptake as percent of the injected dose per gram makes sense from a drug delivery standpoint, it does not easily allow the reader to determine what percent of the total administered dose accumulates in the lymph node. For instance, in animals such as mice with very small lymph nodes that weigh only a fraction of a gram, the accumulated doses can be as high as 100% of the injected dose per gram, but considering that a mouse lymph node weighs only 0.01 g this is only 1% of the injected dose (1% ID). Had this same fraction of the injected dose accumulated in a rat lymph node that weighs approximately 0.1 g, this dose would have been only 10% of the injected dose per gram; in a human with a lymph node with a weight of 1 g, it would have been only 1% ID per gram. It is very important to keep these species differences in mind when interpreting the previous literature and it would be best if all investigators would report their research not only in terms of dose per gram but also in terms of percent of the total injected dose delivered to the lymph node. From a pharmacologic standpoint, percent dose per gram may be considered correct, but it is highly unlikely that the percent dose per gram results in mice would translate into humans. Based on our experience, it is much more likely that the percent that clears from the injected site will always be approximately the same in each species and the percent that accumulates in the lymph nodes, no matter what its weight in grams, will also be approximately the same.

For example, in one study, researchers reported that subcutaneously injected liposomes without a lymph node targeting mechanism had much higher concentration in the lymph nodes on the side of the SC injection compared to the lymph nodes of the opposite side that did not receive the injection (35). Twenty four hours after SC injection, 57.9% of the ID per gram of tissue was found in the inguinal node on the side of the injection (ipsilateral side) versus only 0.48% ID per gram of tissue in the lymph node of the opposite side of the injection (contralateral side) at 24 hours. Here again, this study was carried out in mice and the percentage ID per gram is somewhat deceptive due to the fact that mice have very small lymph nodes. This probably represents no more that 1% to 3% of the total injected dose accumulating in the lymph node that drained from the SC site. The important point in this article is unchanged. There was more than a 100-fold increased amount of liposomes deposited on the side of SC injection compared with the lymph node on the other side (35).

Calculation of Lymph-Node Retention

Using scintigraphic imaging, detailed studies have been carried out to assess the effect of liposome size and surface modifications on movement from the SC site of injection as well as the retention of the liposomes in the lymph node (33). Scintigraphic imaging makes it possible to quantitatively determine the percent of subcutaneously injected activity that clears from the injection site and the percent that accumulates in the lymph node. Our group has developed a method to calculate lymph node retention efficiency (2). The calculation describes the fraction of the liposomes that are cleared from the initial SC site of injection that become trapped and retained in the lymph node. This estimated lymph node retention calculation describes how efficiently the lymph node can retain a particular subcutaneously injected liposome. It also describes what portion of the dose that enters a lymph node and then leaves that primary lymph node and moves to the next lymph node. The calculation requires that the liposome be labeled with a radiotracer and imaged scintigraphically. It is determined by drawing a region of activity around the injection site and determining the total percentage of injected activity that has cleared the injection site. This total cleared activity is then assumed to be the amount that moves through the lymphatic vessels and enters the first lymph node. The activity retained in the lymph node is divided by the total activity cleared from the injection site so that a lymph node retention efficiency can be calculated. With most unmodified liposome preparations, lymph nodes only retain approximately 4% of the liposomes that enter the lymph node. This number is higher with other particles such as unfiltered ^{99m}Tc-sulfur colloid that have a 40% lymph node retention efficiency, but a very poor clearance from the initial injection site (33).

FACTORS INFLUENCING LIPOSOME DELIVERY TO LYMPH NODES

Liposome Size

The fraction of liposomes that are cleared from the SC injection site depends on the size and surface characteristics of a particular liposome formulation. In studying a wide range of liposome sizes from 86 nm liposomes to 520 nm liposome, there was little difference in the ultimate accumulation of liposomes in the first or sentinel lymph node at 24 hour post administration. This lymph node retention ranged from 1.3% to 2.4% of the total administered dose (33). Small liposomes had the greatest clearance from the SC site of injection with small 86 nm liposomes having <40% remaining at the injection site at 24 hours. Larger neutral and negatively charged liposomes had >60% remaining at the initial site of SC injection.

Many factors appear to influence the fraction of the liposomes that are retained at the initial site of SC injection. Liposome diameter appears to be one of the most important factors affecting the clearance of liposomes from the SC site of injection (3). The larger the size of the liposomes that are injected subcutaneously, the greater the fraction of the liposomes that will be retained locally and the less that will enter the lymphatic vessels and have a chance to accumulate in the lymph nodes (3,34).

Much work has been performed evaluating the effect of the size of subcutaneously injected liposomes on lymph node targeting. When small neutral liposomes are injected subcutaneously, more than 60% to 70% of the liposomes will be cleared from the injection site by 24 hours (33,36), with only 30% to 40% of the injected dose remaining at the site of injection. Liposomes larger than 500 nm will have 60% to 80% remaining at the injection site (33,36).

This retention of liposomes in the lymph node is relatively low considering that for most subcutaneously injected liposome preparations more than 50% of the injected liposomes are cleared from the injection site. It appears that the properties of the liposomes that enhance their clearance from the injection site, also decrease their retention in the lymph nodes. The generally low overall lymph node retention by most standard liposome formulations is likely due to their natural lipid composition that probably allows a large percentage of liposomes that enter a lymph node to escape recognition and phagocytosis by macrophages that line the endothelium of the lymph node. This relatively low retention of liposomes in lymph nodes has also been reported by Ousseren et al. (3,36–38).

Even though this retention of liposomes is very low in comparison with the injected dose, liposome uptake can still be quite substantial in terms of drug delivery when considered on a per gram of tissue basis. For example, 1% to 2% of the injection dose in each lymph node represents a total per gram tissue uptake that is generally 30- to 40-fold greater than the liposome uptake that will eventually reach the liver and spleen following the same SC injection. Factors that enhance clearance of liposomes from a local site of SC injection also tend to decrease liposome uptake in the lymph node. For instance, larger liposomes are not cleared from the SC site of injection as readily as smaller liposomes; however, they are better retained in the lymph node. Even though larger liposomes are less well cleared from the injection site, their total retention in the lymph node is similar to other liposomes due to their improved lymph-node retention. This improved lymph-node retention by liposomes that are poorly cleared from the injection site results in liposome retention doses that are approximately equal to liposomes that have improved clearance from the local SC injection site (33).

Total Lipid Dose

The total dose of lipid administered does not appear to have an effect on the percentage of liposomes retained in the lymph node. Lymph-node uptake did not appear to become saturated over a large range of lipid doses administered, ranging from 10 nmol lipid to 10,000 nmol of lipid (36).

Liposome Surface Modification

One of the several liposome surface modifications that have resulted in modestly increased retention of subcutaneously injected liposomes in the lymph node is the use of positively charged lipids in the liposome. Liposomes containing positively charge lipids had approximately two to three times the lymph-node localization (up to 3.6% of the injected dose) as liposomes containing neutral or negatively charged lipids (1.2% of the injected dose) (37). Another method in which the liposomes were coated with the antibody, IgG, has been shown to increase lymph-node localization of liposomes to 4.5% of the injected dose at one hour, but this level decreased to 3% by 24 hours (39). Attaching mannose to the surface of a liposome has also been reported to modestly increase lymph node uptake by threefold compared to control liposomes (40). None of these previously mentioned modifications has resulted in large increases in the percentage of liposomes deposited in the draining lymph nodes, while most of the lymphatically absorbed liposome dose passes through the lymph nodes.

Surface modification of liposomes with polyethylene glycol (PEG) also does not appear to have a very large effect on lymph-node uptake. Small liposomes and those coated with PEG had the greatest clearance from the SC site of injection with small 86-nm PEG-coated liposomes having <40% remaining at the injection site at 24 hours. Larger neutral and negatively charged liposomes had >60% remaining at the initial site of SC injection. However, this smaller amount of large liposomes that were cleared from the injection site was compensated by better retention in the lymph node (33).

Oussoren et al. found that the amount of liposomes that cleared from the injection site was slightly greater with the PEG-coated liposomes (37); however, this improved clearance did not result in improved lymph node retention because the fraction of PEG-liposomes retained by the lymph node is decreased. The slightly improved clearance of PEG-coated liposomes from the SC site of injection was also found by our research group (33).

Effect of Massage on Lymphatic Clearance of Subcutaneously Injected Liposomes

The rate of clearance of liposomes from a s.c injection site can be greatly accelerated with local manual massage (41). Without any mechanical stimulation, subcutaneously injected 200-nm liposomes are usually trapped in the interstitial SC space for a prolonged time. However, after five minutes of manual massage over the s.c injection site, up to 40% of the injected liposomes are cleared from the SC site into the blood via the lymphatic pathway.

Macrophage Phagocytosis

It is generally accepted that liposomes are retained in the lymph node by macrophage phagocytosis. Research findings using liposomes containing colloidal gold appear to support this contention (42). The strong supporting evidence of the role of macrophages in lymph node uptake was provided by a study in which macrophages were temporarily depleted from lymph nodes by prior administration of liposomes containing dichloromethylene diphosphonate (clodronate). Clodronate is toxic to macrophages and previous work has been performed using clodronate encapsulated within liposomes to temporarily deplete macrophages in the liver (43,44). Six days after injection of the clodronate liposomes, small- and large-sized liposomes were also injected subcutaneously. There was a drastic reduction in the uptake of both large and small liposomes in the lymph node following clodronate administration (45).

Fate of Liposomes in Lymph Nodes

Only a few studies have looked at the fate of liposomes once they arrive at the lymph node (3,42). In one study, subcutaneously injected liposomes were found to have accumulated in the subcapsular sinus. Subsequently, liposomes were dispersed throughout the lymph node either by permeation along the sinus or within cells involved in liposome uptake such as macrophages. Once they were in the macrophages, the liposomes were observed to be digested by lysosomes (42).

LIPOSOMES FOR LYMPH-NODE INFECTION THERAPY

Bacterial Disease

Only a few studies have examined the delivery of liposomes to lymph nodes for the treatment of bacterial disease. In one study, liposome encapsulated amikacin was injected subcutaneously, intramuscularly, and intravenously (46). Drug levels in the lymph nodes were studied at various time points following injection. Drug level area under the curve (AUCs) in regional lymph nodes exceeded plasma AUCs by fourfold after SC and intramuscular injection of liposomal amikacin (46). The authors of the study conclude that liposomes encapsulating amikacin have much potential for drug delivery and even suggest that these liposomes could potentially be used for local delivery in perioperative prophylaxis, pneumonias, and intralesional therapy as well as sustained systemic delivery of encapsulated drugs (46).

This effectiveness of liposome encapsulated amikacin following SC injection differs significantly from a previous study in which 400-nm liposome-encapsulated amikacin was administered intravenously. The intravenously administered amikacin encapsulated liposomes were effective against *Mycobacterium avium* intracellulare located in the liver and spleen but they had no effect on the organisms that were located in the lymph nodes (47). It is likely that these intravenously injected liposomes did not accumulate in the lymph nodes to any degree.

Human Immunodeficiency Virus

The use of liposomes to increase the drug delivery to HIV-infected lymph nodes appears promising. Dufresne et al. have investigated liposomes coated with anti-human leukocyte antigen (HLA)-DR Fab' fragments for specifically targeting liposomes to follicular dendritic cells and macrophages within the lymph nodes of mice with the goal of increasing the delivery of antiviral drugs to these cells infected with HIV (48). The uptake of anti-HLA-DR Fab' coated liposomes within lymph nodes was two- to three-fold higher when compared to conventional liposomes, but of more importance is the potential specific delivery of the anti-HLA-DR Fab' liposomes to antigen-presenting cells within the lymph node.

More recently, researchers from this same group have investigated the targeting of lymph nodes with indinavir, a protease inhibitor, encapsulated into immunoliposomes coated with the same anti-HLA-DR Fab' antibody fragment. Mice were injected subcutaneously below the neck with either free indinavir or liposome-encapsulated indinavir. Animals were sacrificed at various times following injection, and tissues collected and analyzed for indinavir drug levels. Drug levels were compared in lymph nodes from the mice receiving the subcutaneously injected free drug and subcutaneously injected liposome-encapsulated drug. Drug levels in the brachial and cervical lymph-nodes were 126 and 69 times greater with the liposome-encapsulated drug compared to the free drug (13). A review of the use of liposomes for delivery drugs to HIV infected lymph nodes has recently been published (15).

LIPOSOMES FOR LYMPH-NODE CANCER TARGETING

Nonliposome Drug Delivery to Lymph Nodes

One of the first studies to investigate the possible use of drugs delivered intralymphatically was performed by Hirnle (1). This study investigated the anticancer drug Bleomycin that was suspended in an oil suspension known as Oil Bleo. This Oil Bleo was injected directly into catheterized lymphatic vessels in dogs. The movement of this agent through the lymph nodes and lymphatic vessels was fairly rapid with peak drug concentrations reaching the blood 15 minutes after intralymphatic administration of Oil Bleo. The drug entering the blood was considered to be spillover from the lymphatic system. Spillover occurred because the drug moved completely through the lymphatic vessels and rejoined the circulation at the thoracic duct. Administering the drug this way required a very tedious catheterization process of the small lymphatic vessels of the extremities. Although drug concentrations were very high in the lymphatic vessels for a fairly short time, the retention of the oil emulsion in the lymphatics was minimal.

Liposomes for Anticancer Lymph-Node Drug Delivery

The investigation of liposomes as a carrier for lymph node drug delivery was first performed by Segal et al. in 1975 (49). Following the intratesticular injection of liposomes encapsulating the anticancer drug actinomycin D, high concentrations of the drug were found in the local lymph nodes.

Subsequently, Hirnle et al. turned to liposomes as an improved carrier for intralymphatically delivered drugs compared with bleomycin emulsions (50). A study in rabbits used liposome-encapsulated bleomycin that was injected directly into the lymphatic vessels of the hindlegs of rabbits. Lymph nodes were removed and measured for bleomycin content at various times following administration. Three days following intralymphatic administration, the drug concentration in the popliteal lymph nodes was $42\,\mu g/g$ of node. Drug deposition and apparent release was sustained over a very long period because concentrations of Bleomycin in the lymph nodes of $0.18\,\mu g/g$ were measured in the popliteal nodes at one month following injection (1).

Further studies were performed by Hirnle with blue dye containing liposomes composed of 80% phosphatidylcholine and 20% cholesterol. The liposomes had a homogeneous size of approximately 170 nm in diameter. The total amount of blue dye injected was 1.6 mg. When the rabbits were sacrificed 28 days later, the retroperitoneal lymph nodes were visually blue and had a concentration of $172\,\mu g$ blue dye per gram of lymph node.

Unfortunately, when these liposomes were administered by direct intra-lymphatic injection in the hindleg of a rabbit, a large fraction of the intact liposomes were found to spillover into the circulation.

Several conclusions were derived from this research with liposomes directly infused into the lymphatic vessels. The amount of drug administered intralymphatically should not exceed that which would be administered intravenously. The limiting factor in administering drugs lymphatically is the amount of the therapeutic agent that moves completely through the lymphatic system and into the circulation through the thoracic duct. The tolerated amount of spillover should be considered with regard to the toxicity of these liposomal agents to the rest of the body. The volume used in humans should remain low, with no more than 4 mL of liposomes being administered into the canulated lymphatic vessels of each leg. It was also suggested that the drug would remain longer in the lymphatics if the patient remains in bed for one day after endolymphatic liposome administration. And most importantly, the lymph nodes will still be filled with measurable amounts of drug a month after injection. Hirnle also introduced the concept that the prolonged retention of anticancer drugs in the lymphatics might be effective for prevention of lymphatic metastasis (1).

Use of Liposomes for Localizing the Sentinel Lymph Node

In the last decade, cancer surgeons have become very interested in methods to definitively localize the sentinel lymph node. The sentinel lymph node is the first lymph node that receives lymphatic drainage from the site of a primary tumor. The sentinel node is much more likely to contain metastatic tumor cells than other lymph nodes in the same region. It is believed that the initial draining lymph node (sentinel node) of a tumor may reflect the status of the tumor's spread to the remaining lymphatic bed. Localization of the sentinel lymph node and its close histological assessment following its removal from the body was initially developed as a prognostic indicator in patients with malignant melanoma (51). If no cancer cells are found in the sentinel node on pathologic examination, the prognosis for the patient is greatly improved. After many detailed studies validating the effectiveness of this approach for patient prognosis and as a method to guide future therapy of melanoma patients, this technique has begun to be applied in other cancers, particularly breast cancer. Total lymphadenectomy procedures are being replaced by intraoperative lymphatic mapping and sentinel lymph node biopsy (52).

Several techniques for identification of the sentinel lymph node have been investigated that use liposome-based systems to enhance the ability of the surgeon to detect the sentinel node. As described previously, Hirnle et al. encapsulated patent blue dye within liposomes for potential use in localizing the sentinel lymph node during surgery (50). This group later performed a study in humans in which blue liposomes were injected directly into the lymphatic vessels of the foot of a patient prior to retroperitoneal staging-lymphadenectomy (53). The lymph nodes were well stained with blue dye and were readily visualized at the time of the surgery performed 24 hours following the intralymphatic injection of the blue liposomes. This group has also investigated blue liposomes for sentinel node detection in the pig. The blue liposomes were found to provide greater intensity blue staining that lasted for a longer duration than free unencapsulated blue dye (54).

Plut et al. also have developed a liposome formulation containing blue dye that can be radiolabeled with ^{99m}Tc (55). The use of the liposome nanoparticle to provide a visual identification and tracking of the liposomes through the lymphatic channels along with the ability to trace the preparation using standard radiation detection instrumentation provides the surgeon with an improved radiolabeled compound for lymphoscintigraphy and intraoperative sentinel lymph node identification. This method also demonstrates the versatility of nanoparticles to carry multiple diagnostic tracers in the same nanoparticle.

Avidin Biotin-Liposome Lymph-Node Targeting Method

With standard liposome formulations, only a small fraction of the liposomes injected subcutaneously is retained in each lymph node encountered, so that the majority of the dose that clears from the injection site returns to the systemic circulation (3,33,36). The relatively low retention of liposomes in lymph nodes led our group to search for new ways to improve liposome retention in lymph nodes. This research resulted in a new method of increasing lymph node retention of subcutaneously injected liposomes (2). This lymph node targeting method utilizes the high affinity ligands, biotin and avidin. Biotin is a naturally occurring cofactor and avidin is a protein derived from eggs. Avidin and biotin have an extremely high affinity for each other. Avidin has four receptor sites for biotin associated with each molecule. These four receptor sites permit the binding of multiple biotin molecules that causes aggregation of liposomes that have biotin on their surface. Following their SC injection, the avidin and the biotin-liposomes move into the lymphatic vessels.

The precise mechanism of liposome accumulation in the lymph nodes with the avidin/biotin-liposome method is not definitely known. It was originally hypothesized that the biotin-liposomes that are migrating through the lymphatic vessels meet with the avidin resulting in an aggregate that becomes trapped in the lymph nodes. Subsequent research suggests that an alternative possibility may be more likely (56,57). This alternative hypothesis is that the positively charged avidin becomes bound to negatively charged endothelial cells in the lymph nodes and the biotin-liposomes become bound by these avidin molecules attached to the endothelial surface. It is possible that both processes are occurring, however research with intracavitary avidin/biotin-liposome systems suggests that the second possibility may be more likely (56,57).

This in vivo nanoassembly of biotin-liposome/avidin aggregates mimics processes that occur naturally in the body such as the aggregation of platelets and the aggregation of infectious agents by antibodies. The biotin-liposome/avidin system has promising potential for application in therapeutic agent delivery to lymph nodes. It can be applied not only to SC targeting of lymph nodes but also to intracavitary lymph node targeting. Scintigraphic imaging of liposomes labeled with ^{99m}Tc, labeled in a stable fashion, has greatly aided the determination of the proper concentration of avidin and biotin and could be used to develop similar targeting methodology with other nanocarriers (2).

As an extension of the avidin/biotin-liposome lymph node targeting system, we have developed a special liposome formulation that contains both encapsulated blue dye and ^{99m}Tc as a potential system for localizing the sentinel lymph node, visually as well as scintigraphically and/or with a gamma probe (34). Potential advantages of this system over the current methods are that it can be performed anytime from one hour to one day before the surgery is planned because the lymph nodes are stained blue for a prolonged time and the sentinel lymph node has the highest concentration of liposomes. Using this method, a separate blue dye injection just prior to surgery would not be necessary.

Methods

Biotin-liposomes encapsulating glutathione and patent blue violet dye were prepared using extrusion through polycarbonate filters to form small unilamellar liposomes (100 nm). The lipid composition of the biotin-liposomes was 58:39:1:2 molar ratio (total lipid) of distearoyl phosphatidylcholine: cholesterol:N-biotinoyl distearoyl phosphoethanolamine:alpha-tocopherol. Liposomes were prepared in a laminar hood under aseptic conditions and liposomal size was monitored using a Brookhaven particle size analyzer. Liposomes were labeled with ^{99m}Tc as described using a lipophilic chelator HMPAO kit and ^{99m}Tc-pertechnetate. Twenty female Fisher-344 rats were inoculated subcutaneously with one million tumor cells [13762-MAT-B(III)] in the mammary fat pad. After 11 to 18 days of tumor growth, the rats were divided into two groups (9–11/gp). The experimental group was injected SC with 50 μL of blue ^{99m}Tc-biotin-liposomes (13 MBq; 0.037 M phospholipid conc.) at the top of the tumor and avidin (50 μL, 5 mg/mL) was injected SC at the axillary region. Control group was only injected with the ^{99m}Tc-biotin-liposomes at the top of the tumor. Static scintigraphic images at 0, 2, 4, 6, 12, and 22 hours, were acquired in 64 × 64 Word Image Matrix using a gamma camera interfaced to an image analysis computer. After imaging, the animals were sacrificed by cervical dislocation. Tumors and tissues were harvested, weighed, and counted for radioactivity. The percentage of injected dose (% ID) per organ was calculated by comparison with a standard aliquot of the radioactive material used.

Results

At 22 hours following administration of the liposomes, the biodistribution results showed that nodal retention of liposomes was significantly higher in the experimental group receiving the avidin versus the control group that did not receive the avidin (approximately 30-fold more). These results are illustrated in the images in Figure 2. The animals receiving the avidin have significant retention in the axillary lymph node while the animals that did not receive avidin had minimal retention in the axillary nodes. Image analysis for the control group indicated that the liposomes on the side that did not receive the avidin passed through the lymph node with minimal retention in this axillary node and most of the dose of liposomes that left the injection site passed through the lymph nodes. The sentinel lymph node proven to contain cancer and receiving the avidin had approximately 2.5 times more liposome uptake compared to those that did not have cancer metastasis. It appears that the presence of the tumor actually increased the uptake of the ^{99m}Tc-biotin-liposomes in the sentinel node. A photograph of an animal that had metastatic cancer and received avidin is shown in Figure 3. The dark staining of the axillary lymph node is easily visualized. When shown in a color photograph, this lymph node is blue. The blue dye encapsulated into the liposomes permitted easy visual identification of the sentinel node (even after 22 hours of injection).

This study demonstrates that it is possible to use the avidin/biotin-liposome system to target lymph nodes that contain metastatic cancer cells. Uptake in lymph nodes was more than 30 times greater in the rats receiving the avidin compared to those that did not receive the avidin. This study

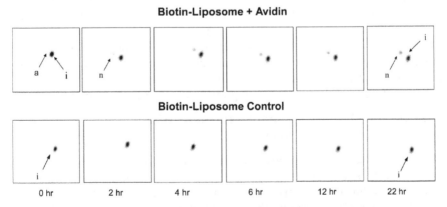

Biotin-Liposome + Avidin

Biotin-Liposome Control

0 hr   2 hr   4 hr   6 hr   12 hr   22 hr

Figure 2 Scintigraphic images of lymphatic clearance at various times from baseline to 22 hours following subcutaneous injection of the ^{99m}Tc-biotin-liposomes into the chest wall surround the breast cancer. The animal in the top panel received avidin while the animal in the lower panel did not. The uptake in the lymph node can be easily visualized in the animal that received avidin. This lymph node contained metastatic cancer cells. *Abbreviations*: a, avidin; i, injection site; n, lymph node.

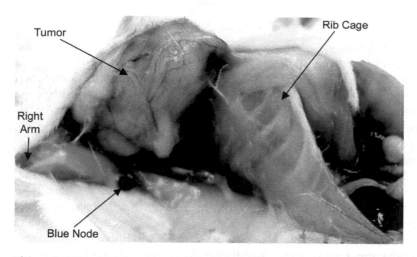

Figure 3 Photograph taken during necropsy showing the sentinel lymph node that is stained blue by the blue dye in the biotin-liposomes. This node appears dark on the black and white images but it is actually blue.

demonstrates the potential of this system for delivery of drugs to lymph nodes containing cancer.

Examination of blue lymph nodes by light microscopy reveals that the liposomes tend to be deposited predominantly in the outer cortex of the lymph node, however this is not always the case as lymph nodes can be completely stained, depending on the concentration and timing of the avidin and biotin-liposomes. Similar to Hirnle et al., we have observed that lymph nodes can be blue stained by visual observation for more than a week following SC injection. The prolonged retention and slow release observed with blue biotin-liposomes demonstrates the potential of this system for the delivery and sustained release of drugs in the lymph nodes. Clinical studies would be required to determine whether the biotin-liposome/ avidin system is effective in targeting the sentinel node in humans.

Intraoperative Therapy for Positive Tumor Margins and Treatment of Lymph Nodes

One possible use of therapeutic liposomes is to target residual tumor in the intraoperative situation. In many cases, the surgeon is unable to remove all of the cancer during surgery so that the margins of the resected tumor are positive. This generally means that there is cancer remaining at the operative site that severely compromises patient survival. This positive margin can frequently be determined during the operation. Therapeutic liposomes that target residual tumor could be injected in the region of the positive tumor margin to sterilize the surgical margin of tumor cells. Because the liposomes will drain through the lymph nodes, they would also have the potential to treat micrometastasis in those nodes. These liposomes could contain a therapeutic radionuclide for radiotherapy or a chemotherapeutic drug or a combination of both. Our group has previously demonstrated that liposomes can be labeled with therapeutic beta particle emitting rhenium-186 or rhenium-188 (58). These beta particles travel 2 mm in the case of rhenium-186 and 4 mm in rhenium-188. Intraoperatively applied liposomes could, therefore, provide an additional tool for the surgeon, particularly when the margins of the tumor are positive.

Even when the margins of the tumor are negative, frequently there is reoccurrence of cancer in the local region or in the nodes that drain from the local region. Cancer surgeons spend many hours of each surgery carefully uncovering and removing lymph nodes in the region of the tumor, while being careful not to damage other critical vessels and nerves. Although these surgeries are very long, it is not always possible to find and remove all of the lymph nodes in the local region of the tumor. Removal of distant lymph nodes that also receive lymph drainage from the tumor is usually not possible. The application of therapeutic liposomes intraoperatively could provide an additional tool to treat micrometastasis in lymph nodes with the goal of decreasing local reoccurrences. Extensive clinical trials would have to be performed to determine the effectiveness of this approach. Effective treatment of lymph nodes draining from a tumor could decrease the need for tedious surgical removal of lymph nodes. One possible method to ensure good lymph-node targeting of liposomes in the intraoperative situation would be to use the avidin/biotin-liposome lymph-node targeting system to ensure trapping of the particles in the lymph nodes that drain from the tumor. This methodology would also limit the spillover of radiotherapeutic liposomes out of the lymphatic vessels and into the bloodstream.

CONCLUSIONS

The delivery of liposomes to lymph nodes for therapeutic purposes has much promise. Significant progress has been made in understanding the various processes involved in liposome delivery and in the development of potential systems for targeting liposomes to lymph nodes. Lymph-node delivery appears to have much promise for improving cancer and infectious disease therapy, treatment of autoimmune disease, as well as improvements in vaccine systems.

REFERENCES

1. Hirnle P. Liposomes for drug targeting in the lymphatic system. Hybridoma 1997; 16(1):127–132.
2. Phillips WT, Klipper R, Goins B. Novel method of greatly enhanced delivery of liposomes to lymph nodes. J Pharmacol Exp Ther 2000; 295(1):309–313.
3. Oussoren C, Storm G. Liposomes to target the lymphatics by subcutaneous administration. Adv Drug Deliv Rev 2001; 50(1–2):143–156.
4. Duzgunes N, Simoes S, Slepushkin V, et al. Enhanced inhibition of HIV-1 replication in macrophages by antisense oligonucleotides, ribozymes and acyclic

nucleoside phosphonate analogs delivered in pH-sensitive liposomes. Nucleosides Nucleotides Nucleic Acids 2001; 20(4–7):515–523.

5. Swartz MA. The physiology of the lymphatic system. Adv Drug Deliv Rev 2001; 50(1–2):3–20.
6. Swartz MA, Skobe M. Lymphatic function, lymphangiogenesis, and cancer metastasis. Microsc Res Tech 2001; 55(2):92–99.
7. Hawley A, Davis S, Illum L. Targeting of colloids to lymph nodes: influence of lymphatic physiology and colloidal characteristics. Adv Drug Delivery Rev 1995; 17:129–148.
8. Porter CJ. Drug delivery to the lymphatic system. Crit Rev Ther Drug Carrier Syst 1997; 14(4):333–393.
9. Porter CJ, Charman WN. Transport and absorption of drugs via the lymphatic system. Adv Drug Deliv Rev 2001; 50(1–2):1–2.
10. Hanahan D, Weinberg RA. The hallmarks of cancer. Cell 2000; 100(1):57–70.
11. Torabi M, Aquino SL, Harisinghani MG. Current concepts in lymph node imaging. J Nucl Med 2004; 45(9):1509–1518.
12. Busby JE, Evans CP. Old friends, new ways: revisiting extended lymphadenectomy and neoadjuvant chemotherapy to improve outcomes. Curr Opin Urol 2004; 14(5):251–257.
13. Gagne JF, Desormeaux A, Perron S, et al. Targeted delivery of indinavir to HIV-1 primary reservoirs with immunoliposomes. Biochim Biophys Acta 2002; 1558(2):198–210.
14. Cohen OJ, Pantaleo G, Lam GK, et al. Studies on lymphoid tissue from HIV-infected individuals: implications for the design of therapeutic strategies. Springer Semin Immunopathol 1997; 18(3):305–322.
15. Desormeaux A, Bergeron MG. Lymphoid tissue targeting of anti-HIV drugs using liposomes. Methods Enzymol 2005; 391:330–351.
16. El Setouhy M, Ramzy RM, Ahmed ES, et al. A randomized clinical trial comparing single- and multi-dose combination therapy with diethylcarbamazine and albendazole for treatment of bancroftian filariasis. Am J Trop Med Hyg 2004; 70(2):191–196.
17. Dixon TC, Meselson M, Guillemin J, et al. Anthrax. N Engl J Med 1999; 341(11):815–826.
18. Jernigan JA, Stephens DS, Ashford DA, et al. Bioterrorism-related inhalational anthrax: the first 10 cases reported in the United States. Emerg Infect Dis 2001; 7(6):933–944.
19. Barakat LA, Quentzel HL, Jernigan JA, et al. Fatal inhalational anthrax in a 94-year-old Connecticut woman. JAMA 2002; 287(7):863–868.
20. Ilgazli A, Boyaci H, Basyiglt I, et al. Extrapulmonary tuberculosis: clinical and epidemiologic spectrum of 636 cases. Arch Med Res 2004; 35(5):435–441.
21. Bermudez LE. Use of liposome preparation to treat mycobacterial infections. Immunobiology 1994; 191(4–5):578–583.
22. Zinkernagel RM, Ehl S, Aichele P, et al. Antigen localisation regulates immune responses in a dose- and time-dependent fashion: a geographical view of immune reactivity. Immunol Rev 1997; 156:199–209.
23. Schijns VE. Induction and direction of immune responses by vaccine adjuvants. Crit Rev Immunol 2001; 21(1–3):75–85.
24. Johansen P, Haffner AC, Koch F, et al. Direct intralymphatic injection of peptide vaccines enhances immunogenicity. Eur J Immunol 2005; 35(2):568–574.

25. Gregoriadis G, Bacon A, Caparros-Wanderley W, et al. A role for liposomes in genetic vaccination. Vaccine 2002; 20(suppl 5):B1–B9.
26. Maloy KJ, Erdmann I, Basch V, et al. Intralymphatic immunization enhances DNA vaccination. Proc Natl Acad Sci USA 2001; 98(6):3299–3303.
27. Osborne MP, Richardson VJ, Jeyasingh K, et al. Radionuclide-labelled liposomes—a new lymph node imaging agent. Int J Nucl Med Biol 1979; 6(2):75–83.
28. Patel HM, Boodle KM, Vaughan-Jones R. Assessment of the potential uses of liposomes for lymphoscintigraphy and lymphatic drug delivery. Failure of 99m-technetium marker to represent intact liposomes in lymph nodes. Biochim Biophys Acta 1984; 801(1):76–86.
29. Love WG, Amos N, Williams BD, et al. Effect of liposome surface charge on the stability of technetium (99mTc) radiolabelled liposomes. J Microencapsul 1989; 6(1):105–113.
30. Andersen AR, Friberg H, Lassen NA, et al. Serial studies of cerebral blood flow using 99Tcm-HMPAO: a comparison with 133Xe. Nucl Med Comm 1987; 8(7):549–557.
31. Phillips WT, Rudolph AS, Goins B, et al. A simple method for producing a technetium-99m-labeled liposome which is stable in vivo. Int J Rad Appl Instrum B 1992; 19(5):539–547.
32. Goins BA, Phillips WT. Radiolabelled liposomes for imaging and biodistribution studies. In: Torchilin VP, Weissig V, eds. Liposomes: A Practical Approach. Oxford: Oxford University Press, 2003:319–336.
33. Phillips WT, Andrews T, Liu H, et al. Evaluation of [(99m)Tc] liposomes as lymphoscintigraphic agents: comparison with [(99m)Tc] sulfur colloid and [(99m)Tc] human serum albumin. Nucl Med Biol 2001; 28(4):435–444.
34. Phillips WT, Klipper R, Goins B. Use of (99m)Tc-labeled liposomes encapsulating blue dye for identification of the sentinel lymph node. J Nucl Med 2001; 42(3):446–451.
35. Harrington KJ, Rowlinson-Busza G, Syrigos KN, et al. Pegylated liposomes have potential as vehicles for intratumoral and subcutaneous drug delivery. Clin Cancer Res 2000; 6(6):2528–2537.
36. Oussoren C, Zuidema J, Crommelin DJ, et al. Lymphatic uptake and biodistribution of liposomes after subcutaneous injection. II. Influence of liposomal size, lipid compostion and lipid dose. Biochim Biophys Acta 1997; 1328(2):261–272.
37. Oussoren C, Storm G. Lymphatic uptake and biodistribution of liposomes after subcutaneous injection. III. Influence of surface modification with poly(ethylene glycol). Pharm Res 1997; 14(10):1479–1484.
38. Oussoren C, Velinova M, Scherphof G, et al. Lymphatic uptake and biodistribution of liposomes after subcutaneous injection. IV. Fate of liposomes in regional lymph nodes. Biochim Biophys Acta 1998; 1370(2):259–272.
39. Mangat S, Patel HM. Lymph node localization of non-specific antibody-coated liposomes. Life Sci 1985; 36(20):1917–1925.
40. Wu MS, Robbins JC, Bugianesi RL, et al. Modified in vivo behavior of liposomes containing synthetic glycolipids. Biochim Biophys Acta 1981; 674(1):19–29.
41. Trubetskoy VS, Whiteman KR, Torchilin VP, et al. Massage-induced release of subcutaneously injected liposome-encapsulated drugs to the blood. J Control Release 1998; 50(1–3):13–19.
42. Velinova M, Read N, Kirby C, et al. Morphological observations on the fate of liposomes in the regional lymph nodes after footpad injection into rats. Biochim Biophys Acta 1996; 1299(2):207–215.

43. Van Rooijen N, Kors N, vd Ende M, et al. Depletion and repopulation of macrophages in spleen and liver of rat after intravenous treatment with liposome-encapsulated dichloromethylene diphosphonate. Cell Tissue Res 1990; 260(2):215–222.
44. Van Rooijen N, Sanders A. Liposome mediated depletion of macrophages: mechanism of action, preparation of liposomes and applications. J Immunol Methods 1994; 174(1–2):83–93.
45. Oussoren C, Storm G. Role of macrophages in the localisation of liposomes in lymph nodes after subcutaneous administration. Int J Pharm 1999; 183(1):37–41.
46. Fielding RM, Moon-McDermott L, Lewis RO. Bioavailability of a small unilamellar low-clearance liposomal amikacin formulation after extravascular administration. J Drug Target 1999; 6(6):415–426.
47. Duzgunes N, Perumal VK, Kesavalu L, et al. Enhanced effect of liposome-encapsulated amikacin on *Mycobacterium avium—M. intracellulare* complex infection in beige mice. Antimicrob Agents Chemother 1988; 32(9):1404–1411.
48. Dufresne I, Desormeaux A, Bestman-Smith J, et al. Targeting lymph nodes with liposomes bearing anti-HLA-DR Fab′ fragments. Biochim Biophys Acta 1999; 1421(2):284–294.
49. Segal AW, Gregoriadis G, Black CD. Liposomes as vehicles for the local release of drugs. Clin Sci Mol Med 1975; 49(2):99–106.
50. Hirnle P, Harzmann R, Wright JK. Patent blue V encapsulation in liposomes: potential applicability to endolympatic therapy and preoperative chromolymphography. Lymphology 1988; 21(3):187–189.
51. Morton DL, Wen DR, Wong JH, et al. Technical details of intraoperative lymphatic mapping for early stage melanoma. Arch Surg 1992; 127(4):392–399.
52. Focht SL. Lymphatic mapping and sentinel lymph node biopsy. AORN J 1999; 69(4):802–809.
53. Pump B, Hirnle P. Preoperative lymph-node staining with liposomes containing patent blue violet. A clinical case report. J Pharm Pharmacol 1996; 48(7):699–701.
54. Dieter M, Schubert R, Hirnle P. Blue liposomes for identification of the sentinel lymph nodes in pigs. Lymphology 2003; 36(1):39–47.
55. Plut EM, Hinkle GH, Guo W, et al. Kit formulation for the preparation of radioactive blue liposomes for sentinel node lymphoscintigraphy. J Pharm Sci 2002; 91(7):1717–1732.
56. Medina LA, Calixto SM, Klipper R, et al. Avidin/biotin-liposome system injected in the pleural space for drug delivery to mediastinal lymph nodes. J Pharm Sci 2004; 93(10):2595–2608.
57. Medina LA, Klipper R, Phillips WT, et al. Pharmacokinetics and biodistribution of [111In]-avidin and [99mTc]-biotin-liposomes injected in the pleural space for the targeting of mediastinal nodes. Nucl Med Biol 2004; 31(1):41–51.
58. Bao A, Goins B, Klipper R, et al. 186Re-liposome labeling using 186Re-SNS/S complexes: in vitro stability, imaging, and biodistribution in rats. J Nucl Med 2003; 44(12):1992–1999.

14

Intracellular Delivery of Therapeutic Oligonucleotides in pH-Sensitive and Cationic Liposomes

Nejat Düzgüneş

Department of Microbiology, Arthur A. Dugoni School of Dentistry, University of the Pacific, San Francisco, California, U.S.A.

Sérgio Simões

Laboratory of Pharmaceutical Technology, Faculty of Pharmacy, and Center for Neuroscience and Cell Biology, University of Coimbra, Coimbra, Portugal

Montserrat Lopez-Mesas

Chemical Engineering Department, EUETIB-Universitat Politecnica de Catalunya, Barcelona, Spain

Maria C. Pedroso de Lima

Department of Biochemistry, Faculty of Sciences and Technology, and Center for Neuroscience and Cell Biology, University of Coimbra, Coimbra, Portugal

INTRODUCTION

Antisense oligonucleotide drugs and other similar drugs have much to offer as potential therapeutic agents, and as tools in genomics and proteomics. These drugs may be useful in the treatment of viral infections, cancer, restenosis, rheumatoid arthritis, and allergic disorders. The current clinical status of antisense therapeutics is shown in Table 1. Despite the extensive work with antisense oligonucleotides to date, the only antisense-based

Table 1 Current Clinical Status of Antisense Oligonucleotides

Compound	Company	Target	Indication	Clinical status	Study type
Vitravene®	ISIS Novartis AG	CMV	CMV-retinitis (AIDS)	Licensed	Monotherapy
Genasense®	Genta	Bcl-2	Solid tumors/leukemia	Phase III	Combination
Alicaforsen ISIS-2302	ISIS	ICAM-1	Crohn's disease	Phase III	
Affinitak	ISIS Eli Lilly	PKCα	Solid tumors	Phase III	Combination
ISIS-2503	ISIS	Ha-Ras	Solid tumors	Phase II	Combination
ISIS-5132	ISIS	c-Raf-1	Solid tumors	Phase II	Monotherapy
GEM-231	Hybridon	PKA-R1α	Solid tumors	Phase II	Combination
INX-3280	Lynx Therapeutics Inex	c-Myc	Restenosis/solid tumors	Phase II completed	
GEM-92	Hybridon	Gag region of HIV-1	AIDS	Phase I completed	
LR-3001	Inex Genta	c-Myb	CML	Phase I/II[a]	Monotherapy
ISIS-14803	ISIS	HCV	Hepatitis C	Phase I/II	
OL(1)p53		p53	Leukemia	Phase I	Monotherapy
GTI-2040	Lorus	R2 subunit of RR	Solid tumors	Phase II	Combination
GTI-2501	Lorus	R1 subunit of RR	Solid tumors	Phase I	Monotherapy

MG98	MethylGene MGI Pharma	DNA MeTase	Solid tumors	Phase II	Monotherapy
AP12009	Antisense Pharma	TGF-beta 2	Glioma	Phase II[b]	Monotherapy
OGX-011	ISIS OncoGenex	Clusterin	Solid tumors	Phase I/II	Monotherapy
ATL-1101	Antisense Therapeutics	IGF-1R	Psoriasis	Phase I	
ATL-1102	Antisense Therapeutics	VLA4 (CD 49d)	Multiple sclerosis	Phase II	
ISIS 113715	ISIS	PTB-1B	Diabetes type II	Phase II	
ISIS 301012	ISIS	Apo B100	Cardiovascular disease	Phase I	
LY 2181308	ISIS Eli Lilly	Survivin	Solid tumors	Phase I	

[a]Genta obtained the Orphan Drug status in the treatment of chronic myeloid leukemia in March 2005.

[b]Antisense Pharma GmbH obtained Orphan Drug status in the treatment of malignant glioma.

Abbreviations: RR, ribonucleotide reductase; ICAM-1, intercellular adhesion molecule-1; MeTase, methyl transferase; CMV, cytomegalovirus; Ha-RAS, p21 ras gene product from Harvey murine sarcoma virus; PKA- R1alpha, type 1 alpha regularory subunit of the cAMP-dependent protein kinase; HCV, hepatitis C virus; TGF, transforming growth factor; IGF- 1R, insulin-like growth factor-1 receptor; PTB-1B, protein tyrosine phosphatase 1B; CML, chronic myelogenous leukemia; PKC, protein kinase C.

Source: From Refs. 1, 2.

product that is approved for use is Vitravene (fomivirsen) against cytomega-lovirus (CMV) retinitis (1,2).

Oligonucleotide-based therapeutic agents bind to target nucleic acids or transcription factors, inhibiting the eventual synthesis of proteins at the transcription and translation stages. The five major classes of oligonucleotides are the following:

1. Antisense oligodeoxynucleotides that are short segments of DNA or modified DNA (such as phosphorothioate oligonucleotides) that hybridize to complementary sequences of target DNA or RNA. They act primarily by binding to target RNAs by Watson–Crick hybridization, and either inhibit the translation of the RNA or induce its degradation (3–5).
2. Triple-helix–forming oligonucleotides that recognize sequences in double-stranded DNA, thereby blocking gene expression or inducing mutations and recombination (6,7).
3. Ribozymes that recognize and cleave specific target mRNAs, with turnover (8–11).
4. DNA and RNA decoys that are designed to specifically bind transcription factors or to sequester regulatory proteins, respectively, thus competitively inhibiting gene expression (12).
5. Short interfering RNA (siRNA) that is produced from double-stranded RNA via the enzyme Dicer and are incorporated into the RNA-inducing silencing complex, which then facilitates the recognition and cleavage of the target mRNA (10,13–15).

CELLULAR ENTRY OF OLIGONUCLEOTIDES

Plain uncomplexed oligonucleotides are thought to bind to cell surface receptors, but these receptors are not well characterized (16,17). Proteins ranging in size 20–143 kDa are thought to be involved in binding, especially at lower oligonucleotide concentrations (18). Oligonucleotides are then internalized via endocytosis, and the endosomes formed subsequently can be visualized by confocal laser microscopy by the internalization of fluorescently labelled molecules (3). Internalized oligonucleotides have to be released from the endosome into the cytosol before the endosome contents are destined to degradation in lysosomes. Certain drug carriers, including peptides, polymers, and dendrimers, can be utilized to facilitate the destabilization of the endosome membrane. Both pH-sensitive and cationic liposomes are particularly useful in this respect (5,19). Fusion of cationic liposomes [containing a cationic cholesterol (Chol) derivative and complexed to plasmids] with endosomes has been reported (20), and may be one of the mechanisms whereby oligonucleotides are released into the cytoplasm.

The mechanisms by which oligonucleotides are transported to the nucleus and through the nuclear membrane are not well understood.

Microinjection of naked oligonucleotides into the cytoplasm leads to rapid accumulation of the macromolecules within the nucleus, probably due to its relatively small size (3,21–23). Lappalainen et al. (24) have shown that, after release from intracellular vesicles, oligonucleotides localize to the perinuclear area, but that the nuclear membrane constitutes a barrier. In the case of plasmid DNA, nuclear entry can be facilitated by nuclear localization signal peptides (25–27), or via the importin α and β proteins (28). Phosphorothioate oligonucleotides were shown to shuttle between the nucleus and the cytoplasm (29). The shuttling process was inhibited by chilling and adenosine triphosphate (ATP) depletion, which was saturable and thus possibly carrier mediated. Being sensitive to treatment with wheat-germ agglutinin, it was likely to be mediated by the nuclear pore complex.

Relatively high doses of free oligonucleotides are necessary to achieve therapeutic effects. Degradation of oligonucleotides in biological milieu and their rapid clearance from the circulation are additional disadvantages of free oligonucleotides. Oligonucleotides may also have undesired non–sequence-specific effects, and thus the delivery of free oligonucleotides in large doses may be problematic. In free form, they may also not be able to reach certain target tissues. Therefore, the delivery of oligonucleotides in pH-sensitive or cationic liposomes may be particularly advantageous.

pH-SENSITIVE LIPOSOMES

The concept of pH-sensitive liposomes emerged from the observation that certain enveloped viruses infect cells following acidification of the endosomal lumen to infect cells, and from the knowledge that some pathological tissues (tumors, inflamed, and infected tissue) have a more acidic environment compared to normal tissues. Although pH-sensitive liposomes are stable at physiological pH, they destabilize under acidic conditions, leading to the release of their aqueous contents (30–32). In addition, they appear to destabilize or fuse with the membranes of endosomes in which they are internalized, enabling even macromolecular liposome contents to enter the cytoplasm (33,34).

The response to acidic pH can be facilitated by a variety of molecules (35–38), including fusogenic peptides incorporated in the lipid bilayer (39–43), pH-sensitive lipids (44–46) and pH-sensitive polymers on the surface of liposomes (47–49). The combination of phosphatidylethanolamine (PE) or its derivatives with molecules with a protonatable group (e.g., carboxylic group) that acts as a stabilizer of PE membranes at neutral pH, is the most commonly used composition. PE has a minimally hydrated and small headgroup that occupies a lower volume compared to the hydrocarbon chains, and can be imagined to have a cone shape, in contrast to the cylinder shape exhibited by phospholipids such a phosphatidylcholine (PC). Strong intermolecular interactions between the amino and phosphate groups of neighboring polar headgroups, along with the cone shape, facilitate the formation of an inverted hexagonal phase at temperatures above a critical temperature (T_H) characteristic of the species of PE (50,51). These properties preclude the preparation of liposomes composed solely of PE or its derivatives under physiological conditions of pH, ionic strength, and temperature. Several conditions tend to facilitate the formation of liposomes composed mostly of PE (52): (i) PE can be mixed with other phospholipids, including the zwitterionic PC, and the net negatively charged phosphatidylglycerol or phosphatidylserine (PS). These lipids decrease the intermolecular interactions between the polar headgroups of PE and increase the hydration layer of the membrane. (ii) High pH (≥ 9.0) confers a net negative charge on PE molecules, due to deprotonation of the amino groups, decreases the intermolecular interactions between the polar headgroups, and increases the hydration layer. (iii) Amphiphilic molecules containing a protonatable acidic group that is negatively charged at physiological pH can be incorporated alongside PE in the liposome membrane. These molecules not only cause electrostatic repulsion between bilayers, but also disrupt the strong interactions between PE headgroups, thereby allowing the formation of bilayer structures and liposomes at physiological pH and temperature (32,53). With this approach, stable liposomes are formed at physiological pH, while at mildly acidic pH the carboxyl groups of the amphiphiles are protonated and their stabilizing effect on PE bilayers is diminished. PE molecules then tend to revert to their inverted hexagonal phase and thus cause liposome destabilization.

Following binding to cells, the liposomes are internalized through the endocytotic pathway. Liposomes are retained in early endosomes that mature into late endosomes. The potential of pH-sensitive liposomes lies in their ability to undergo destabilization at this stage, thus preventing their degradation at the lysosomal level and consequently increasing access to the cytosolic or nuclear targets (35,54). Although non–pH-sensitive liposomes [e.g., containing PC instead of dioleoylphosphatidylethanolamine (DOPE)] are internalized as extensively as pH-sensitive immunoliposomes, their capacity to mediate cytoplasmic delivery of the encapsulated molecules is significantly lower (34,55). This observation suggests that fusion or destabilization of liposomes induced by acidification of the endosomal lumen represents the most important stage in the process of intracellular delivery (Fig. 1).

Studies involving the incubation of cells with lysosomotropic agents (e.g., ammonium chloride or chloroquine) that prevents endosome acidification demonstrate that the efficacy of pH-sensitive liposomes depends on the pH decrease upon endosome maturation. Different molecular mechanisms by which the liposomes release their contents into the cytoplasm have been proposed: (i) destabilization of pH-sensitive liposomes triggers the destabilization of the endosomal membrane, most likely through pore formation, leading to cytoplasmic delivery of their contents; (ii) upon liposome

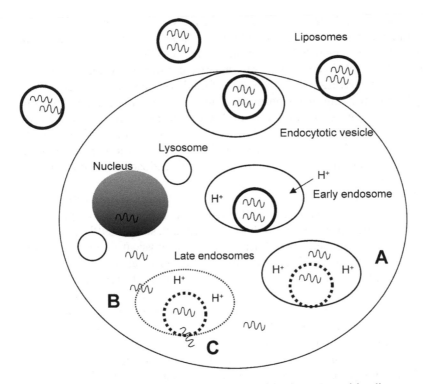

Figure 1 Intracellular delivery of oligonucleotides by pH-sensitive liposomes. The liposomes are internalized by endocytosis after binding to cell surface receptors. The lumen of resulting endosomes is acidified by the action of a H^+-ATPase. The liposomes destabilize at acidic pH, the threshold pH being determined by the composition of the liposomes. The liposomes in the figure have been designed ("programmed") to destabilize at the lower pH achieved in late endosomes. In case **A**, the encapsulated oligonucleotides are released into the endosome lumen, but the endosome is not destabilized, and thus the contents are trapped in the endosome. In case **B**, the endosome membrane is also destabilized due to the structural transformation of the pH-sensitive liposomes, enabling the cytoplasmic entry of the oligonucleotides. Alternatively (case **C**), the liposomes may undergo fusion with the endosome membrane, and release their contents directly into the cytoplasm. Some of the oligonucleotides can diffuse into the nucleus.

destabilization the encapsulated molecules diffuse to the cytoplasm through the endosomal membrane; and (iii) fusion between the liposome and the endosomal membranes, leading to cytoplasmic delivery of their contents (34,35,54,56). The fusogenic properties of PE associated with its tendency to form an inverted hexagonal phase under certain conditions favor hypotheses (i) and (iii). The fusogenic properties of the liposomes do not always correlate with their efficacy in mediating intracellular delivery. Although aggregation, release of contents, and lipid intermixing are observed at low pH with DOPE:cholesteryl hemisuccinate (CHEMS) liposomes, no intermixing of aqueous contents takes place (57), but these liposomes are efficient in delivering their encapsulated contents into cultured cells (58). Divalent cations may also play a role in delivery by pH-sensitive liposomes. PE:oleic acid (OA) liposomes undergo fusion in the presence of millimolar concentrations of Ca^{2+} or Mg^{2+}, and the rate of fusion under acidic conditions is enhanced significantly in the presence of $2\,mM\ Ca^{2+}$(32). Cytoplasmic delivery of calcein by DOPE:CHEMS liposomes is inhibited in the presence of ethylenediamine tetraacetic acid (EDTA) (58), indicating that divalent cations participate in the destabilization of pH-sensitive liposomes and endosomal membranes, or their fusion with each other.

The efficiency of interaction of pH-sensitive liposomes with cells is dependent on the inclusion of DOPE in their composition, independently of the type of the amphiphilic stabilizer used. In fact, some DOPE-containing liposomes shown to be non–pH-sensitive by biophysical assays, mediated cytoplasmic delivery of their contents as efficiently as well known pH-sensitive formulations (59). Nevertheless, among the different formulations studied, DOPE:CHEMS liposomes had the highest extent of cell association. Results with cells pretreated with metabolic inhibitors or lysosomotropic agents indicate clearly that DOPE-containing liposomes are internalized essentially by endocytosis and that acidification of the endosomes is not the only mechanism involved in the destabilization of the liposomes inside the cell (59). Although some of the liposomes tested had similar abilities to deliver calcein, the delivery of higher molecular weight molecules was highest when encapsulated in pH-sensitive DOPE:CHEMS liposomes compared to other DOPE-containing liposomes (60).

DELIVERY OF THERAPEUTIC OLIGONUCLEOTIDES IN pH-SENSITIVE LIPOSOMES

The first use of pH-sensitive liposomes to mediate intracellular delivery of antisense oligonucleotides was reported by Ropert et al. (56,61). The oligonucleotide was against the *env* gene mRNA of Friend retrovirus. A significant inhibition of viral replication was observed when the antisense oligonucleotides were delivered in liposomes composed of DOPE:OA:Chol, compared to that observed with non–pH-sensitive liposomes.

Selvam et al. (62) reported the sequence-specific suppression of HIV-1 replication in H9 cells and peripheral blood lymphocytes by a 20-mer anti-*rev* antisense phosphorothioate oligonucleotides, using DOPE:OA:Chol (45:10:35 mol%) liposomes targeted to cell surface CD4 via covalently coupled monoclonal antibodies. HIV-1 replication was reduced by 85% in antisense immunoliposome-treated cells, whereas either empty immunoliposomes or immunoliposomes containing scrambled *rev* phosphorothioate oligonucleotide sequences did not inhibit HIV-1 replication.

The authors' laboratory has investigated the potential of pH-sensitive liposomes to mediate intracellular delivery of antisense oligonucleotides and ribozymes against HIV-1 sequences. They used a 15-mer phosphorothioate oligonucleotide against the Rev-responsive element (RRE) of HIV-1. The macromolecular drug was encapsulated in pH-sensitive CHEMS/DOPE (4:6) liposomes and added to cultures of primary human macrophages infected with HIV-1$_{BaL}$. Virus production was assessed by the production of the viral core protein p24 in the culture medium, using an enzyme-linked immunosorbent assay (ELISA). The encapsulated oligonucleotide inhibited virus replication by 91%, while the free (unencapsulated) oligonucleotide was not effective (Fig. 2). A nonspecific oligonucleotide encapsulated in pH-sensitive liposomes inhibited HIV infection by 53%. Nonspecific inhibition of HIV infection of lymphocyte cell lines by free phosphorothioate oligonucleotides has been reported by other laboratories (63–65).

Similar results were obtained with a 38-mer chimeric ribozyme complementary to the 5'-long terminal repeat region of HIV-1. Virus production was inhibited by 88% when HIV-infected macrophages were treated with the ribozyme encapsulated in pH-sensitive liposomes, whereas the free ribozyme caused a decrease of only 10% (60).

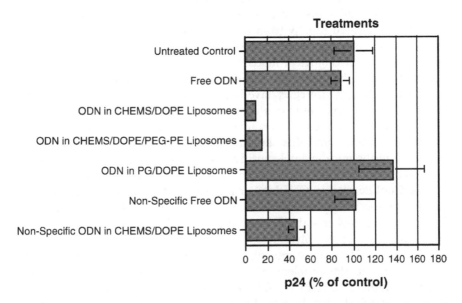

Figure 2 Inhibition of HIV-1 production in human macrophages by free and liposome-encapsulated anti-Rev-responsive element 15-mer phosphorothioate oligodeoxynucleotide at a concentration of 3 μM. The viral p24 values are given as the percentage of untreated controls. *Abbreviations*: ODN, oligodeoxyribonucleotide; CHEMS, cholesteryl hemisuccinate; DOPE, dioleoylphosthatidylethanolamine; PEG–PE, poly(ethylene glycol)-phosphatidylethanolamine; PG, phosphatidylglycerol.

Ponnappa et al. (66) encapsulated antisense phosphorothioate oligonucleotides (TJU-2755) against tumor necrosis factor (TNF)-α in pH-sensitive liposomes that were administered intravenously into rats (1–2 mg/kg body weight). A significant inhibition of TNF-α production was observed, leading to a 65% to 70% reduction in plasma levels of TNF-α, compared with controls. The authors concluded that the oligonucleotide TJU-2755 encapsulated in pH-sensitive liposomes can be used to effectively reduce endotoxin-mediated production of TNF-α in macrophages in vivo and thus may be of value in attenuating or preventing macrophage-mediated liver injury. In a subsequent study, they showed that lipopolysaccharoid (LPS)-induced serum TNF-α was reduced by 54%, and LPS-induced liver injury was reduced by 60% when the antisense oligonucleotide was administered in pH-sensitive liposomes at a fivefold lower dose compared to the free oligonucleotide (67).

DELIVERY OF OLIGONUCLEOTIDES IN STERICALLY STABILIZED pH-SENSITIVE LIPOSOMES WITH PROLONGED CIRCULATION

Although they are able to deliver therapeutic macromolecules to target cells, classical pH-sensitive liposomes are destabilized in plasma at physiological pH (68) and cleared rapidly from the circulation, accumulating in the liver and spleen (36,69–71). The tendency of pH-sensitive liposomes to aggregate in the presence of biological fluids may be the underlying reason for their accumulation in the lungs (69). To improve the biostability of pH-sensitive liposomes, different strategies have been reported, including the incorporation of a third component to confer stability to the lipid bilayer. Inclusion of Chol in DOPE:OA liposome formulations resulted in a significant increase in plasma stability, and this stabilizing effect was not accompanied by a reduction in the liposome pH-sensitivity (72). The use of other amphiphilic stabilizers that were shown to be resistant to extraction by albumin, including the Chol derivative, CHEMS, or lipids with double acyl chains such as dipalmitoylsuccinylglycerol (DPSG), result in the formation of pH-sensitive liposomes with higher stability in biological fluids (73,74).

The presence of poly(ethylene glycol)-distearoylphosphatidylethanol-amine (PEG-PE) in liposomes of various non–pH-sensitive compositions could overcome the problem of rapid removal from circulation by the reticuloendothelial system (75–78). Prolonged circulation time and "pH" sensitivity would be highly desirable for the delivery of therapeutic macromolecules to cells in vivo. Monosialoganglioside (GM$_1$) was shown to confer relatively prolonged residence in circulation to pH-sensitive liposomes composed of DOPE and DPSG (70). The incorporation of GM$_1$ enhanced the retention of the liposome-associated radioactive tracer (^{125}I-tyraminylinulin) in the blood up to two hours. Nevertheless, both control and GM$_1$-containing liposomes were almost completely cleared from circulation at later times (5 and 24 hours).

The incorporation of PEG (2000)-DSPE into the membrane of pH-sensitive liposomes conferred stability on the liposomes and enabled them to circulate for prolonged periods (71). Liposomes composed of DOPE: CHEMS:PEG-PE had $t_{1/2}$ similar to that of control sterically stabilized liposomes ($t_{1/2} = 11.1 \pm 0.6$ hours) and to that of "stealth" compositions described by others (79,80). In contrast, DOPE:CHEMS liposomes were cleared rapidly from the circulation (0.5 hours). In addition, pH-sensitive sterically stabilized liposomes delivered significantly higher amounts of an encapsulated water-soluble marker to the cells of the reticuloendothelial system than any of the sterically stabilized liposomes described by others, or control non–pH-sensitive liposomes used in the studies mentioned above.

The low pH-dependent release of encapsulated charged fluorophores was inhibited with this modification of the liposome membrane; nevertheless, the ability of these liposomes to facilitate the intracellular delivery of their aqueous contents remained unaltered. This was demonstrated by a flow cytometry assay involving dual fluorescence labeling of the liposomes (71). This lack of correlation between biophysical assays measuring liposome destabilization and the efficacy of the liposomes to deliver their contents intracellularly, suggests that the mechanisms by which pH-sensitive liposomes mediate intracellular delivery of their contents are not simply governed by the pH-dependent release of their contents in endosomes.

Other laboratories have also shown that liposomes composed of DOPE, DPSG, and PEG-DSPE (up to 5%) are pH-sensitive, plasma stable, and have a long circulation time in the blood (81). These liposomes were able to release an entrapped marker rapidly in tumor tissue homogenates, where the pH is lower than normal healthy tissues, in contrast to non–pH-sensitive dipalmitoyl-PC/Chol/PEG-DSPE liposomes.

Because sterically stabilized pH-sensitive liposomes circulate for prolonged periods in the bloodstream and may localize in lymph nodes after intravenous or subcutaneous injection as shown for plain sterically stabilized liposomes (82,83), they may be useful for the delivery of antisense molecules to lymph nodes where active HIV replication takes place (84,85). The anti-RRE oligonucleotide encapsulated in sterically stabilized pH-sensitive DOPE:CHEMS:PEG-PE liposomes inhibited HIV replication in infected macrophages, but to a slightly lower extent than that in regular pH-sensitive liposomes (60).

OLIGONUCLEOTIDE DELIVERY BY CATIONIC LIPOSOMES

Cationic liposomes were first introduced by Felgner et al. (86) as gene delivery vehicles and have been used both in cell culture and in vivo (87–91). Cationic liposome–DNA complexes ("lipoplexes") can protect DNA or RNA from inactivation or degradation in biological milieu, do not induce specific immunity, and are safer than viral vectors (91–93). Large-scale production of liposomes is easy compared to that of viruses expressing particular antisense oligonucleotide or ribozyme sequences. Liposomes are also highly versatile.

The plasma membrane is an obvious barrier to the entry of oligonucleotides. Complexation with cationic liposomes is a potentially useful method to overcome this barrier (94,95), and to confer improved stability against nucleases (96). Nevertheless, improvement of the cellular uptake of oligonucleotides does not necessarily translate into functional delivery into the cytoplasm and the nucleus, as evidenced by our studies on ribozyme delivery to HIV-infected cells (95). The latter observation points to the importance of facilitating the escape of the oligonucleotide from the endosome into the cytoplasm and the nucleus. If the oligonucleotides cannot escape the endosomes, they will be targeted to lysosomes and degraded (97). Once in the cytoplasm, oligonucleotides are able to diffuse into the nucleus through nuclear pores (21). This aspect constitutes a difference between oligonucleotides and plasmid DNA because the latter neither diffuse readily in the cytoplasm nor can they enter the nucleus efficiently. The inclusion of potentially fusogenic lipids in the cationic liposomes may facilitate the cytoplasmic delivery of oligonucleotides. One proposal for the mechanism of this process is that the flip-flop of the PS in the cytoplasmic leaflet of the endosomal membrane (most likely mediated by the fusogenic lipid) to the lumenal side may facilitate the dissociation of the oligonucleotide from the cationic lipids while the membranes of the cationic liposomes and the endosome fuse (98). The dissociation of the cationic lipid–oligonucleotide complex upon interaction with liposomes that mimic the endosome membrane has been investigated using fluorescence resonance energy transfer (99). These studies showed that the presence of DOPE in the cationic liposome membrane is an important factor in the dissociation of the nucleic acid from the liposomes. Supporting these in vitro observations, experiments by Marcusson et al. (100) have indicated that oligonucleotides dissociate from their cationic lipid carriers before entering the nucleus.

Oligonucleotides incorporated into cationic liposomes composed of dioctadecyl amidoglycyl spermidine and DOPE during liposome preparation were taken up by HeLa cells, released from endocytotic vesicles, and localized in the cytoplasm and the nucleus (101). This "DLS lipoplex" system was much more effective in inhibiting HIV-1 replication in chronically infected MOLT-3 cells than the free phosphorothioate oligonucleotides. However, control (e.g., random and sense sequence) oligonucleotides also had significant inhibitory effects in this system. Levigne et al. (101) have suggested that siRNA and decoy oligonucleotides may be more promising than antisense oligonucleotides in the therapy of HIV infections. In a system involving the complexation of DLS liposomes with oligonucleotides, the latter were localized in the cytoplasm, the perinuclear region, and nucleoli of KS-Y1 Kaposi's sarcoma cells, as well as HepG2 hepatoma cells and human monocyte–derived macrophages (101). Colony formation by KS-Y1 cells was inhibited by 93% using an antisense oligonucleotide against vascular endothelial growth factor (VEGF) at 1 µM when administered via DLS liposomes, and by 60% at 100 nM. Free oligonucleotides and control oligonucleotides in DLS liposomes caused only 4% to 5% inhibition. Intratumoral injection of the DLS-anti-VEGF oligonucleotides also inhibited tumor growth and caused tumor necrosis.

Some nuclear delivery of oligonucleotides was observed in human cervical epithelial CaSki cells, with liposomes composed of dimethyl-dioctadecyl-ammonium bromide and DOPE (2:5), under conditions where the complexes were positively charged (24). Although liposomes composed of polycationic 2,3-dioleoyloxy-*N*-[2(sperminecarboxamido)ethyl]-*N,N*-dimethyl-1-propanaminium trifluoroacetate and DOPE (3:1) could deliver oligonucleotides into the cytoplasm, nuclear delivery was not observed. The same liposomes, however, facilitated both cytosolic and nuclear delivery of oligonucleotides in C6 glioma cells, at an optimal lipid/DNA charge ratio of 1:1 (102). The combination of cationic 1,2-dioleoyl-3-(trimethylammonium) propane (DOTAP)/Chol liposomes with the cationic polymer polyethylenimine enhanced the inhibitory effect of an antisense oligonucleotide against the p53 tumor suppressor protein in HepG2 and hepatoma 2.2.15 cells, compared to the use of liposomes alone (103).

Liposomes composed of cationic cholesteryl-3-β-carboxyamidoethylene-*N*-hydroxyethylamine and DOPE (3:2) were shown to deliver an antisense oligonucleotide against the mRNA of the antiapoptotic protein Bcl-2 into the nuclei of HeLa cells (104). The oligonucleotide induced significant apoptosis, whereas a scrambled oligonucleotide was essentially ineffective. The extent of apoptosis was much greater than that achieved by 3-*N*-(dimethyl amino ethyl) carbamate (DC-Chol):DOPE (2:3) liposomes in this system.

Another vector for oligonucleotide delivery is influenza virus envelopes reconstituted in DOTAP liposomes (105). This system was used to deliver an oligonucleotide antisense to L-*myc* in human small cell lung cancer cells and resulted in inhibition of thymidine incorporation in the picomolar range, whereas micromolar concentrations of free oligonucleotides were necessary to achieve inhibition in previous studies (106). Antisense oligonucleotides can also be delivered via the "hemagglutinating virus of Japan (HVJ)-liposome" method, in which the oligonucleotides are first encapsulated in PS:PC:Chol (1:4:8:2) liposomes, and the liposomes are allowed to interact with UV-inactivated HVJ (Sendai virus) (107,108). HVJ-liposomes-mediated delivery of antisense oligonucleotides against cyclin B1 and CDC2 kinase-encoding genes into cells lining the rat carotid artery mediated partial inhibition in neointima formation (107). Systemic delivery of antiapolipoprotein E oligonucleotides in these liposomes resulted in mice deficient in the production of this protein (107).

Antisense phosphorothioate oligonucleotides complementary to the influenza A virus RNA polymerase component *PB2* gene complexed to the cationic liposome reagent, Tfx-10, were administered intravenously to infected mice, and prolonged significantly the mean survival time and increased the overall survival rates of the animals (109). The liposome-complexed oligonucleotide also inhibited viral growth in the lungs and reduced pulmonary pathology.

Cationic liposomes can also have nonspecific effects on treated cells, such as inhibitory effects on the transcription of vascular cell adhesion molecule-1 (110) or on the viability of HIV-infected cells (111). The cytotoxicity of cationic lipid-oligonucleotide complexes have to be ascertained and may limit their usefulness in some cases (111,112). Another disadvantage of cationic liposome–oligonucleotide complexes is that intravenous injection can result in their rapid uptake by the lungs and liver, making delivery to other organs problematic (113,114).

One method that may be able to overcome the deleterious effects of the net positively charged cationic liposome–oligonucleotide complexes is to "coat" the complexes formed in an organic phase with a monolayer of neutral or zwitterionic lipids (115,116). An alternative approach, similar to that employed for pH-sensitive liposomes described above, is the use of PEGylated lipids to confer steric stabilization to cationic liposome–oligonucleotide complexes. DOTAP:DOPE:PEG-PE liposomes retained their structure after complexation with oligonucleotides, were relatively stable in serum, and enhanced oligonucleotide uptake by the breast cancer cells SKBR3 and MCF-7 (117). Oligonucleotides can also be entrapped in cationic liposomes containing PEG-lipids using 25% to 40% (v/v) ethanol (118).

TARGETING OF ANTISENSE OLIGONUCLEOTIDES

Oligonucleotide-liposomes can be targeted to specific cells via covalently coupled antibodies to cell surface markers. Neutral lipid-coated cationic liposomes encapsulating antisense oligonucleotides against c-*myb* and targeted to disialoganglioside, GD_2, were much more effective in inhibiting the growth of neuroblastoma cells compared to nontargeted liposomes and free oligonucleotides (119). These liposomes also had significant antitumor effects in vivo, but part of this effect could be attributed to the immunostimulatory effect of CpG sequences in the oligonucleotide (120). GD_2-targeted coated cationic liposomes containing anti-c-*myc* oligonucleotides inhibited melanoma cell proliferation, the development of microscopic metastases, and tumor growth (121).

Coated cationic liposomes encapsulating antisense oligonucleotides against the mRNA of the P-glycoprotein can be targeted to multidrug-resistant B-cell lymphoma cells via the attachment of anti-CD19 antibodies to the liposomes (122). Coated cationic liposomes targeted to scavenger receptors via aconitylated human serum albumin accumulated in liver endothelial cells and could downregulate intercellular adhesion molecule-1 mRNA in the macrophage cell line J774 (123).

Although antibody coupling to PEG-DSPE has the advantage that the targeting ligand protrudes out from the liposome surface, the inhibitory effect of this polymeric lipid on transfection (124,125) should be taken into consideration. Thus, it may be necessary to incorporate exchangeable PEG-lipids into the cationic vector (5,124,126,127). Utilizing such a system termed "programmable fusogenic vesicles" with exchangeable PEG-ceramides, Hu et al. (127) showed a relatively modest reduction in *bcl-2* mRNA levels in 518A2 melanoma cells, whereas control oligonucleotides with reverse polarity resulted in an increase in mRNA levels. PEGylated cationic liposomes targeted to HER-2 and complexed with an oligonucleotide antisense to the Bcl-2 protein reduced protein expression by about 50% in BT-474 breast carcinoma cells, while the nontargeted liposomes were not effective (117).

In another targeting approach, Rodriguez et al. (128) utilized biotinylated antibodies to p185/HER-2, which is overexpressed on breast cancer cells, and associated them with cationic liposomes containing streptavidin-DOPE:DOTAP complexed with an antisense oligonucleotide directed toward the translational start site of dihydrofolate reductase RNA. As an alternative, they associated streptavidin to biotinylated antibody and biotinylated oligonucleotide that was complexed with DOTAP. The immunoliposomes were more toxic to SKBR3 cells that overexpress p185 than the antisense oligonucleotide in the absence of the antibody. Antisense oligonucleotides against HER-2 mRNA complexed with folate-targeted liposomes inhibited cell growth and HER-2 expression in SCC-2CP head and neck tumor cells, induced apoptosis, and increased the sensitivity of the cells to chemotherapeutic agents (129).

CONCLUDING REMARKS

Antisense oligonucleotides and similar macromolecular drugs are promising investigational drugs, some of which are in clinical trials for the treatment of diseases ranging from cancer to restenosis, viral infections, and multiple sclerosis (Table 1). Due to the relatively high doses of free oligonucleotides necessary for treatment, their degradation and rapid clearance from the circulation, their potential nonspecific effects, and limitations in their ability to reach certain target tissues, it may be advantageous to deliver them in pH-sensitive or cationic liposomes. These carrier systems can be stabilized by the inclusion of PEGylated lipids in their membranes, preventing their destabilization in serum and rapid clearance by the reticuloendothelial system. Cationic liposomes can also be coated with neutral lipids to prevent their rapid interaction with serum components. Both types of liposomes can be targeted to cell surface receptors via antibodies or ligands, to facilitate specific uptake by cancer cells or other types of cells. The versatility of these carrier systems and their "programmable" feature are likely to enhance the therapeutic value of antisense oligonucleotides, ribozymes, siRNAs, and triple-helix forming oligonucleotides.

REFERENCES

1. Orr RM, O'Neill CF. Patent review: Therapeutic applications for antisense oligonucleotides. Curr Opin Mol Ther 2000; 2:325–331.
2. Orr RM, Dorr FA. Clinical studies of antisense oligonucleotides for cancer therapy. Methods Mol Med 2005; 106:85–111.
3. Akhtar S, Hughes MD, Khan A, et al. The delivery of antisense therapeutics. Adv Drug Deliv Rev 2000; 44:3–21.
4. Crooke ST. Antisense strategies. Curr Mol Med 2004; 4:465–487.
5. Shi F, Hoekstra D. Effective intracellular delivery of oligonucleotides in order to make sense of antisense. J Control Rel 2004; 97:189–209.
6. Praseuth D, Guieysse AL, Hélène C. Triple helix formation and the antigene strategy for sequence-specific control of gene expression. Biochim Biophys Acta 1999; 1489:181–206.
7. Seidman MM, Glazer PM. The potential for gene repair via triple helix formation. J Clin Invest 2003; 112:487–494.
8. Rossi JJ. The application of ribozymes to HIV infection. Curr Opin Mol Ther 1999; 1:316–322.
9. Schubert S, Kurreck J. Ribozyme- and deoxyribozyme-strategies for medical applications. Curr Drug Targets 2004; 5:667–681.
10. Scanlon KJ. Anti-genes: siRNA, ribozymes and antisense. Curr Pharm Biotechnol 2004; 5:415–420.
11. Peracchi A. Prospects for antiviral ribozymes and deoxyribozymes. Rev Med Virol 2004; 14:47–64.
12. Tomita N, Ogihara T, Morishita R. Transcription factors as molecular targets: molecular mechanisms of decoy ODN and their design. Curr Drug Targets 2003; 4:603–608.
13. Sontheimer EJ. Assembly and function of RNA silencing complexes. Nat Rev Mol Cell Biol 2005; 6:127–138.
14. Campbell TN, Choy FY. RNA interference: past, present and future. Curr Issues Mol Biol 2005; 7:1–6.
15. Bennink JR, Palmore TN. The promise of siRNAs for the treatment of influenza. Trends Mol Med 2004; 10:571–574.
16. Wagner RW, Matteucci MD, Lewis JG, Gutierrez AJ, Moulds C, Froechler BC. Antisense gene inhibition by oligonucleotides containing C-5 propyne pyrimidines. Science 1993; 260:1510–1513.
17. Crooke RM, Graham MJ, Cooke ME, Crooke ST. In vitro pharmacokinetics of phosphorothioate antisense oligonucleotides. J Pharmacol Exp Ther 1995; 275:923–937.
18. Beltinger C, Saragovi HU, Smith RM, et al. Binding, uptake, and intracellular trafficking of phosphorothioate-modified oligodeoxynucleotides. J Clin Invest 1995; 95:1814–1823.

19. Fattal E, Couvreur P, Dubernet C. "Smart" delivery of antisense oligonucleotides by anionic pH-sensitive liposomes. Adv Drug Deliv Rev 2004; 56:931–946.
20. Noguchi A, Furuno T, Kawaura C, Nakanishi M. Membrane fusion plays an important role in gene transfection mediated by cationic liposomes. FEBS Lett 1998; 433:169–173.
21. Chin DJ, Green GA, Zon G, Szoka FC Jr., Straubinger RM. Rapid nuclear accumulation of injected oligodeoxyribonucleotides. New Biol 1990; 2:1091–1100.
22. Dokka S, Rojanasakul Y. Novel non-endocytic delivery of antisense oligonucleotides. Adv Drug Deliv Rev 2000; 44:35–49.
23. Pichon C, Roufaï MB, Monsigny M, Midoux P. Histidylated oligolysines increase the transmembrane passage and the biological activity of antisense oligonucleotides. Nucl Acids Res 2000; 28:504–512.
24. Lappalainen K, Miettinen R, Kellokoski J, Jääskeläinen I, Syrjanen S. Intracellular distribution of oligonucleotides delivered by cationic liposomes: light and electron microscopic study. J Histochem Cytochem 1997; 45:265–274.
25. Aronsohn AI, Hughes JA. Nuclear localization signal peptides enhance cationic liposome-mediated gene therapy. J Drug Target 1997; 5:163–169.
26. Sebestyén MG, Ludtke JJ, Bassik MC, et al. DNA vector chemistry: the covalent attachment of signal peptides to plasmid DNA. Nat Biotechnol 1998; 16:80–85.
27. Escriou V, Carrière M, Scherman D, Wils P. NLS bioconjugates for targeting therapeutic genes to the nucleus. Adv Drug Deliv Rev 2003; 55:295–306.
28. Carrière M, Escriou V, Savarin A, Scherman D. Coupling of importin beta binding peptide on plasmid DNA: transfection efficiency is increased by modification of lipoplex's physico-chemical properties. BMC Biotechnol 2003; 3:1–11.
29. Lorenz P, Misteli T, Baker BF, Bennett CF, Spector DL. Nucleocytoplasmic shuttling: a novel in vivo property of antisense phosphorothioate oligodeoxynucleotides. Nucl Acids Res 2000; 28:582–592.
30. Ellens H, Bentz J, Szoka FC. pH-induced destabilization of phosphatidylethanolamine-containing liposomes: role of bilayer contact. Biochemistry 1984; 23:1532–1538.
31. Düzgüneş N, Straubinger RM, Papahadjopoulos D. pH-dependent membrane fusion. J Cell Biol 1983; 97:178a.
32. Düzgüneş N, Straubinger RM, Baldwin PA, Friend DS, Papahadjopoulos D. Proton-induced fusion of oleic acid/phosphatidylethanolamine liposomes. Biochemistry 1985; 24:3091–3098.
33. Straubinger RM, Düzgüneş N, Papahadjopoulos D. pH-sensitive liposomes: enhanced cytoplasmic delivery of encapsulated macromolecules. J Cell Biol 1983; 97:109a.
34. Straubinger RM, Düzgüneş N, Papahadjopoulos D. pH-sensitive liposomes mediate cytoplasmic delivery of encapsulated macromolecules. FEBS Lett 1985; 179:148–154.
35. Düzgüneş N, Straubinger RM, Baldwin PA, Papahadjopoulos D. pH-sensitive liposomes: introduction of foreign substances into cells. In: Wilschut J, Hoekstra D, eds. Membrane Fusion. New York: Marcel Dekker, 1991:713–730.
36. Torchilin VP, Zhou F, Huang L. pH-sensitive liposomes. pH-sensitive liposomes. J Liposome Res 1993; 3:201–255.

37. Drummond DC, Zignani M, Leroux J. Current status of pH-sensitive liposomes in drug delivery. Prog Lipid Res 2000; 39:409–460.
38. Venugopalan P, Jain S, Sankar S, Singh P, Rawat A, Vyas SP. pH-sensitive liposomes: mechanism of triggered release to drug and gene delivery prospects. Pharmazie 2002; 57:659–671.
39. Parente RA, Nadasdi L, Subbarao NK, Szoka FC Jr. Association of a pH-sensitive peptide with membrane vesicles: role of amino acid sequence. Biochemistry 1990; 29:8713–8719.
40. Ishiguro R, Matsumoto M, Takahashi S. Interaction of fusogenic synthetic peptide with phospholipid bilayers: orientation of the peptide alpha-helix and binding isotherm. Biochemistry 1996; 35:4976–4983.
41. Bailey AL, Monck MA, Cullis PR. pH-induced destabilization of lipid bilayers by a lipopeptide derived from influenza hemagglutinin. Biochim Biophys Acta 1997; 1324:232–244.
42. Nir S, Nicol F, Szoka FC Jr. Surface aggregation and membrane penetration by peptides: relation to pore formation and fusion. Mol Membr Biol 1999; 16: 95–101.
43. Turk MJ, Reddy JA, Chmielewski JA, Low PS. Characterization of a novel pH-sensitive peptide that enhances drug release from folate-targeted liposomes at endosomal pHs. Biochim Biophys Acta 2002; 1559:56–68.
44. Anderson VC, Thompson DH. Triggered release of hydrophilic agents from plasmalogen liposomes using visible light or acid. Biochim Biophys Acta 1992; 1109:33–42.
45. Drummond DC, Daleke DL. Synthesis and characterization of N-acylated, pH-sensitive "caged" aminophospholipids. Chem Phys Lipids 1995; 75:27–41.
46. Reddy JA, Low PS. Enhanced folate receptor mediated gene therapy using a novel pH-sensitive lipid formulation. J Control Release 2000; 64:27–37.
47. Leroux J, Roux E, Le Garrec D, Hong K, Drummond DC. N-isopropylacrylamide copolymers for the preparation of pH-sensitive liposomes and polymeric micelles. J Control Rel 2001; 72:71–84.
48. Roux E, Francis M, Winnik FM, Leroux JC. Polymer based pH-sensitive carriers as a means to improve the cytoplasmic delivery of drugs. Int J Pharm 2002; 242:25–36.
49. Mizoue T, Horibe T, Maruyama K, et al. Targetability and intracellular delivery of anti-BCG antibody-modified, pH-sensitive fusogenic immunoliposomes to tumor cells. Int J Pharm 2002; 237:129–137.
50. Cullis PR, De Kruijff B. Lipid polymorphism and the function role of lipids in biological membranes. Biochim Biophys Acta 1979; 559:399–420.
51. Seddon JM, Cevc G, Marsh D. Calorimetric studies of the gel-fluid (L_β-L_α) and lamellar-inverted hexagonal phase (L_α-H_{II}) transition in dialkyl-and diacyl-phosphatidylethanolamine. Biochemistry 1983; 22:1280–1289.
52. Simões S, Fonseca C, Pedroso de Lima MC, Düzgüneş N. pH-sensitive liposomes: from biophysics to therapeutic applications. In: Arshady R, Kono K, eds. Trigger-Sensitive Assemblies and Particulates. London: Citus Books. In press.
53. Lai MZ, Düzgüneş N, Szoka FC. Effect of replacement of the hydroxyl group of cholesterol and tocopherol on the thermotropic behavior of phospholipid membranes. Biochemistry 1985; 24:1646–1653.

54. Collins D. pH-sensitive liposomes as tools for cytoplasmic delivery. In: Philippot JR, Schuber F, eds. Liposomes as Tools in Basic Research and Industry. Boca Raton: CRS Press, 1995:201–214.
55. Ho RJY, Rouse BT, Huang L. Target-sensitive immunoliposomes: preparation and characterization. Biochemistry 1986; 25:5500–5506.
56. Ropert C, Malvy C, Couvreur P. Inhibition of the Friend retrovirus by antisense oligonucleotides encapsulated in liposomes: mechanism of action. Pharm Res 1993; 10:1427–1433.
57. Ellens H, Bentz J, Szoka FC. H^+- and Ca^{2+}-induced fusion and destabilization of liposomes. Biochemistry 1985; 24:3099–3106.
58. Chu CJ, Dijkstra J, Lai MZ, Hong K, Szoka FC. Efficiency of cytoplasmic delivery by pH-sensitive liposomes to cells in culture. Pharm Res 1990; 7: 824–834.
59. Simões S, Slepushkin V, Düzgüneş N, Pedroso de Lima MC. On the mechanisms of internalization and intracellular delivery mediated by pH-sensitive liposomes. Biochim Biophys Acta 2001; 1515:23–37.
60. Düzgüneş N, Simões S, Slepushkin V, et al. Enhanced inhibition of HIV-1 replication in macrophages by antisense oligonucleotides, ribozymes and acyclic nucleoside phosphonate analogs delivered in pH-sensitive liposomes. Nucleos Nucleot Nucl Acids 2001; 20:515–523.
61. Ropert C, Lavignon M, Dubernet C, Couvreur P, Malvy C. Oligonucleotides encapsulated in pH sensitive liposomes are efficient toward Friend retrovirus. Biochem Biophys Res Commun 1992; 183:879–885 [Erratum in: Biochem Biophys Res Commun 1993; 192:982].
62. Selvam MP, Buck SM, Blay RA, Mayner RE, Mied PA, Epstein JS. Inhibition of HIV replication by immunoliposomal antisense oligonucleotide. Antiviral Res 1996; 33:11–20.
63. Lisziewicz J, Sun D, Klotman M, Agrawal S, Zamecnik P, Gallo R. Specific inhibition of human immunodeficiency virus type 1 replication by antisense oligonucleotides: an in vitro model for treatment. Proc Natl Acad Sci USA 1992; 89:11,209–11,213.
64. Zelphati O, Imbach J-L, Signoret N, Zon G, Rayner B, Leserman L. Antisense oligonucleotides in solution or encapsulated in immunoliposomes inhibit replication of HIV-1 by several different mechanisms. Nucl Acids Res 1994; 22:4307–4314.
65. Weichold FF, Lisziewicz J, Zeman RA, et al. Antisense phosphorothioate oligodeoxynucleotides alter HIV type 1 replication in cultured human macrophages and peripheral blood mononuclear cells. AIDS Res Hum Retroviruses 1995; 11:863–867.
66. Ponnappa BC, Dey I, Tu GC, et al. In vivo delivery of antisense oligonucleotides in pH-sensitive liposomes inhibits lipopolysaccharide-induced production of tumor necrosis factor-alpha in rats. J Pharmacol Exp Ther 2001; 297:1129–1136.
67. Ponnappa BC, Israel Y, Aini M, et al. Inhibition of tumor necrosis factor-alpha secretion and prevention of liver injury in ethanol-fed rats by antisense oligonucleotides. Biochem Pharmacol 2005; 69:569–577.
68. Liu D, Huang L. Interaction of pH-sensitive liposomes with blood components. J Liposome Res 1994; 4:121–141.

69. Connor J, Huang L. pH-sensitive immunoliposomes: an efficient and target-specific carrier for antitumor drugs. Cancer Res 1986; 46:3431–3435.
70. Liu D, Huang L. pH-sensitive, plasma stable liposomes with relatively prolonged residence in circulation. Biochim Biophys Acta 1990; 1022:348–354.
71. Slepushkin VA, Simões S, Dazin P, et al. Sterically stabilized pH-sensitive liposomes, intracellular delivery of aqueous contents and prolonged circulation in vivo. J Biol Chem 1997; 272:2382–2388.
72. Liu D, Huang L. Small, but not large, unilamellar liposomes composed of dioleoylphosphatidylethanolamine and oleic acid can be stabilized by human plasma. Biochemistry 1989; 28:7700–7707.
73. Collins D, Litzinger DC, Huang L. Structural and functional comparisons of phosphatidylethanolamine and three different diacylsuccinylglycerol. Biochim Biophys Acta 1990; 1025:234–242.
74. Chu CJ, Szoka FC Jr. pH-sensitive liposomes. J Liposome Res 1994; 4:361–395.
75. Blume G, Cevc G. Liposomes for the sustained drug release in vivo. Biochim Biophys Acta 1990; 1029:91–97.
76. Klibanov AL, Maruyama K, Torchilin VP, Huang L. Amphipathic polyethyleneglycols effectively prolong the circulation time of liposomes. FEBS Lett 1990; 268:235–237.
77. Papahadjopoulos D, Allen TM, Gabizon A, et al. Sterically stabilized liposomes: improvements in pharmacokinetics and antitumor therapeutic efficacy. Proc Natl Acad Sci USA 1991; 88:11,460–11,464.
78. Senior J, Delgado C, Fisher D, Tilcock C, Gregoriadis G. Influence of surface hydrophilicity of liposomes on their interaction with plasma protein and clearance from the circulation: studies with poly(ethylene glycol)-coated vesicles. Biochim Biophys Acta 1991; 1062:77–82.
79. Woodle MC, Matthay KK, Newman MS, et al. Versatility in lipid compositions showing prolonged circulation with sterically stabilized liposomes. Biochim Biophys Acta 1992; 1105:193–200.
80. Bakker-Woudenberg IAJM, Lokerse AF, Ten Kate MT, Storm G. Enhanced localization of liposomes with prolonged blood circulation time in infected lung tissue. Biochim Biophys Acta 1992; 1138:318–326.
81. Hong MS, Lim SJ, Oh YK, Kim CK. pH-sensitive, serum-stable and long-circulating liposomes as a new drug delivery system. J Pharm Pharmacol 2002; 54:51–58.
82. Allen TM, Hansen CB, Guo LS. Subcutaneous administration of liposomes: a comparison with the intravenous and intraperitoneal routes of injection. Biochim Biophys Acta 1993; 1150:9–16.
83. Oussoren C, Storm G. Liposomes to target the lymphatics by subcutaneous administration. Adv Drug Deliv Rev 2001; 50:143–156.
84. Pantaleo G, Graziosi C, Demarest JF. HIV infection is active and progressive in lymphoid tissue during the clinically latent stage of disease. Nature 1993; 362:355–358.
85. Pantaleo G, Fauci AS. New concepts in the immunopathogenesis of HIV infection. Annu Rev Immunol 1995; 13:487–412.
86. Felgner PL, Gadek TR, Holm M, et al. Lipofection: a highly efficient, lipid-mediated DNA-transfection procedure. Proc Natl Acad Sci USA 1987; 84:7413–7417.

87. Düzgüneş N, Felgner PL. Intracellular delivery of nucleic acids and transcription factors by cationic liposomes. Methods Enzymol 1993; 221: 303–306.
88. Nabel GJ, Nabel EG, Yang ZY, et al. Direct gene transfer with DNA-liposome complexes in melanoma: expression, biologic activity, and lack of toxicity in humans. Proc Natl Acad Sci USA 1993; 90:11,307–11,311.
89. Gao X, Huang L. Cationic liposome-mediated gene transfer. Gene Ther 1995; 2:710–722.
90. Templeton NS, Lasic DD. New directions in liposome gene delivery. Mol Biotechnol 1999; 11:175–180.
91. Simões S, Pires P, Düzgüneş N, Pedroso de Lima MC. Cationic liposomes as gene transfer vectors: barriers to successful application in gene therapy. Curr Opin Mol Therapeut 1999; 1:147–157.
92. Clark PR, Hersh EM. Cationic lipid-mediated gene transfer: current concepts. Curr Opin Mol Ther 1999; 1:158–176.
93. Düzgüneş N, Simões S, Pires P, Pedroso de Lima MC. Gene delivery by cationic liposome-DNA complexes. In: Dumitriu S, ed. Polymeric Biomaterials. New York: Marcel Dekker, 2002:943–958.
94. Bennett CF, Chiang M-Y, Chan H, Shoemaker JEE, Mirabelli CK. Cationic lipids enhance cellular uptake and activity of phosphorothioate antisense oligonucleotides. Mol Pharmacol 1992; 41:1023–1033.
95. Konopka K, Rossi JJ, Swiderski P, Slepushkin VA, Düzgüneş N. Delivery of anti-HIV-1 ribozyme into HIV-infected cells via cationic liposomes. Biochim Biophys Acta 1998; 1372:55–68.
96. Lappalainen K, Urtti A, Söderling E, Jääskeläinen I, Syryänen K, Syryänen S. Cationic liposomes improve stability and intracellular delivery of antisense oligonucleotides into CaSki cells. Biochim Biophys Acta 1994; 1196: 201–208.
97. Jääskeläinen I, Lappalainen K, Honkakoski P, Urtti A. Requirements for delivery of active antisense oligonucleotides into cells with lipid carriers. Methods Enzymol 2004; 387:210–230.
98. Zelphati O, Szoka FC Jr. Mechanism of oligonucleotide release from cationic liposomes. Proc Natl Acad Sci USA 1996; 93:11,493–11,498.
99. Jääskeläinen I, Sternberg B, Mönkkönen J, Urtti A. Physicochemical and morphological properties of complexes made of cationic liposomes and oligonucleotides. Int J Pharm 1998; 167:191–203.
100. Marcusson EG, Bhat B, Manoharan M, Bennett CF, Dean NM. Phosphorothioate oligodeoxyribonucleotides dissociate from cationic lipids before entering the nucleus. Nucl Acids Res 1998; 26:2016–2023.
101. Levigne C, Lunardi-Iskandar Y, Lebleu B, Thierry AR. Cationic liposomes/lipids for oligonucleotide delivery: application to the inhibition of tumorigenicity of Kaposi's sarcoma by vascular endothelial growth factor antisense oligonucleotides. Methods Enzymol 2004; 387:189–210.
102. Islam A, Handley SL, Thompson KK, Akhtar S. Studies on uptake, subcellular trafficking and efflux of antisense oligodeoxynucleotides in glioma cells using self-assembling cationic lipoplexes as delivery systems. J Drug Target 2000; 7:373–382.

103. Lee C-H, Ni Y-H, Chen C-C, Chou C-K, Chang F-H. Synergistic effect of polyethyleneimine and cationic liposomes in nucleic acid delivery to human cancer cells. Biochim Biophys Acta 2003; 1611:55–62.

104. Noguchi S, Hirashima N, Furuno T, Nakanishi M. Remakable induction of apoptosis in cancer cells by a novel cationic liposome complexed with a bcl-2 antisense oligonucleotide. J Control Rel 2003; 88:313–320.

105. Waelti ER, Glück R. Delivery to cancer cells of antisense L-myc oligonucleotides incorporated in fusogenic, cationic-lipid-reconstituted influenza-virus envelopes (cationic virosomes). Int J Cancer 1998; 77:728–733.

106. Dosaka-Akita H, Akie K, Hiroumi H, Kinoshita I, Kawakami Y, Murakami A. Inhibition of proliferation by L-myc antisense DNA for the translational initiation site in human small cell lung cancer. Cancer Res 1995; 55: 1559–1564.

107. Morishita R, Gibbons GH, Kaneda Y, Ogihara T, Dzau VJ. Pharmacokinetics of antisense oligodeoxyribonucleotides (cyclin B1 and CDC 2 kinase) in the vessel wall in vivo: enhanced therapeutic utility for restenosis by HVJ-liposome delivery. Gene 1994; 149:13–19.

108. Morishita R, Gibbons GH, Kaneda Y, Zhang L, Ogihara T, Dzau VJ. Apolipoprotein E-deficient mice created by systemic administration of antisense oligodeoxynucleotides: a new model for lipoprotein metabolism studies. J Endocrinol 2002; 175:475–485.

109. Mizuta T, Fujiwara M, Hatta T, et al. Antisense oligonucleotides directed against the viral RNA polymerase gene enhance survival of mice infected with influenza A. Nat Biotechnol 1999; 17:583–587.

110. Maus U, Rosseau S, Mandrakas N, et al. Cationic lipids employed for antisense oligodeoxynucleotide transport may inhibit vascular cell adhesion molecule-1 expression in human endothelial cells: a word of caution. Antisense Nucl Acid Drug Dev 1999; 9:71–80.

111. Konopka K, Pretzer E, Felgner PL, Düzgüneş N. Human immunodeficiency virus type-1 (HIV-1) infection increases the sensitivity of macrophages and THP-1 cells to cytotoxicity by cationic liposomes. Biochim Biophys Acta 1996; 1312:186–196.

112. Kanamaru T, Takagi T, Takakura Y, Hashida M. Biological effects and cellular uptake of c-myc antisense oligonucleotides and their cationic liposome complexes. J Drug Target 1998; 5:235–246.

113. Litzinger DC, Brown JM, Wala I, et al. Fate of cationic liposomes and their complex with oligonucleotide in vivo. Biochim Biophys Acta 1996; 1281: 139–149.

114. Bennett CF, Zuckerman JE, Kornbrust D, Sasmor H, Leeds JM, Crooke ST. Pharmacokinetics in mice of a [^{3}H]-labeled phosphorothioate oligonucleotide formulated in the presence and absence of a cationic lipid. J Control Rel 1996; 41:121–130.

115. Stuart DD, Semple SC, Allen TM. High efficiency entrapment of antisense oligonucleotides in liposomes. Methods Enzymol 2004; 387:171–188.

116. Stuart DD, Allen TM. A new liposomal formulation for antisense oligonucleotides with small size, high incorporation efficiency and good stability. Biochim Biophys Acta 2000; 1463:219–229.

117. Meyer O, Kirpotin D, Hong K, et al. Cationic liposomes coated with polyethylene glycol as carriers for oligonucleotides. J Biol Chem 1998; 273: 15,621–15,627.
118. Maurer N, Wong KF, Stark H, et al. Spontaneous entrapment of polynucleotides upon electrostatic interaction with ethanol-destabilized cationic liposomes. Biophys J 2001; 80:2310–2326.
119. Pagnan G, Stuart DD, Pastorino F, et al. Delivery of c-myb antisense oligodeoxynucleotides to human neuroblastoma cells via disialoganglioside GD_2-targeted immunoliposomes: antitumor effects. J Natl Cancer Inst 2000; 92:253–261.
120. Brignole C, Marimpietri D, Pagnan G, et al. Neuroblastoma targeting by c-myb-selective antisense oligonucleotides entrapped in anti-GD_2 immunoliposome: immune cell-mediated anti-tumor activities. Cancer Lett 2005; 228:181–186. (PMID: 15936140).
121. Pastorino F, Brignole C, Marimpietri D, et al. Targeted liposomal c-myc antisense oligodeoxynucleotides induce apoptosis and inhibit tumor growth and metastases in human melanoma models. Clin Cancer Res 2003; 9:4595–4605.
122. Stuart DD, Kao GY, Allen TM. A novel, long-circulating, and functional liposomal formulation of antisense oligodeoxynucleotides targeted against MDR1. Cancer Gene Ther 2000; 7:466–475.
123. Bartsch M, Weeke-Klimp AH, Hoenselaar EP, et al. Stabilized lipid coated lipoplexes for the delivery of antisense oligonucleotides to liver endothelial cells in vitro and in vivo. J Drug Target 2004; 12:613–621.
124. Pedroso de Lima MC, Simões S, Pires P, Faneca H, Düzgüneş N. Cationic lipid-DNA complexes in gene delivery: from biophysics to biological applications. Adv Drug Deliv Rev 2001; 47:277–294.
125. Shi F, Wasungu L, Nomden A, et al. Interference of poly(ethylene glycol)-lipid analogues with cationic lipid-mediated delivery of oligonucleotides: role of lipid exchangeability and non-lamellar transitions. Biochem J 2002; 366: 333–341.
126. Wheeler JJ, Palmer L, Ossanlou M, et al. Stabilized plasmid-lipid particles: construction and characterization. Gene Ther 1999; 6:271–281.
127. Hu Q, Shew CR, Bally MB, Madden TD. Programmable fusogenic vesicles for intracellular delivery of antisense oligodeoxynucleotides: enhanced cellular uptake and biological effects. Biochim Biophys Acta 2001; 1514:1–13.
128. Rodriguez M, Coma S, Noe V, Ciudad CJ. Development and effects of immunoliposomes carrying an antisense oligonucleotide against DHFR RNA and directed toward human breast cancer cells overexpressing HER2. Antisense Nucl Acid Drug Dev 2002; 12:311–325.
129. Rait ASA, Pirollo KF, Ulick D, Cullen K, Chang EH. HER-2-targeted antisense oligonucleotides results in sensitization of head and neck cancer cells to chemotherapeutic agents. Ann NY Acad Sci 2003; 1002:78–89.

15

Liposomes for Intracavitary and Intratumoral Drug Delivery

William T. Phillips and Beth A. Goins

Department of Radiology, The University of Texas Health Science Center at San Antonio, San Antonio, Texas, U.S.A.

Ande Bao

Department of Radiology and Otolaryngology, University of Texas Health Science Center at San Antonio, San Antonio, Texas, U.S.A.

Luis A. Medina

Institute of Physics, Universidad Nacional Autonoma de Mexico, Mexico City, Mexico

INTRODUCTION

The local application of liposomes into body cavities or directly into body tissues such as solid tumors has many potential therapeutic advantages. These advantages can exceed the well-known improvement in distribution and pharmacokinetics observed for encapsulated agents delivered intravenously (1). The pharmacokinetic behavior of a liposome encapsulated-therapeutic agent that has been injected locally is greatly altered in comparison to the intracavitary administration of the same agent as an unencapsulated free drug. For example, when free drugs are injected into a body cavity, they are generally cleared very rapidly from that cavity by direct absorption through membranes that line the cavity (2). This situation is very different in comparison with administration of the same drug encapsulated within a liposome. The liposome-encapsulated drug is prevented from passing through the lining of the cavity and must be cleared from the cavity by passage through the lymphatic system (3). This lymphatic clearance results in a prolonged retention of the therapeutic agent in the body cavity, increasing the possibility of achieving higher, more sustained drug levels in targeted tissue as the liposomes slowly degrade. By attaching ligands to the surface of liposomes administered intracavitarily, they can be targeted to specific cells or structures located within the cavity.

Similar advantages are also obtained when liposomes are injected directly into targeted tissue. Liposomes injected directly into tissues are well retained locally, however this retention is not absolutely fixed, thus providing the possibility of local diffusion within the tissue. This local diffusion can prove advantageous compared to a free unencapsulated agent that is often rapidly absorbed directly through blood capillaries at the site of injection. Following direct injection into tissue, a portion of the injected liposomes are retained for a very long time in the local region where they slowly degrade and release therapeutic agents at high concentrations. Another portion of the directly injected liposomes are cleared from the local tissue by the lymphatic system, providing the opportunity to deliver therapeutic agents to the lymphatic vessels and lymph nodes, which are frequently affected by the same disease process that affects the local tissue. These draining lymph nodes are often difficult targets for intravenously injected free drugs. An example of this is illustrated by the situation in which solid tumors frequently metastasize and spread tumor cells to the lymph nodes that receive lymphatic drainage from that tumor. It is these lymph nodes where cancer tends to reoccur after the primary solid tumor has been surgically removed.

Local applications of liposomes include direct injection into tissues and injection into intracavitary sites whose fluid is cleared through the lymphatics. Many different local applications of liposomes have been investigated and are too numerous to cover adequately in one review. This chapter will review local applications of liposomes with particular emphasis on recent developments in this field. For the purposes of limiting the scope of this review, this chapter will focus on the use of liposomes administered through two different intracavitary routes, intraperitoneal and intrapleural, and the direct local administration of liposomes into solid tumors for the local treatment of cancer.

INTRACAVITARY ADMINISTRATION OF LIPOSOME ENCAPSULATED THERAPEUTIC AGENTS

Intracavitary sites that have been investigated as potential applications for liposome drug delivery include the pleural space surrounding the lungs (4,5), the peritoneal space surrounding the intestines (3,6), the articular cavity of the joints (7,8), and the central spinal fluid surrounding the brain

(9–12). Local administration of liposomes into the central spinal fluid has been investigated for the purposes of treating neoplastic lymphomatous meningitis, cerebral ischemia, gene transfection of the brain, and induction of prolonged-analgesia (9–12). Liposomes that are injected directly into body cavities appear to have minimal retention at the focal site of their injection as they disperse freely throughout the whole cavity (6). These intra-cavitarily administered liposomes either become associated with targeted tissue in the cavity or drain into lymphatic vessels where they can be trapped in lymph nodes that may contain the same disease process that affects the cavity. A portion of the liposome-encapsulated drug that clears from the cavity is able to pass through the lymphatics, return to the blood circulation, and circulate as if it had been injected intravenously. These liposomes returning to the circulation still have a chance to accumulate in pathologic tissue through either a targeted receptor mechanism or nonspecifically by the enhanced permeability and retention mechanism (13).

Intraperitoneal Drug Delivery

There has been a long-standing interest in the local delivery of pharmaceutical and biologic agents into the peritoneum for the treatment of peritoneal diseases (14–16). Many disease processes spread by dissemination through the peritoneum. For instance, dissemination of cancer cells throughout the peritoneum is a very common manifestation of ovarian and gastric cancer (17). When the cancer cells spread throughout the peritoneum, they are frequently trapped in lymph nodes that receive peritoneal fluid drainage (18).

Cancers that originate primarily in the peritoneum also have a high incidence of lymph node metastasis. The incidence of pelvic and para-aortic lymph node metastases was similar among women with two different types of primary peritoneal cancer, primary peritoneal carcinoma and peritoneal serous papillary carcinoma (72.7% vs. 66.6%, $p = 0.701$, 72.7% vs. 48.1%, $p = 0.172$, respectively) (19). Based on this frequent occurrence of lymph node metastasis in these primary peritoneal cancers, investigators have suggested that pelvic and para-aortic lymphadenectomy should be considered among women with these primary peritoneal cancers in whom the primary tumor can be optimally cytoreduced (20).

The basic goal of intraperitoneal drug administration is to increase the local drug concentration and the duration of drug exposure to a peritoneal disease process while decreasing systemic drug toxicity. Although several recent investigations have examined the efficacy of intraperitoneally delivered antibiotics, therapeutic radionuclides and genes (21–23), the majority of the research in this area has been conducted with intraperitoneally administered chemotherapeutic agents for the treatment of peritoneal carcinomatosis and ovarian cancer (24). Intraperitoneal drug delivery is currently considered a viable approach for the treatment of ovarian cancer (24–26).

Studies in which free drugs are administered into the peritoneum have shown survival benefits in ovarian cancer patients (24). Although most intraperitoneally delivered unencapsulated free drugs are rapidly cleared from the peritoneal fluid without entering the lymphatic system, direct intraperitoneal administration of drugs can achieve much higher peak concentrations in the peritoneal fluid compared to the same drug administered intravenously (20-fold higher for cisplatin and carboplatin to as high as 1000-fold for taxol) (24–26). Although these drug levels quickly equilibrate with plasma after termination of the peritoneal infusion (27), transiently elevated peritoneal drug levels provide a significant therapeutic advantage. These elevated drug levels have led many investigators to be enthusiastic about intraperitoneal drug administration for treatment of ovarian cancer (24,28). Unfortunately, rapid clearance of these free drugs from the peritoneum diminishes the advantages derived from the intraperitoneal infusion procedure. Studies with chemotherapeutic agents administered intraperitoneally have yielded encouraging results for the treatment of ovarian cancer (2,25). A recent consensus statement from specialists in the field of ovarian cancer recommends that intraperitoneal therapy with chemotherapy and/or biological agents be pursued as a legitimate area of research (25,29). Intraperitoneal chemotherapy has also been suggested as a future direction for ovarian cancer research (30).

The normal pathway of drug clearance from the peritoneum is either through direct absorption across the peritoneal membrane or by drainage into the lymphatic system through absorption by the diaphragmatic stomata (31). These diaphragmatic stomata are fairly large. Studies have shown that these large stoma can absorb red blood cells from the peritoneal fluid (32,33). Most intraperitoneally delivered drugs are rapidly cleared from the peritoneal fluid. In a clinical study of free cisplatin administered by continuous hyperthermic peritoneal perfusion, only 27% of the administered cisplatin remained in the peritoneal fluid at the end of a 90-minute infusion (27). Most of the cisplatin dose rapidly entered the systemic circulation by direct absorption through the peritoneal membrane.

Intraperitoneal Liposome Administration for Cancer Therapy

One approach to prolong the retention of intraperitoneally administered drugs is to encapsulate the drug within a liposome (34–36). Administration of liposomes encapsulating therapeutic agents directly into the peritoneum increases and prolongs the concentration of drug in the peritoneum. Because liposomes are cleared from the peritoneum through lymphatic drainage, the delivery of therapeutic agents to the lymph nodes that filter lymph fluid that drains from the peritoneum can also be greatly increased. This lymphatic drainage is commonly taken as the passage of cancer cell invasion and these lymph nodes are one of the most common types of tumor metastasis.

The encapsulation of drugs in liposomes for intraperitoneal administration has several potential advantages. First, direct local toxicity of the chemotherapeutic agent may be attenuated because of encapsulation of the drug inside the protective lipid bilayer of the liposome. The dose-limiting toxicity of many intraperitoneally administered drugs is due to abdominal pain from direct peritoneal irritation (2). Liposome encapsulation of drugs has already been shown to have reduced local toxicity compared to free drug following accidental injection directly into tissue (37).

Second, the encapsulated drug is blocked from rapid direct absorption through the peritoneal lining, resulting in increased time for the liposome-encapsulated drug to reach tumor cells, while the encapsulated drug is cleared through the lymphatics. Many studies have clearly demonstrated that the pharmacokinetics of liposome-encapsulated drugs administered intraperitoneally are very different from the same nonencapsulated drug administered intraperitoneally (34,36,38). Slow removal of liposomes from the peritoneal cavity appears to provide a sustained release of drugs from the liposomes into the peritoneal cavity. In one study in which liposomes encapsulating the drug cefoxitin were administered intraperitoneally, the release of cefoxitin from the liposome complex was estimated to be well in excess of the maximum inhibitory concentration (21). Similar findings were also described in a model of peritoneally disseminated cancer in which intraperitoneally administered doxorubicin encapsulated in liposomes was considered to be slowly released in the abdominal cavity from gradually degrading liposomes (38).

Liposome-encapsulated agents are cleared from the peritoneum by movement through the diaphragmatic stomata into the lymphatic vessels and then into the blood. Ellens et al. (39) reported that 19% of the intraperitoneally administered liposomes were detected in the blood at two hours, with 7% in the liver and 4% in the spleen, indicating fairly rapid clearance of liposomes from the peritoneal cavity through the lymphatics into the systemic circulation. This clearance is much slower than most free drugs that are directly absorbed through the peritoneal lining. In another study of intraperitoneally administered liposomes, 30% of the liposomes reached the liver by six hours (36). Allen et al. has also demonstrated that liposomes labeled with iodine-125 (^{125}I) eventually had a tissue distribution that was equivalent to that of intravenously injected liposomes (40). Medina et al. also demonstrated that intraperitoneally administered liposomes had a distribution similar to an intravenous administration with only minimal retention in the mediastinal lymph nodes (0.6% ID) (41).

A third advantage of intraperitoneally administered liposomes is that increased abdominal and mediastinal lymph node targeting is possible because liposome-encapsulated drugs are cleared through the lymphatic vessels with at least a portion of the administered drug being deposited in the lymph nodes, where it degrades and is slowly released from the liposomes in high concentration (3,35).

Effect of Liposome Size on Peritoneal Retention

The effect of liposome size has been evaluated as a method to increase the retention of liposome-encapsulated drugs within the peritoneum. It appears that making liposomes larger does not increase the retention of liposome-encapsulated drugs in the peritoneum or within lymph nodes that receive drainage from the peritoneum. Hirono and Hunt have performed a detailed study on the effect of liposome size ranging from 48 to 720 nm on subsequent distribution after intraperitoneal administration (3). In their studies, 50% to 60% of the intraperitoneal dose of liposomes of varying sizes encapsulating carbon-14 (^{14}C) labeled-sucrose cleared from the peritoneum by five hours in all liposomes studied. The greatest amount of ^{14}C-sucrose (~40%) appeared in the urine after administration of the largest 720-nm liposomes. The authors speculated that the large 460- and 720-nm liposomes were unstable in the peritoneum so that they rapidly released their encapsulated ^{14}C-sucrose.

It appears that simply increasing the size of the liposomes, in and of itself, is not sufficient to result in increased peritoneal and lymph node retention because particles as large as erythrocytes have been demonstrated to readily drain from the peritoneum by passing through the diaphragmatic stomata and into the bloodstream. When chromium-51 labeled red blood cells were injected into the peritoneal cavity of sheep, 80% of the chromium-51 labeled red blood cells were returned to blood circulation by six hours after administration (32).

Intraperitoneal Delivery of Liposome-Encapsulated
Therapeutic Agents

In the last decade, a moderate amount of research has been performed to investigate intraperitoneally administered liposome-encapsulated anticancer chemotherapeutic agents (42–44). These studies have demonstrated prolonged retention time of the liposome-encapsulated chemotherapeutic agents in the peritoneum. These studies support the hypothesis that there is a marked pharmacological advantage for the treatment of intraperitoneal malignancies by encapsulating the intraperitoneally administered chemotherapeutic agent in a liposome (42,44). These studies demonstrated a prolongation of the mean retention time of liposome-encapsulated agents in the peritoneum following intraperitoneal administration without compromising the systemic distribution of the drug. The investigators suggested that prolonged retention in the peritoneum might result in a significant enhancement of the therapeutic efficacy of the liposome drug against malignancies confined to the peritoneal cavity compared to the intraperitoneal administration of nonencapsulated drug (44).

Other studies demonstrate an improved toxicity profile. For instance, encapsulation of paclitaxel in a liposome that is administered intraperitoneally has been shown to decrease toxicity while retaining equal efficacy for the treatment of intraperitoneal P388 leukemia (43). It is likely that the reduced toxicity results from decreased local toxicity of encapsulated-paclitaxel compared to the free drug. In humans, the dose limiting toxicity from intraperitoneal administration of paclitaxel was severe abdominal pain, which was thought to be due to direct toxicity from either the paclitaxel or the ethanol/polyethoxylated castor oil delivery vehicle (16).

Intraperitoneal Delivery of Liposomes for Gene Transfection

Intraperitoneal delivery has also shown promise for liposome gene transfection with novel cationic lipid containing liposomes (45). These cationic liposomes contained luciferase and beta-galactosidase genes that served as reporter genes. Intraperitoneally administered liposomal gene delivery for peritoneal disseminated ovarian cancer in nude mice was performed using a stable chloramphenicol acetyl transferase (CAT)-expressing ovarian cancer cell line (OV-CA-2774/CAT), which permitted quantification of the exact tumor burden in various organs. Intraperitoneal gene delivery to these disseminated ovarian cancer cells was excellent with gene transfection appearing to be specific to intraperitoneal ovarian cancer cells. The 3 beta [L-ornithinamide-carbamoyl]-cholesterol (O-Chol):DNA lipoplex appeared to offer potential advantages over other commercial transfection reagents due to its high level of gene expression in vivo; its reduced susceptibility to serum inhibition; and its highly selective transfection into tumor cells. These results suggest that the O-Chol:DNA lipoplex is a promising tool for intraperitoneal gene therapy for patients with peritoneal disseminated ovarian cancer (45).

Avidin/Biotin-Liposome System for Intraperitoneal and Lymph-Node Drug Delivery

Few of the above previously described studies with intraperitoneally administered liposomes have focused on the fact that intraperitoneally administered liposomes clear from the peritoneum by passing through the lymphatic vessels that provide an opportunity to deliver therapeutic agents to these lymph nodes. The liposomes pass through and are partially trapped in lesser or greater degrees by the lymph nodes that drain from the peritoneum. These lymph nodes frequently contain cancer metastasis because intraperitoneally disseminated cancer cells follow the same pathways of lymphatic fluid clearance.

Although intraperitoneally administered conventional liposomes are more slowly cleared from the peritoneal fluid than free drug, the clearance of liposomes from the peritoneum through the lymphatic system still remains fairly rapid (peritoneal clearance half-life of one to six hours) (34,38,41) and the retention of liposomes in individual lymph nodes receiving lymph fluid draining from the peritoneum is relatively low ($<1\%$ ID per lymph node) (34). This low lymph-node retention occurs because the majority of intraperitoneally administered liposomes return to the systemic circulation by passing through abdominal and mediastinal lymph nodes. This minimal retention of conventional liposomes in lymph nodes has been previously described for liposomes administered subcutaneously (46,47).

Our group has developed an avidin/biotin-liposome system to increase the retention of intraperitoneally administered liposome encapsulated drugs within the peritoneum and the lymph nodes that receive drainage of peritoneal lymphatic fluid (6). This lymph-node targeting method utilizes the high-affinity ligands, biotin and avidin. Biotin is a naturally occurring cofactor and avidin is a protein derived from eggs. Avidin and biotin have an extremely high affinity for each other. Avidin has four biotin-binding sites. These four receptor sites permit the binding of multiple biotin molecules that cause aggregation of liposomes that have biotin on their surface. This system is described in more detail in Volume III, Chapter 13, which is specifically dedicated to lymph-node delivery following subcutaneous injection.

This system has potential as a delivery system for the local treatment of intraperitoneal and intralymphatic disease processes by greatly increasing the retention of drugs in the peritoneum and in the lymph nodes that receive lymphatic drainage from the peritoneum. When liposomes that have biotin attached to their surfaces are administered into the peritoneum and followed with administration of avidin, the retention of liposomes in the peritoneum is greatly increased.

The interaction of biotin-liposomes with avidin apparently results in aggregation of the liposomes within the peritoneum. This aggregation greatly alters the distribution of liposomes and results in a greatly prolonged retention of liposomes in the peritoneum as well as an increased accumulation and retention of liposomes in lymph nodes receiving drainage from the peritoneum.

Methodology

Preparation of biotin-liposomes containing blue dye: It is very useful for tracking liposomes after intracavitary injection to encapsulate blue dye for visual identification. This method supplements the ability to label liposomes with technetium-99m (^{99m}Tc) for noninvasive imaging. The liposomes used for this purpose are comprised of a 50.5:45:2.5:2 molar ratio (total lipid) of distearoyl phosphatidylcholine (DSPC):cholesterol:N-biotinoyl distearoyl phosphoethanolamine:α-tocopherol. Liposomes are prepared in a laminar flow hood using aseptic conditions as previously described (48). A dried film of lipid ingredients in chloroform is formed by rotary evaporation and vacuum desiccation for at least four hours. The dried lipid film is rehydrated in 300 mM sucrose in sterile water and lyophilized overnight. The resultant lyophilized powder is then rehydrated with 200 mM reduced glutathione (GSH) and 10 mg/mL patent blue dye in Dulbecco's phosphate buffered saline (PBS), pH 6.3, at a final total lipid concentration of 120 μmol/mL. Immediately before extrusion, the lipid suspension is diluted to 40 μmol/mL with 100 mM GSH and 10 mg/mL blue dye in PBS (pH 6.3) containing 150 mM sucrose, and extruded through a series (2 μ, two passes; 400 nm, two passes; 100 nm, five passes) of polycarbonate filters at 55°C. Extruded liposomes are washed three times in PBS, pH 6.3, containing 75 mM sucrose and centrifuged at 45,000 rpm for 45 minutes in an ultracentrifuge to remove any unencapsulated sucrose, GSH and blue dye. The final liposome pellet is reconstituted in 300 mM sucrose/PBS to a total lipid concentration of approximately 60 μmol/mL, and stored at 4°C until needed.

Labeling of biotin-liposomes containing blue dye with technetium-99m: Liposomes are labeled with ^{99m}Tc as previously described (49). A commercial kit of the lipophilic chelator, hexamethylpropyleneamine oxime (HMPAO), is reconstituted with 5 mL of saline containing 370 MBq of ^{99m}Tc-pertechnetate. An aliquot (1 mL) of ^{99m}Tc-HMPAO is added to a concentrated suspension of liposomes encapsulating GSH and blue dye, and incubated at room temperature for 30 minutes. Labeling efficiencies are determined from the ^{99m}Tc activity associated with the ^{99m}Tc-liposomes before and after Sephadex G-25 column separation with a dose calibrator. For three separate labeling experiments, the labeling efficiency was 92.8% ± 2.4%.

Study with Intraperitoneal Avidin/Biotin-Liposomes

The biodistribution of biotin-liposomes was compared with that of rats receiving an intraperitoneal injection of ^{99m}Tc-labeled biotin-liposomes and avidin with rats that received the ^{99m}Tc-labeled biotin-liposomes alone. The rats that received intraperitoneal avidin in addition to the ^{99m}Tc-biotin-liposomes had only a minimal percentage of the injected dose (% ID) of liposomes that reached the systemic circulation by 24 hours and a low % ID was found in the spleen, blood and liver at 24 hours (<9% ID combined). In contrast, control animals, administered only the biotin-liposomes without the avidin, had 23% ID in the spleen, 14% ID in the blood, and 9.8% ID in the liver. Figure 1 demonstrates the very different distribution of the ^{99m}Tc-biotin-liposomes in a rat that received the avidin compared with a control rat that received only the ^{99m}Tc-biotin liposomes. The observations from this study suggest that the avidin/biotin-liposome methodology would enhance the reservoir-like effect previously observed for standard liposome formulations by blocking rapid lymphatic transit of liposomes from the peritoneum to the systemic circulation. The interaction of biotin-liposomes with avidin apparently results in aggregation of the liposomes in the peritoneum. This aggregation greatly alters the distribution

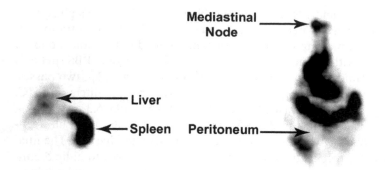

Figure 1 Scintigraphic images obtained 24 hours after intraperitoneal administration of ^{99m}Tc-biotin liposomes in a control rat (*left panel*) are compared with those of a rat that received ^{99m}Tc-biotin liposomes followed by administration of avidin (*right panel*). The biodistribution of the ^{99m}Tc-biotin liposomes is very different in the rat that received the avidin. This rat has significant retention of liposomes in the peritoneal space and in abdominal and mediastinal lymph nodes. The control rat has significant uptake in liver and spleen, a distribution of liposomes similar to that observed for intravenously delivered liposomes after 24 hours.

of liposomes and appears to result in a greatly prolonged retention of liposomes in the peritoneum as well as an increased accumulation and retention of liposomes in lymph nodes receiving drainage from the peritoneum. The lymph nodes in the abdomen and in the mediastinum in the rats that received the avidin also had greatly increased uptake of the biotin-liposomes.

The liposome biodistribution in control animals that did not receive the avidin was similar to previous reports with standard liposome formulations that were administered intraperitoneally (36,39–41). This retention of liposomes in the rats receiving the avidin should result in increased release of a liposome-encapsulated drug in the peritoneal fluid and in the lymph nodes receiving lymphatic drainage from the peritoneum. Delivery of liposome-encapsulated drugs using this method should provide sustained local release of drug within the peritoneum and the lymph nodes draining the peritoneum as the liposomes degrade or become phagocytosed by macrophages. This delivery system could also attenuate systemic drug toxicities by greatly reducing the movement of drug into the systemic circulation by either passage through the lymphatic vessels and lymph nodes, or through direct absorption through the peritoneal membrane.

The potential for treatment of micrometastasis in lymph nodes secondary to lymphatic dissemination with this avidin/biotin liposome method is also great. For example, liposome retention in mediastinal lymph nodes as demonstrated in this study could be efficacious in ovarian cancer therapy as metastasis to mediastinal and other lymph nodes are not uncommon findings in ovarian cancer at autopsy (18).

The investigations with intraperitoneal avidin/biotin liposomes have been in normal animals. Further studies need to be carried out in models of intraperitoneal cancer metastasis using delivery of therapeutic agents with this avidin/biotin-liposome methodology. In summary, the intraperitoneal avidin/biotin-liposome delivery method described has potential as a delivery system for the local treatment of intraperitoneal and intralymphatic disease processes by increasing the retention of drugs in the peritoneum and in the lymph nodes that receive lymphatic drainage from the peritoneum.

Potential Use of Liposome Therapeutic Agents for Prophylaxis Against Recurrent Peritoneal Cancer

An important potential application of the intraperitoneal delivery of liposomes that carry anticancer agents is in the prophylaxis of peritoneal carcinomatosis. Because 50% of patients with malignant gastrointestinal or gynecological diseases experience peritoneal carcinomatosis shortly after local curative resection, there is a great interest in delivering intraperitoneal chemotherapy during the perioperative period (2,50). One study found that the intraperitoneal administration of the chemotherapeutic agents, cisplatin and mitomycin, prevented perioperative peritoneal carcinomatosis in a rat model (51,52). The rats receiving cisplatin did, however, experience severe, local toxicity with bleeding into the peritoneum and toxic necrotic reactions of the colon.

Liposomes encapsulating anticancer agents could potentially be used for this type of perioperative prophylactic chemotherapy. The potential for treatment of micrometastasis in lymph nodes is also great.

Intrapleural Administration of Liposomes for Cancer Therapy

The pleural space is the region between the mesothelium of the parietal pleura, which surrounds and covers the inner surface of the thoracic cage, mediastinum, and diaphragm; and the visceral pleura, which covers the entire surface of the lung. Openings between mesothelial cells in the diaphragm and in the dorso-caudal part of the thorax-called stomata are the exit points for pleural liquid, protein, and cells that are removed from the pleural space (53,54). The stomata communicates directly with lymphatic lacunae, which drain into lymphatic channels that finally drain into the mediastinal nodes (53,55). Some of these lymphatic channels move through the diaphragm and share the same lymphatic vessels as the lymphatic fluid that is removed from the diaphragm.

Very few studies have examined the potential of the intrapleural route for the administration of liposome encapsulated therapeutic agents for any type of therapy. In one pioneering study, Perez-Soler et al. investigated the intrapleural administration of a liposome-entrapped chemotherapeutic platinum compound in patients with malignant pleural effusions secondary to lung cancer, malignant pleural mesothelioma and ovarian cancer (5). In this study, a lipophilic noncross-resistant platinum compound formulated in large multilamellar liposomes (1–3 μm in diameter) was administered intrapleurally into patients with free flowing malignant pleural effusions. Twentyone patients were treated with escalating doses of this liposome-encapsulated platinum compound by intrapleural administration over 30 minutes every 21 days. Considering the very poor prognosis of this disease, the results were very promising. In one of these patients with malignant pleural mesothelioma, the pleural effusion disappeared without evidence of recurrence for 19 months, and in six patients (three adenocarcinoma of the lung, two with malignant pleural mesothelioma, and one ovarian carcinoma), the pleural effusion was reduced by >50% for 5+, 10+, 18+, 8, 5+, and 2+ months, respectively. Plasma pharmacokinetic studies showed that the absorption of this liposomal platinum compound from the pleural cavity was rapid during the first two hours, with levels becoming steady or increasing slowly between 6 and 24 hours after administration. The maximum tolerated dose of the intrapleural liposomal platinum was 50% higher than the maximum tolerated dose after IV administration. The absorption of liposomal platinum into the systemic circulation was much slower than that of the free cisplatin compound. The therapeutic advantages of intrapleural liposome administration compared to the intravenous administration of the same liposomal formulation and the free drug included the following: (i) a favorable depot effect, (ii) lack of systemic toxicity, and (iii) control of the pleural effusion in three of five patients with malignant pleural effusion, a disease similar to ovarian carcinoma in that it tends to remain confined to a body cavity. In initial studies, myelosuppresion was the dose-limiting toxicity that probably indicated that the liposomal agent returned fairly rapidly from the pleural space back into the circulation through the lymphatic system. In subsequent studies, when the dose was kept below the maximum tolerated value, no myelosuppression was observed.

Avidin/Biotin Liposomes Method For Intrapleural and Mediastinal Lymph-Node Drug Delivery

Our group has investigated the avidin/biotin-liposome system as a method to prolong drug delivery and increase liposomal drug retention in the pleural cavity and the mediastinal nodes that receive drainage from this cavity using a rat model (4,41). The avidin/biotin-liposome system can greatly prolong the retention of liposomes in the pleural space and also greatly increase liposome trapping in the mediastinal nodes (4). Mediastinal nodes are important therapeutic targets. Mediastinal nodes are involved as centers of incubation and dissemination in several diseases including lung cancer, tuberculosis, and anthrax (56–58). Treatment and control of these diseases is hard to accomplish because of the limited access of drugs to mediastinal nodes using common pathways of drug delivery. Also, the anatomical location of mediastinal nodes represents a difficult target for external beam irradiation due to its close proximity to major vessels and the heart.

Methodology

^{99m}Tc-biotin liposomes containing blue dye were prepared as previously described for the intraperitoneal studies in Section on "Preparation of Biotin-Liposomes Containing Blue Dye." These liposomes were injected into the pleural space using the following technique. Anesthetized rats were shaved in the lateral left chest. An incision of approximately 8 mm was made through the skin, then the fascia was dissected away and a small incision was made through the external oblique muscle layer, the latissimus dorsi and the serratus layers. Using fine scissors a nick was made in the intercostal layers. The intercostal layers were punctured using a flat tipped needle stub (19 gauge ~ 4.5 mm in length). To confirm penetration and to prevent damage to the underlying lungs, a 1 mL tuberculin syringe was fitted to the 19 gauge luer hub and 0.1 mL of air was injected into the pleural space. When successfully placed, the air will enter the pleural space without resistance. If resistance was encountered, the 19 gauge stub was removed and reintroduced again. The material was then injected using a flat tipped 23 gauge needle stub ($\sim$20 mm in length) inserted through the 19 gauge needle stub.

Study with Intrapleural Avidin/Biotin-Liposomes

Studies were performed by injecting ^{99m}Tc-biotin-liposomes containing blue dye into the pleural space followed two hours later by an injection of avidin. This approach was the reverse of the sequence used with the intraperitoneal studies in which avidin was injected after the biotin-liposomes (6,48).

By 22 hours after injection, good retention (15.7% ID/mediastinal nodes; 515 % ID/g) of liposomes was achieved in the mediastinal nodes with the avidin/biotin-liposome system. The scintigraphic images that visually demonstrate the mediastinal node uptake are shown in Figure 2. The images demonstrate the high uptake of liposomes in the mediastinal nodes. In the absence of avidin, liposomes were minimally retained in the nodes (<1.0% ID/organ; 36% ID/g). The specific targeting of a liposome-encapsulated drug to mediastinal lymph nodes could result in a prolonged targeted sustained depot-like delivery of high drug concentrations to these nodes while the liposomes are slowly degraded and metabolized by phagocytic cells located within these nodes. In the study by Medina et al., evidence of prolonged retention and sustained release of liposome-encapsulated agent in the mediastinal lymph nodes is provided by the continued blue staining of lymph nodes for 22 hours (4). The very high retention of liposomes in

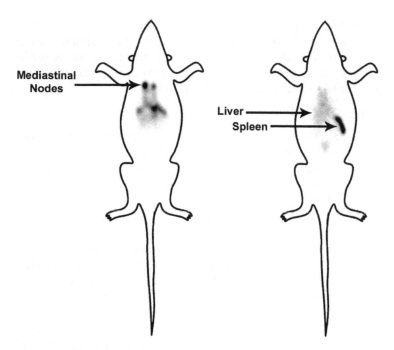

Figure 2 Scintigraphic images at 22 hours following intrapleural administration of ^{99m}Tc-biotin liposomes. The rat in the left panel received intrapleural avidin two hours before administration of the liposomes while the rat in the right panel did not receive avidin. Note the high accumulation of liposomes in the mediastinal nodes in the rat the received the prior injection of avidin. The control rat had a biodistribution of liposomes that resembled an intravenous administration.

the pleural space and in the mediastinal lymph nodes suggests that this delivery methodology could be used for treatment of disease processes that involve this space or affect the mediastinal lymph nodes. Future experiments using intrapleural injection of the avidin/biotin-liposome system to target drugs to mediastinal nodes should be pursued.

Diaphragmatic Targeting with Avidin/Biotin-Liposome System

It was serendipitously discovered that when ^{99m}Tc-biotin-liposomes containing blue dye were injected into the peritoneal cavity and avidin was simultaneously injected into the pleural cavity surrounding the lungs, the liposomes aggregated strongly in the diaphragm as well as in the mediastinal nodes. This accumulation in the diaphragm occurred when the avidin draining from the pleural space into the diaphragmatic lymphatics encountered the biotin-liposomes draining from the peritoneal space causing the liposomes to aggregate within the lymphatic vessels of the diaphragm. A scintigraphic

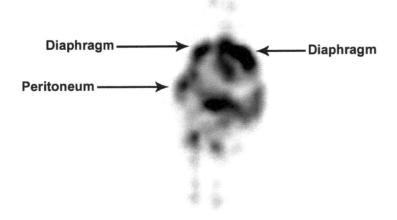

Figure 3 Scintigraphic image demonstrating the high diaphragm uptake in rats that received an injection of avidin in the pleural space and ^{99m}Tc-biotin-liposomes in the peritoneal space.

image of this diaphragm and mediastinal node accumulation is shown in Figure 3. The scintigraphic image shows the intense activity of linear uptake in the region of the diaphragm as well as uptake in the mediastinal nodes. At necropsy, blue dye containing biotin-liposomes accumulate in the linear lymphatic vessels coursing through the diaphragm. This study confirms the fact that in the rat, the pleural lymphatic drainage pathway and the peritoneal lymphatic drainage pathway share the same lymphatic vessels in the diaphragm.

One potential application of diaphragmatic drug delivery is for treatment of mesothelioma. Mesothelioma is a cancer of the diaphragm that generally has a very poor prognosis that has not changed with any attempted therapies including surgery, chemotherapy and radiation (59).

POTENTIAL OF LIPOSOMES FOR INTRATUMORAL THERAPY

Intratumoral Drug Therapy

The direct injection of therapeutic agents into solid tumors has received significant recent interest (60–63). This interest has been stimulated by recent awareness of the difficulty in delivering drugs into the higher-pressure central core of a solid tumor (64). Even though this central region of the solid tumor is ischemic, it has been shown to frequently contain viable tumor cells. Although few studies have evaluated the intratumoral distribution of traditional chemotherapeutic agents in solid tumors of humans, it is likely that the majority of drug that reaches a solid tumor following intravenous administration accumulates in the periphery of a solid tumor with minimal drug reaching the inner core of the tumor. This would explain the difficulty in treating solid tumors even though individual cells have shown to be responsive to a particular chemotherapeutic agent. It also explains why there has been much more success in treating non-solid tumors such as leukemia and lymphomas (65).

A variety of different therapeutic agents have been proposed for treatment of tumors by direct intratumoral injection. For instance, direct injection of nanoparticles into solid tumors has been investigated as a method of delivering genes into tumors (66). This approach has also been applied in combination with external physical modalities. Magnetic nanoparticles have been directly injected into a solid tumor and exposed to alternating current as a new type of thermal ablation of solid tumors (63).

Liposomes for Intratumoral Delivery of Therapeutic Agents

Although the direct injection of liposomes into tissues has been investigated for a variety of purposes, including the treatment of infections with antimicrobial agents and administration of local anesthetics (67,68), liposomes have also received much recent attention for local delivery into tumors (69,70). Liposomes appear to offer significant advantages for direct intratumoral administration, including excellent biocompatibility and an improved intratumoral biodistribution compared to unencapsulated free drugs (70). Liposomes appear to diffuse to some degree through the interstitial space of the tumor along primitive and chaotic lymph vessels within the tumor. As demonstrated in Figure 4, the neutral DSPC:cholesterol liposomes labeled with ^{99m}Tc-N,N-bis(2-mercaptoethyl)-N',N'-diethylethylenediamine (BMEDA) appear to diffuse more readily through the solid tumor compared to the free unencapsulated ^{99m}Tc-BMEDA radiopharmaceutical. The degree of diffusion may depend on the characteristics of the particular liposome injected. This improved biodistribution associated with liposome encapsulation compared to unencapsulated drugs should result in improved solid cancer therapy due to a more homogeneous distribution throughout the tumor. In spite of the fact that liposomes appear to diffuse within solid tumors to a certain degree, liposomes can still be well retained within the tumor. When free drug is injected intratumorally, it appears to be absorbed directly into the blood supply of the tumor with less diffusion throughout the tumor so that there is a less homogeneous distribution throughout the tumor following the intratumoral injection of a free drug as compared with intratumoral injection of liposomes. Also, depending on the nature of the free drug, intratumorally injected free drug is likely to be cleared from the tumor more rapidly than liposome-encapsulated drugs due to the direct absorption into the tumor blood capillaries. The magnitude

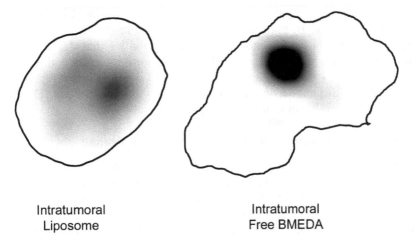

Intratumoral
Liposome

Intratumoral
Free BMEDA

Figure 4 Scintigraphic images of excised head and neck tumors removed from rats at necropsy at 44 hours after intratumoral administration of ^{99m}Tc-liposomes or the free radiopharmaceutical, ^{99m}Tc-N,N-bis(2-mercaptoethyl)-N',N'-diethyl-ethylene-diamine (BMEDA). Note the wider diffusion of liposomes throughout the tumor compared to the more local focus of uptake of the free ^{99m}Tc-BMEDA.

of difference between liposome-encapsulated agents and free agents with regard to intratumoral distribution may depend on the specific characteristics of the free drug and the liposome formulation. The methods of intratumoral administration that are still in development will also be a factor in the ultimate intratumoral distribution of agents.

Even with this improved local diffusion associated with liposomes compared with free drug, obtaining a homogeneous distribution throughout the solid tumor with intratumoral administration of liposomes still remains a challenge that will require new approaches to improving homogeneity of the injected dose throughout the tumor. The development of new tools and methods to study intratumoral distribution in tumors that are the size of those commonly encountered in humans will also be required. One approach is to modify the injection method by such methods as using multiple sites of injections within the solid tumor (71). This approach has been recently applied in the case of gene delivery with nanoparticles. Another treatment method is the use of beta-emitting therapeutic isotopes attached to liposomes that are administered intratumorally. The beta particles penetrate millimeter distances away from the radiolabeled liposome enabling the beta-emitting liposomes to deliver therapy to regions of the solid tumor that the liposomes themselves cannot reach.

An important significant advantage of liposomes for use in intratumoral injection is that a certain portion of the injected dose appears to clear through the tumor by moving into the lymphatic vessels. By moving into the lymphatic vessels, liposomes have the chance to deliver anticancer therapy to the sentinel lymph node and other lymphatics that drain from the tumor. Therefore, it is possible that the intratumoral injection with liposomes would not only treat the tumor but also could potentially treat lymph nodes that receive drainage from the tumor such as the sentinel node. These lymph nodes frequently contain metastatic cancer cells (72).

Liposome Pharmacokinetics after Intratumoral Administration

Studies of liposome intratumoral pharmacokinetics have been stimulated by attempts to use liposomes as gene carriers. Clinical trials using cationic liposomes carrying E1A gene were performed to treat squamous cell carcinoma using an intratumoral injection technique (73,74). Pharmacokinetic studies have indicated that the size and surface charge of liposomes have a significant effect on their in vivo intratumoral distribution (75,76).

Increasing the liposome diameter and adding a positive surface charge to the liposomes slowed liposome clearance from injection site compared with smaller-sized and neutral liposomes, respectively. At two hours after intratumoral injection, about 70% and 90% of injected dose remained in the tumor for a 254.0 ± 5.1 nm neutral liposome and a 125.0 ± 29.4 nm cationic liposome, respectively (76). Based on their observation of intratumorally administered cationic liposomes, Nomura et al. stated that there is a need to improve the control of the cationic liposome complexes to ensure better distribution throughout the tumor (76). Biodistribution of ^{111}In-labeled pegylated liposomes via intratumoral or subcutaneous injection techniques has also shown that liposomes have excellent potential as vehicles for intratumoral and subcutaneous drug delivery (70).

Rhenium-Labeled Liposomes for Intratumoral Radionuclide Therapy

Our group has developed a novel method of labeling liposomes with a radionuclide of rhenium. These radiotherapeutic liposomes have potential for intravenous, intracavitary, and direct intratumoral administration. This method uses BMEDA to post-load either ^{99m}Tc, rhenium-188 ([188]Re) or rhenium-186 ([186]Re) into liposomes (72).

Rhenium Isotopes

One of the significant advantages of rhenium-labeled liposomes that carry therapeutic beta particles is the short-range field effect that they have due to the millimeter range of beta particle penetration (2 mm average beta particle penetration for [186]Re and 4 mm average penetration for [188]Re) (77). This length of penetration is adequate to treat a large number of cancer cells in the region of the radiolabeled liposome, but not so far as to cause extensive damage to normal tissue. The 2- to 4-mm range of beta particle emission penetration with the rhenium-186/188 isotopes compares favorably with the most common clinically used radiotherapeutic isotope, iodine-131 (^{131}I), which has a shorter 1 mm average beta particle penetration. The 4-mm beta particle penetration with ^{188}Re provides an extensive treatment field around the injected liposome while still limiting the dose to normal structures. This field effect of the beta particles can compensate to some degree for a heterogeneous distribution of the liposomes within tumors. The liposome has to reach only within a 4-mm vicinity of the cancer cells (77).

Both rhenium isotopes, ^{186}Re and ^{188}Re, emit gamma photons at 10% and 15.5% of beta emissions, respectively. These are ideal ratios of beta to gamma emissions permitting localization of the rhenium liposomes within the body by use of single photon emission computed tomography (SPECT). A higher number of gamma emissions would deliver an excessive dose outside the local region of the tumor, as is the case for ^{131}I, which has a one to one ratio of beta to gamma photons. The photon emission energy of both rhenium isotopes is in the range of the photon energy of ^{99m}Tc (140 keV) so that the radiolabeled liposomes can be tracked through the body. Many therapeutic radioisotopes are pure beta emitters so that it is more difficult to track their distribution in the body. As a transitional element, ^{186}Re/^{188}Re also has many other advantages over most heavy metal radiotherapeutic isotopes, such as yttrium-90, because it has almost no affinity for bone uptake. It shares this characteristic with ^{99m}Tc as both radioisotopes tend to be cleared through the kidney, while most heavy metal beta-emitting radioisotopes have a high affinity for bone. This high bone accumulation can deliver a high radiation dose to bone marrow cells that are very sensitive to radiation. This occurs when the radioisotope becomes separated from its liposome and radioisotope chelator following metabolism of the liposomes and the chelating molecules in the body.

Previous theoretical dosimetry studies have addressed the potential use of radiotherapeutic liposomes for treatment of tumors via intravenous injection (78–80). In addition to these intravenous investigations, our group has investigated the potential use of rhenium-liposomes for intratumoral therapy (72). There are some significant advantages of using intratumoral delivery route for rhenium-liposomes compared with intravenous injection, such as the much lower radiation dose delivered to liver, spleen, kidney and other normal tissues, and the potential of simultaneous targeting of metastatic lymph nodes that drain from the region of the tumor (48).

^{99m}Tc to Track Liposome Distribution

^{99m}Tc-liposomes can be used to preevaluate the suitability of using ^{186}Re/^{188}Re-liposomes to treat a tumor. This is because the same chemistry is used to label liposomes with the diagnostic isotope, ^{99m}Tc, as the therapeutic rhenium isotopes. The likely dose distribution from the rhenium-liposomes can be calculated by performing SPECT/computed tomographic (CT) images of the ^{99m}Tc-liposome distribution in order to determine the potential dose distribution of the rhenium-liposomes (81,82).

We have performed studies with ^{99m}Tc-labeled liposomes to assess the potential intratumoral administration of radiolabeled liposomes. In these studies, prolonged tumor retention and very high tumor-to-normal tissue ratio of ^{99m}Tc-activity were observed. ^{99m}Tc-liposomes were injected intratumorally into a head and neck tumor in a rat model using the same methodology for labeling liposomes with radiotherapeutic rhenium as shown in Figure 4. ^{99m}Tc-liposomes had good tumor retention with 47.6% to 65.7% of injected activity still in tumors at 44 hours after injection, while unencapsulated ^{99m}Tc-BMEDA cleared from tumors quickly with only 37.1 % of injected activity remaining in tumors at two hours and 19.4% at 44 hours.

Potential for Combination of Radionuclide Therapy with Chemotherapy

Using combinations of different therapies that are coencapsulated within the same liposome might result in improved cancer therapy. For example, therapeutic radionuclides can be coencapsulated within liposomes that also contain standard anticancer agents. The pH gradient loading mechanism that is used to load doxorubicin into these liposomes can be used to label commercially available liposomes such as Doxil® with rhenium. Anthracyclines are radiosensitizing drugs so that combining direct intratumoral administration of anthracyclines with radiotherapeutic rhenium agents within the same liposome could potentially have synergistic properties. Potentially, the beta particles could improve the intratumoral drug distribution of the liposome encapsulated drug. The therapeutic beta particles that have an average penetration of 4 mm and a maximum penetration of >1 cm could cause defects within the solid tumor that would permit increased penetration of the encapsulated anticancer drug when the liposome degrades. Our group has already shown that the labeling of Doxil with radionuclides is highly feasible (83). We have previously shown that Doxil can be easily labeled with ^{99m}Tc using the same chemistry that is used to label liposomes with radiotherapeutic rhenium agents (83).

Recent Progress in Image Guided Therapy

The recent progress in imaging technology makes the potential use of intratumoral administration of cancer therapeutic agents more feasible than ever. Imaging can be used to sensitively diagnosis the occurrence and location of cancer with physiologic imaging using positron emission tomography (PET) and fluorine-18 fluorodeoxyglucose. This imaging modality detects the location of solid cancers that can be superimposed on CT imaging that is acquired simultaneously with combined PET/CT imaging cameras (84).

The CT camera can be used to place a needle directly into the cancer located in almost any region of the body. Many different possible uses of imaging could eventually be applied in the clinical setting. For instance, SPECT that images agents such as ^{99m}Tc could be used to determine the precise distribution of the liposomes within the tumor to ensure total coverage of the tumor with therapeutic agents. Following complete coverage of the tumor, PET imaging could again be used to determine therapeutic effectiveness of the solid tumor treatment as soon as one week after administration of the intratumoral therapy. Extended serial follow-up surveillance imaging could be performed with PET imaging to sensitively detect cancer reoccurrence [85]. Retreatment with the same local procedures should always be possible with this local intratumoral treatment methodology because most of the therapy is limited to the pathologic tissue. This highly targeted therapy differs from the traditional radiation treatment methods using external beam radiation because external beam radiation delivers higher levels of radiation to normal tissues in addition to the tumor target. This radiation delivered to normal tissues precludes the possibility of retreatment if the tumor reoccurs in the same region of the body.

CONCLUSION

The local delivery of liposomes into body cavities and directly into tumors appears to have much promise. These local delivery methods take advantage of the special properties of liposomes to improve the pharmacokinetics of encapsulated therapeutic agents. Recent developments in imaging technology permit the monitoring of therapeutic liposomes within the body as well as their accurate placement within tumors or cavities of the body.

REFERENCES

1. Abraham SA, Waterhouse DN, Mayer LD, et al. The liposomal formulation of doxorubicin. Meth Enzymol 2005; 391:71–97.
2. Markman M. Intraperitoneal drug delivery of antineoplastics. Drugs 2001; 61(8):1057–1065.
3. Hirano K, Hunt CA. Lymphatic transport of liposome-encapsulated agents: effects of liposome size following intraperitoneal administration. J Pharm Sci 1985; 74(9):915–921.
4. Medina LA, Calixto SM, Klipper R, et al. Avidin/biotin-liposome system injected in the pleural space for drug delivery to mediastinal lymph nodes. J Pharm Sci 2004; 93(10):2595–2608.
5. Perez-Soler R, Shin DM, Siddik ZH, et al. Phase I clinical and pharmacological study of liposome-entrapped NDDP administered intrapleurally in patients with malignant pleural effusions. Clin Cancer Res 1997; 3(3):373–379.

6. Phillips WT, Medina LA, Klipper R, et al. A novel approach for the increased delivery of pharmaceutical agents to peritoneum and associated lymph nodes. J Pharmacol Exp Ther 2002; 303(1):11–16.
7. Bonanomi MH, Velvart M, Stimpel M, et al. Studies of pharmacokinetics and therapeutic effects of glucocorticoids entrapped in liposomes after intraarticular application in healthy rabbits and in rabbits with antigen-induced arthritis. Rheumatol Int 1987; 7(5):203–212.
8. Tomita T, Hashimoto H, Tomita N, et al. In vivo direct gene transfer into articular cartilage by intraarticular injection mediated by HVJ (Sendai virus) and liposomes. Arthritis Rheum 1997; 40(5):901–906.
9. Bomgaars L, Geyer JR, Franklin J, et al. Phase I trial of intrathecal liposomal cytarabine in children with neoplastic meningitis. J Clin Oncol 2004; 22(19): 3916–3921.
10. Takanashi Y, Ishida T, Kirchmeier MJ, et al. Neuroprotection by intrathecal application of liposome-entrapped fasudil in a rat model of ischemia. Neurol Med Chir (Tokyo) 2001; 41(3):107–113.
11. Zou LL, Huang L, Hayes RL, et al. Liposome-mediated NGF gene transfection following neuronal injury: potential therapeutic applications. Gene Ther 1999; 6(6):994–1005.
12. Grant GJ, Piskoun B, Bansinath M. Intrathecal administration of liposomal neostigmine prolongs analgesia in mice. Acta Anaesthesiol Scand 2002; 46(1): 90–94.
13. Brannon-Peppas L, Blanchette JO. Nanoparticle and targeted systems for cancer therapy. Adv Drug Deliv Rev 2004; 56(11):1649–1659.
14. Weisberger AS, Levine B, Storaasli JP. Use of nitrogen mustard in treatment of serous effusions of neoplastic origin. J Am Med Assoc 1955; 159:1704–1707.
15. Dedrick RL, Myers CE, Bungay PM, et al. Pharmacokinetic rationale for peritoneal drug administration in the treatment of ovarian cancer. Cancer Treat Rep 1978; 62(1):1–11.
16. Markman M, Rowinsky E, Hakes T, et al. Phase I trial of intraperitoneal taxol: a Gynecologic Oncology Group study. J Clin Oncol 1992; 10(9):1485–1491.
17. Morice P, Joulie F, Rey A, et al. Are nodal metastases in ovarian cancer chemoresistant lesions? Analysis of nodal involvement in 105 patients treated with preoperative chemotherapy. Eur J Gynaecol Oncol 2004; 25(2):169–174.
18. Montero CA, Gimferrer Jm, Baldo X, et al. Mediastinal metastasis of ovarian carcinoma. Eur J Obstet Gynecol Reprod Biol 2000; 91(2):199–200.
19. Eltabbakh GH, Mount SL. Lymphatic spread among women with primary peritoneal carcinoma. J Surg Oncol 2002; 81(3):126–131.
20. Eltabbakh GH, Piver MS, Hempling RE, et al. Clinical picture, response to therapy, and survival of women with diffuse malignant peritoneal mesothelioma. J Surg Oncol 1999; 70(1):6–12.
21. Kresta A, Shek PN, Odumeru J, et al. Distribution of free and liposome-encapsulated cefoxitin in experimental intra-abdominal sepsis in rats. J Pharm Pharmacol 1993; 45(9):779–783.
22. Reimer DL, Kong S, Monck M, et al. Liposomal lipid and plasmid DNA delivery to B16/BL6 tumors after intraperitoneal administration of cationic liposome DNA aggregates. J Pharmacol Exp Ther 1999; 289(2):807–815.

23. Meredith RF, Alvarez RD, Partridge EE, et al. Intraperitoneal radioimmuno-chemotherapy of ovarian cancer: a phase I study. Cancer Biother Radiopharm 2001; 16(4):305–315.
24. Markman M. Intraperitoneal antineoplastic drug delivery: rationale and results. Lancet Oncol 2003; 4(5):277–283.
25. Alberts DS, Markman M, Armstrong D, et al. Intraperitoneal therapy for stage III ovarian cancer: a therapy whose time has come! J Clin Oncol 2002; 20(19): 3944–3946.
26. Markman M. Intraperitoneal hyperthermic chemotherapy as treatment of peritoneal carcinomatosis of colorectal cancer. J Clin Oncol 2004; 22(8):1527; author reply 1529.
27. Cho HK, Lush RM, Bartlett DL, et al. Pharmacokinetics of cisplatin administered by continuous hyperthermic peritoneal perfusion (CHPP) to patients with perito-neal carcinomatosis. J Clin Pharmacol 1999; 39(4):394–401.
28. Conti M, De Giorgi U, Tazzari V, et al. Clinical pharmacology of intraperitoneal cisplatin-based chemotherapy. J Chemother 2004; 16(suppl 5):23–25.
29. Berek JS, Markman M, Blessing JA, et al. Intraperitoneal alpha-interferon alter-nating with cisplatin in residual ovarian carcinoma: a phase II Gynecologic Oncology Group study. Gynecol Oncol 1999; 74(1):48–52.
30. Kaye SB. Future directions for the management of ovarian cancer. Eur J Cancer 2001; 37(suppl 9):S19–S23.
31. Zakaria ER, Simonsen O, Rippe A, et al. Transport of tracer albumin from peri-toneum to plasma: role of diaphragmatic, visceral, and parietal lymphatics. Am J Physiol 1996; 270(5 Pt 2):H1549–H1556.
32. Yuan ZY, Rodela H, Hay JB, et al. 51Cr-RBCs and 125I-albumin as markers to estimate lymph drainage of the peritoneal cavity in sheep. J Appl Physiol 1994; 76(2):867–874.
33. Flessner MF, Parker RJ, Sieber SM. Peritoneal lymphatic uptake of fibrinogen and erythrocytes in the rat. Am J Physiol 1983; 244(1):H89–H96.
34. Parker RJ, Hartman KD, Sieber SM. Lymphatic absorption and tissue disposition of liposome-entrapped [14C]adriamycin following intraperitoneal administration to rats. Cancer Res 1981; 41(4):1311–1317.
35. Parker RJ, Priester ER, Sieber SM. Comparison of lymphatic uptake, metabolism, excretion, and bio distribution of free and liposome-entrapped [14C]cytosine beta-D-arabinofuranoside following intraperitoneal administration to rats. Drug Metab Dispos 1982; 10(1):40–46.
36. Rosa P, Clementi F. Absorption and tissue distribution of doxorubicin entrapped in liposomes following intravenous or intraperitoneal administration. Pharmacology 1983; 26(4):221–229.
37. Madhavan S, Northfelt DW. Lack of vesicant injury following extravasation of liposomal doxorubicin. J Natl Cancer Inst 1995; 87(20):1556–1557.
38. Sadzuka Y, Hirota S, Sonobe T. Intraperitoneal administration of doxorubicin encapsulating liposomes against peritoneal dissemination. Toxicol Lett 2000; 116(1–2):51–59.
39. Ellens H, Morselt H, Scherphof G. In vivo fate of large unilamellar sphingomyelin-cholesterol liposomes after intraperitoneal and intravenous injection into rats. Biochim Biophys Acta 1981; 674(1):10–18.

40. Allen TM, Hansen CB, Guo LS. Subcutaneous administration of liposomes: a comparison with the intravenous and intraperitoneal routes of injection. Biochim Biophys Acta 1993; 1150(1):9–16.
41. Medina LA, Klipper R, Phillips WT, et al. Pharmacokinetics and bio distribution of [111In]-avidin and [99mTc]-biotin-liposomes injected in the pleural space for the targeting of mediastinal nodes. Nucl Med Biol 2004; 31(1):41–51.
42. Daoud SS. Combination chemotherapy of human ovarian xenografts with intraperitoneal liposome-incorporated valinomycin and cis-diamminedichloroplatinum(II). Cancer Chemother Pharmacol 1994; 33(4):307–312.
43. Sharma A, Sharma US, Straubinger RM. Paclitaxel-liposomes for intracavitary therapy of intraperitoneal P388 leukemia. Cancer Lett 1996; 107(2):265–272.
44. Vadiei K, Siddik ZH, Khokher AR, et al. Pharmacokinetics of liposome-entrapped cis-bis-neodecanoato-trans-R,R-1,2-diaminocyclohexane platinum(II) and cisplatin given i.v. and i.p. in the rat. Cancer Chemother Pharmacol 1992; 30(5):365–369.
45. Lee MJ, Cho SS, You JR, et al. Intraperitoneal gene delivery mediated by a novel cationic liposome in a peritoneal disseminated ovarian cancer model. Gene Ther 2002; 9(13):859–866.
46. Oussoren C, Storm G. Lymphatic uptake and bio distribution of liposomes after subcutaneous injection: III. Influence of surface modification with poly (ethyleneglycol). Pharm Res 1997; 14(10):1479–1484.
47. Phillips WT, Andrews T, Liu H, et al. Evaluation of [(99m)Tc] liposomes as lymphoscintigraphic agents: comparison with [(99m)Tc] sulfur colloid and [(99m)Tc] human serum albumin. Nucl Med Biol 2001; 28(4):435–444.
48. Phillips WT, Klipper R, Goins, B. Novel method of greatly enhanced delivery of liposomes to lymph nodes. J Pharmacol Exp Ther 2000; 295(1):309–313.
49. Phillips WT, Rudolph AS, Goins B, et al. A simple method for producing a technetium-99m-labeled liposome that is stable in vivo. Int J Rad Appl Instrum B 1992; 19(5):539–547.
50. Jonas S, Weinrich M, Tullius SG, et al. Microscopic tumor cell dissemination in gastric cancer. Surg Today 2004; 34(2):101–106.
51. Hribaschek A, Kuhn R, Pross M, et al. Prophylaxis of peritoneal carcinomatosis in experimental investigations. Int J Colorectal Dis 2001; 16(5):340–345.
52. Hribaschek A, Ridwelski K, Pross M, et al. Intraperitoneal treatment using taxol is effective for experimental peritoneal carcinomatosis in a rat model. Oncol Rep 2003; 10(6):1793–1798.
53. Wang NS. Anatomy and physiology of the pleural space. Clin Chest Med 1985; 6(1):3–16.
54. Shinohara H. Distribution of lymphatic stomata on the pleural surface of the thoracic cavity and the surface topography of the pleural mesothelium in the golden hamster. Anat Rec 1997; 249(1):16–23.
55. Sahn SA. State of the art. The pleura. Am Rev Respir Dis 1988; 138(1):184–234.
56. Grinberg LM, Abramova FA, Yampolskaya OV, et al. Quantitative pathology of inhalational anthrax I: quantitative microscopic findings. Mod Pathol 2001; 14(5):482–495.
57. Ilgazli A, Boyaci H, Basyigit I, et al. Extrapulmonary tuberculosis: clinical and epidemiologic spectrum of 636 cases. Arch Med Res 2004; 35(5):435–441.

58. Jackson PJ, Hugh-Jones ME, Adair DM, et al. PCR analysis of tissue samples from the 1979 Sverdlovsk anthrax victims: the presence of multiple Bacillus anthracis strains in different victims. Proc Natl Acad Sci USA 1998; 95(3): 1224–1229.
59. Hughes RS. Malignant pleural mesothelioma. Am J Med Sci 2005; 329(1):29–44.
60. Currier MA, Adams LC, Mahller YY, et al. Widespread intratumoral virus distribution with fractionated injection enables local control of large human rhabdomyosarcoma xenografts by oncolytic herpes simplex viruses. Cancer Gene Ther 2005; 12(4):407–416.
61. Duncan IC, Fourie PA, Alberts AS. Direct percutaneous intratumoral bleomycin injection for palliative treatment of impending quadriplegia. Am J Neuroradiol 2004; 25(6):1121–1123.
62. Duvillard C, Benoit L, Moretto P, et al. Epinephrine enhances penetration and anticancer activity of local cisplatin on rat sub-cutaneous and peritoneal tumors. Int J Cancer 1999; 81(5):779–784.
63. Hilger I, Hiergeist R, Hergt R, et al. Thermal ablation of tumors using magnetic nanoparticles: an in vivo feasibility study. Invest Radiol 2002; 37(10):580–586.
64. Stohrer M, Boucher Y, Stangassinger M, et al. Oncotic pressure in solid tumors is elevated. Cancer Res 2000; 60(15):4251–4255.
65. Pui CH, Cheng C, Leung W, et al. Extended follow-up of long-term survivors of childhood acute lymphoblastic leukemia. N Engl J Med 2003; 349(7):640–649.
66. Gopalan B, Ito I, Branch CD, et al. Nanoparticle based systemic gene therapy for lung cancer: molecular mechanisms and strategies to suppress nanoparticle-mediated inflammatory response. Technol Cancer Res Treat 2004; 3(6):647–657.
67. Fielding RM, Moon-McDermott L, Lewis RO. Bioavailability of a small unilamellar low-clearance liposomal amikacin formulation after extravascular administration. J Drug Target 1999; 6(6):415–426.
68. Grant SA. The Holy Grail: long-acting local anaesthetics and liposomes. Best Pract Res Clin Anaesthesiol 2002; 16(2):345–352.
69. Voges J, Reszka R, Grossman A, et al. Imaging-guided convection-enhanced delivery and gene therapy of glioblastoma. Ann Neurol 2003; 54(4):479–487.
70. Harrington KJ, Rowlinson-Busza G, Syrigos KN, et al. Pegylated liposomes have potential as vehicles for intratumoral and subcutaneous drug delivery. Clin Cancer Res 2000; 6(6):2528–2537.
71. Currier MA, Adams LC, Mahller YY, et al. Widespread intratumoral virus distribution with fractionated injection enables local control of large human rhabdomyosarcoma xenografts by oncolytic herpes simplex viruses. Cancer Gene Ther 2005; 12(4):407–416.
72. Bao A, Goins B, Klipper R, et al. 186Re-liposome labeling using 186Re-SNS/ S complexes: in vitro stability, imaging, and biodistribution in rats. J Nucl Med 2003; 44(12):1992–1999.
73. Ueno NT, Bartholomeusz C, Xia W, et al. Systemic gene therapy in human xenograft tumor models by liposomal delivery of the E1A gene. Cancer Res 2002; 62(22):6712–6716.
74. Villaret D, Glisson B, Kenady D, et al. A multicenter phase II study of tgDCC-E1A for the intratumoral treatment of patients with recurrent head and neck squamous cell carcinoma. Head Neck 2002; 24(7):661–669.

75. Nishikawa M, Hashida M. Pharmacokinetics of anticancer drugs, plasmid DNA, and their delivery systems in tissue-isolated perfused tumors. Adv Drug Deliv Rev 1999; 40(1–2):19–37.
76. Nomura T, Nakajima S, Kawabata K, et al. Intratumoral pharmacokinetics and in vivo gene expression of naked plasmid DNA and its cationic liposome complexes after direct gene transfer. Cancer Res 1997; 57(13):2681–2686.
77. Bao A, Zhao X, Phillips WT, et al. Theoretical study of the influence of a heterogeneous activity distribution on intratumoral absorbed dose distribution. Med Phys 2005; 32(1):200–208.
78. Emfietzoglou D, Kostarelos K, Sgouros G. An analytic dosimetry study for the use of radionuclide-liposome conjugates in internal radiotherapy. J Nucl Med 2001; 42(3):499–504.
79. Kostarelos K, Emfietzoglou D, Papakostas A, et al. Tissue dosimetry of liposome-radionuclide complexes for internal radiotherapy: toward liposome-targeted therapeutic radiopharmaceuticals. Anticancer Res 2000; 20(5A):3339–3345.
80. Kostarelos K, Emfietzoglou D, Papakostas A, et al. Binding and interstitial penetration of liposomes within avascular tumor spheroids. Int J Cancer 2004; 112(4):713–721.
81. Wagner A, Schicho K, Glaser C, et al. SPECT-CT for topographic mapping of sentinel lymph nodes prior to gamma probe-guided biopsy in head and neck squamous cell carcinoma. J Craniomaxillofac Surg 2004; 32(6):343–349.
82. Bao A, Phillips WT, Goins B, et al. Small animal imaging and histopathological studies on head and neck squamous cell carcinoma xenograft in nude rats. Mol Imaging Biol 2005; 7:175.
83. Bao A, Goins B, Klipper R, et al. Direct 99mTc labeling of pegylated liposomal doxorubicin (Doxil) for pharmacokinetic and non-invasive imaging studies. J Pharmacol Exp Ther 2004; 308(2):419–425.
84. Townsend DW, Beyer T. A combined PET/CT scanner: the path to true image fusion. Br J Radiol 2002; 75 Spec No:S24–S30.
85. Barker D, Zagoria RJ, Morton KA, et al. Evaluation of liver metastases after radiofrequency ablation: utility of 18F-FDG PET and PET/CT. Am J Roentgenol 2005; 184(4):1096–1102.

16

The Liposome-Mediated "Macrophage Suicide" Technique: A Tool to Study and Manipulate Macrophage Activities

Nico van Rooijen and Esther van Kesteren-Hendrikx

Department of Molecular Cell Biology, Vrije University Medical Center, Amsterdam, The Netherlands

INTRODUCTION

Macrophages

Macrophages are multifunctional cells. They play a key role in natural and acquired host defense reactions, in homeostasis, and in the regulation of numerous biological processes. Their main tools to achieve these goals are: phagocytosis followed by intracellular digestion and production and release of soluble mediators such as cytokines, chemokines, and nitric oxide. Macrophages can be found as resident cells in all organs of the body and they can be recruited to sites of inflammation. Their immediate precursors are monocytes, which are released in the blood circulation from the bone marrow. After some time, monocytes leave the circulation, cross the barrier formed by the walls of blood vessels, and enter into one of the organs where their final differentiation into mature macrophages will take place.

Depletion of macrophages followed by functional studies in such macrophage-depleted animals forms a generally accepted approach to establish their role in any particular biomedical phenomenon. Early methods for depletion of macrophages were based on the administration of silica, carrageenan, or by various other treatments. However, incompleteness of depletion and even stimulation of macrophages as well as unwanted effects on nonphagocytic cells were obvious disadvantages (1).

For that reason, we have developed a more sophisticated approach, based on the liposome-mediated intracellular delivery of the bisphosphonate clodronate (2,3). In this so-called "macrophage suicide" approach, liposomes are used as a Trojan horse to get the small hydrophilic clodronate molecules into the macrophage.

Liposome-Mediated Depletion of Macrophages

Strong hydrophilic molecules such as the negatively charged bisphosphonate dichloromethylene-bisphosphonate (clodronate) and the positively charged diamidine propamidine can be dissolved in aqueous solutions in substantial concentrations. As a consequence, such molecules can be encapsulated in multilamellar liposomes with a high efficacy (4). Once encapsulated, they cannot easily escape from the liposomes because they are not able to cross their phospholipid bilayers. Leakage remains very low for that reason. After administration of such liposomes in vivo, their natural fate is phagocytosis by macrophages. Once ingested by a macrophage, a liposome will be digested with the help of the lysosomal panel of lytic enzymes, among which are phospholipases that are able to break down the phospholipid bilayers. In this way, the encapsulated molecules are released within the cell.

Because they cannot easily escape from the cell either, owing to the fact that its cell membrane is, in its most basic form, also consisting of phospholipid bilayers, these molecules will be accumulating in the cell as more liposomes are ingested and digested by the macrophage. At a certain intracellular concentration, molecules such as clodronate and propamidine will eliminate the macrophage by initiating its programmed cell death (apoptosis) (5). Reversely, clodronate molecules released from dead macrophages will be rapidly cleared from the circulation by the renal system, because their half-life—when free in the circulation—is in the order of minutes (6). Macrophages can be found in nearly all tissues of the body. By choosing the right route of administration of clodronate liposomes, particular organs or tissues can be depleted of macrophages. In this way, i.e., by creating a macrophage-depleted organ or tissue, macrophage functions can be studied in vivo. Moreover, from a therapeutic perspective, promising results were obtained by the application of clodronate liposomes for suppression of macrophage activity in various models of autoimmune diseases, transplantation, neurological disorders, and gene therapy (7). For more information and specific references, see the "clodronate liposomes" in Ref. 8.

COMPARATIVE ACCESSIBILITY OF MACROPHAGES IN DIFFERENT TISSUES

Administration Routes for Liposomes and Physical Barriers

The extent to which resident macrophage populations in different organs are accessible to single molecules, molecular complexes, or particulate carriers such as liposomes depend on both the position of the macrophages in the tissues and the properties of the molecules or particles. In general, all macrophages can be reached by small molecules if the latter are able to pass the walls of blood vessels, e.g., capillaries, in order to penetrate into the parenchymal tissues. Large molecules, molecular complexes, or particles can reach a macrophage only if there is no physical barrier between the site of injection and the macrophage. Such a barrier can be formed, e.g., by endothelial cells in the wall of blood vessels, by alveolar epithelial cells in the lung, by reticular fibers or collagen fibers in the spleen, or by the presence of densely packed cells such as lymphocytes in the white pulp of the spleen or in the paracortical fields of lymph nodes. By choosing the right administration route for the materials to be injected, this barrier can be kept at a minimum.

The in vivo accessibility of various macrophages to liposomes is the main factor that determines the efficacy of the approach. Both the dose of clodronate liposomes required for depletion of macrophages and the time interval between injection of liposomes and their depletion depend on this accessibility.

Intravenous Administration

Intravenously injected materials can reach macrophages in the liver (Kupffer cells), spleen, and bone marrow. Kupffer cells in the liver sinuses, as well as marginal zone macrophages and red pulp macrophages in the spleen, have a strategic position with respect to large molecular aggregates and particulate materials in the circulation. Liposomes have a nearly unhindered access to these macrophages as concluded from their fast and complete depletion within one day after intravenous injection of clodronate liposomes in mice and rats (9). Obviously, it is a little more difficult for intravenously injected liposomes to reach the marginal metallophilic macrophages in the outer periphery of the white pulp. Depletion of the white pulp macrophages in the periarteriolar lymphocyte sheaths is incomplete emphasizing the barrier formed by the reticulin fiber network and/or the densely packed lymphocytes in the white pulp (10). Also, macrophages in the bone marrow were reached by intravenously injected clodronate liposomes. However, two consecutive injections with a time interval of two days were required to get a nearly complete depletion of macrophages from the bone marrow (11).

Kupffer cells in the liver play a key role in the homeostatic function of the liver. They form the largest population of macrophages in the body, make up 30% of the hepatic nonparenchymal cell population, and have easy access to particulate materials in the circulation. Consequently, a large proportion of all intravenously administered particulate carriers used for drug targeting or gene transfer will be prematurely destroyed before they reach their targets. Therefore, transient blockade of phagocytosis by Kupffer cells might be an important factor to optimize in drug targeting, gene transfer, xenogeneic cell grafting (7), and in some autoantibody-mediated disorders in which macrophages consume the body's own platelets (12) or red blood cells (13). Also, transient suppression of the cytokine-mediated activity of Kupffer cells might have a beneficial effect on various disorders of the liver (14).

Subcutaneous Administration

Subcutaneously injected clodronate liposomes are able to deplete macrophages in the draining lymph nodes of mice and rats. Such liposomes, when e.g., injected in the footpad of mice, led to the depletion of subcapsular sinus lining macrophages and medulla macrophages in the draining popliteal lymph nodes (15). Macrophages in the paracortical fields and those in the follicles were not affected, emphasizing the existence of a barrier formed by reticular fibers and/or densely packed lymphocytes in these lymph node compartments, comparable to that formed in the white pulp of the spleen. After passing the popliteal lymph nodes, the lymph flow is still filtered by consecutive draining of lymph-node stations such as the lumbar lymph nodes (in the mouse). Macrophages in these lymph nodes were partially depleted. It was apparent that only macrophages that directly drained the popliteal lymph nodes had been depleted in those compartments. Whereas the blood flow entering the spleen by the *arteria lienalis* is evenly distributed over the entire spleen, different parts in the lymph nodes are corresponding each with their own draining area and have their own afferent lymph vessels. As a consequence, particles such as liposomes are not equally distributed over all macrophages in the lymph nodes.

Intraperitoneal Administration

Macrophages from the peritoneal cavity and the omentum of the rat were depleted by two consecutive intraperitoneal injections with clodronate liposomes, given at an interval of three days (16). The peritoneal cavity is drained by the parathymic lymph nodes (in rats and mice). After passing these lymph nodes, the lymph flow reaches the blood circulation via the larger lymph vessels such as the *ductus thoracicus*. As a consequence, intraperitoneally injected clodronate liposomes are also able to deplete the macrophages of parathymic lymph nodes and once they arrive in the blood circulation, they may deplete macrophages in liver and spleen. Given the relatively large volume that can be administered via the intraperitoneal route, the total number of macrophages that can be affected is even higher than that affected by intravenous injection.

Intratracheal and Intranasal Administration

Alveolar macrophages form a first line of defense against microorganisms entering the lung via the airways. In contrast to the interstitial macrophages that are separated from the alveolar space by an epithelial barrier, alveolar macrophages which are located in the alveolar space have direct access to liposomes administered via the airways, for instance by intratracheal instillation, intranasal administration or by the application of aerosolized liposomes. The direct access of clodronate liposomes to alveolar macrophages is demonstrated by their ability to eliminate these cells in mice and rats (17). Alveolar macrophages make up about 80% of the total macrophage population in the lung. Given their presence in high numbers and the total mass of lung tissue, they form an important population of macrophages in the body.

Intraventricular Administration in the Central Nervous System

Stereotaxical injection of clodronate liposomes into the fourth ventricle of the central nervous system (CNS) of rats resulted in a complete depletion of perivascular and meningeal macrophages in the cerebellum, cerebrum, and spinal cord of these rats (18). These results confirm that also macrophages in the brain are accessible to liposomes if the latter are administered along the right route.

In other recent studies, it was shown that microglia can be depleted from cultured slices of brain tissue using clodronate liposomes. This approach has been used to demonstrate that, in addition to their phagocytic activity, microglia in the CNS promotes the death of developing neurons engaged in synaptogenesis (19).

Intra-articular Injection in the Synovial Cavity of Joints

Phagocytic synovial lining cells play a crucial role in the onset of experimental arthritis induced with immune complexes or collagen type II. A single intra-articular injection with clodronate liposomes caused the selective depletion of phagocytic synovial lining cells in mice and rats, demonstrating that this administration route allows easy access of liposomes to the macrophages lining the synovial cavity (20). Recent experiments have confirmed that liposomes are also able to reach synovium lining macrophages in men (21).

Local Injection in the Testes

Local injection of a suspension of liposomes can be performed in most organs. However, whether or not the liposomes will be able to diffuse from the injection site over the rest of the tissue will largely depend on the tissue structure. In the testis of rats, a loosely woven tissue structure allows the liposomes to reach most of the testicular macrophages, as demonstrated by the finding that at least 90% of the testicular macrophages can be depleted by clodronate liposomes (22).

SPECIFICITY WITH RESPECT TO MACROPHAGES

Selective Depletion of Phagocytic Cells

Liposomes of more than a few hundred nanometers will not be internalized by nonphagocytic cells. This explains why other cells such as lymphocytes and granulocytes are not depleted by multilamellar clodronate liposomes (23). According to a recent publication, blood monocytes (the precursors of mature resident macrophages) can be depleted by intravenous injection of clodronate liposomes (24). This may explain why in quite a number of studies, clodronate liposomes appeared to affect macrophages in tissues, in spite of the presence of a vascular barrier between liposomes and macrophages (8). In such cases, mature macrophages in these tissues might be prevented from substitution by new ones, because their precursors are killed in the circulation. In this way, the normal turn-over of resident macrophages could be blocked.

Normal dendritic cells (DC), localized in the T-cell areas in the spleen, will not be depleted by the application of clodronate liposomes. However, a particular group of so-called myeloid DC, localized at the border between marginal zone and red pulp, will be depleted as efficacious as were the macrophages (25). This is not surprising, because these cells are able to internalize particles of more than 1 mm. Because macrophages and DC show a considerable overlap in their activities, it remains an open question whether these cells should be considered macrophages or DC.

Uptake of Liposomes by Macrophage Subsets

Although macrophages, in general, seem to prefer liposomes with an overall negative charge, e.g., achieved by incorporation of the anionic phospholipid phosphatidylserine in their bilayers, also neutral and cationic liposomes are rapidly taken up by macrophages. Several modifications of the original liposome formulations, such as the incorporation of amphipathic poly-ethylene glycol conjugates in the liposomal bilayers, have been proposed in order to reduce the recognition and uptake of liposomes by macrophages. Nevertheless, a large percentage of these so-called long-circulating liposomes will still be ingested by macrophages, emphasizing that macrophages form the logical target for all liposomes, irrespective of their surface molecules (26).

Given the fact that macrophages will ingest all types of nonself-macromolecules and particulate materials, it is difficult to achieve specific targeting to only one macrophage subset, e.g., in the spleen. In studies, intended to reveal the conditions for monoclonal antibody-mediated specific targeting of enzyme molecules to marginal metallophilic macrophages in the spleen, we found that highly specific targeting of the enzyme molecules could be achieved only by using monomeric conjugates of the antibody and the enzyme. Larger conjugates lead to their uptake by all macrophage subsets in the spleen (27). As yet, the choice of an administration route for liposomes remains the main approach to achieve some degree of selectivity with respect to macrophage subsets.

TECHNICAL DETAILS

Preparation of Clodronate Liposomes and Control Liposomes

Materials

- 100 mg/mL phosphatidylcholine (egg lectin) solution in chloroform, filtered through 0.2-μm pore filter
- 10 mg/mL cholesterol solution in chloroform, filtered through 0.2-μm pore filter
- 0.7 M clodronate solution in distilled water, pH adjusted to 7.1 to 7.3 with NaOH and filtered through 0.2-μm pore filter
- Chloroform, analytical grade
- Argon gas (or other inert gas, e.g., nitrogen gas)
- Sterile phosphate-buffered saline (PBS) for injection, containing 8.2 g NaCl, 1.9 g Na2HPO4.2H2O, 0.3 g NaH2PO4.2H2O at pH 7.4 per liter
- Rotary evaporator

Method

1. Add 4.30 mL phosphatidylcholine solution to 4.00 mL cholesterol solution in a 0.5 L round-bottom flask.
2. Remove the chloroform by low vacuum (120 mbar) rotary (150 rpm) evaporation at 40°C. At the end, a thin phospholipid film will form against the inside of the flask. Remove the condensed chloroform by aerating the flask three times.
3. Vent the flask with argon gas. Ensure ventilating the whole film and thus removing all remaining chloroform.
4. Disperse the phospholipid film in 20 mL 0.7 M clodronate solution (for clodronate liposomes) or 20 mL PBS (for empty liposomes) by gentle rotation (maximum 100 rpm) at room temperature (RT). Development of foam should be avoided by reducing the speed of rotation.
5. Keep the milky white suspension at RT for about two hours.
6. Shake the solution gently and sonicate it in a waterbath (55 kHz) for three minutes.
7. Keep the suspension at RT for two hours (or overnight at 4°C) to allow swelling of the liposomes. In order to limit the maximum diameter of the liposomes for intravenous injection, the suspension can be filtered using membrane filters with 3.0-μm pores.

8. Before using the clodronate liposomes:

   a. Remove the nonencapsulated clodronate by centrifuging the liposomes at 22,000 × g and 10°C for 60 minutes. The clodronate liposomes will form a white band at the top of the suspension, whereas the suspension itself will be nearly clear.

   b. Carefully remove the clodronate solution under the white band of liposomes with a pipet (about 1% will be encapsulated). Resuspend the liposomes in approximately 45 mL PBS.

9. Wash the liposomes four to five times using centrifugation at 22,000×g and 10°C for 25 minutes. Remove each time the upper solution and resuspend the pellet in approximately 45 mL PBS.

10. Resuspend the final liposome pellet in PBS and adjust to a final volume of 20.0 mL. The suspension should be shaken (gently) before administration to animals or before dispensing, in order to achieve a homogeneous distribution of the liposomes in suspension.

Preparation of Mannosylated Clodronate Liposomes for CNS Research

For some studies in the CNS, e.g., for research on the role of macrophages in experimental allergic encephalomyelitis, a rodent model for multiple sclerosis, clodronate liposomes should be mannosylated (28,29).

Materials

- 1.85 mg/mL *p*-aminophenyl α-d-mannopyranoside (syn.: mannoside) solution in methanol
- 100 mg/mL phosphatidylcholine (egg lectin) solution in chloroform, filtered through 0.2-μm pore filter
- 10 mg/mL cholesterol solution in chloroform, filtered through 0.2-μm pore filter
- 0.7 M clodronate solution in distilled water, pH adjusted to 7.1 to 7.3 with NaOH and filtered through 0.2-μm pore filter
- Chloroform, analytical grade
- Argon gas (or other inert gas, e.g., nitrogen gas)
- Sterile PBS for injection, containing 8.2 g NaCl, 1.9 g Na_2HPO_4. $2H_2O$, 0.3 g $NaH_2PO_4.2H_2O$ at pH 7.4 per liter

Method

1. Add 0.710 mL phosphatidylcholine solution, 2.00 mL manoside solution and 1.08 mL cholesterol solution in a 0.5 L round-bottom flask.

2. Remove the chloroform and methanol by low vacuum (120 mbar) rotary (150 rpm) evaporation at 40°C. At the end, a thin

phospholipid film will form against the inside of the flask. Remove the condensed chloroform by aerating the flask three times.

3. Add 5 mL chloroform and dissolve the lipid film by gentle rotation.
4. Remove the chloroform by low vacuum (120 mbar) rotary (150 rpm) evaporation at 40°C. At the end, a thin phospholipid film will form against the inside of the flask. Remove the condensed chloroform by aerating the flask three times.
5. Vent the flask with argon gas. Ensure ventilating the whole film and thus removing all remaining chloroform.
6. Disperse the phospholipid film in 4 mL 0.7 M clodronate solution (for clodronate liposomes) or 4 mL PBS (for empty liposomes) by gentle rotation (maximum 100 rpm) at RT. Development of foam should be avoided by reducing the speed of rotation.
7. Keep the milky white suspension at RT for about two hours.
8. Shake the solution gently and sonicate it in a waterbath (55 kHz) for three minutes.
9. Keep the suspension at RT for two hours (or overnight at 4°C) to allow swelling of the liposomes.
10. Before using the clodronate liposomes:

    a. Remove the nonencapsulated clodronate by centrifuging the liposomes at 22,000×g and 10°C for 60 minutes. The clodronate liposomes will form a white band at the top of the suspension, whereas the suspension itself will be nearly clear.
    b. Carefully remove the clodronate solution under the white band of liposomes with a pipet (about 1% will be encapsulated). Resuspend the liposomes in approximately 8 mL PBS.

11. Wash the liposomes four to five times using centrifugation at 22,000×g and 10°C for 25 minutes. Remove each time the upper solution and resuspend the pellet in approximately 8 mL PBS.
12. Resuspend the final liposome pellet in PBS and adjust to a final volume of 4.00 mL. The suspension should be shaken (gently) before administration to animals or before dispensing, in order to achieve a homogeneous distribution of the liposomes in suspension.

Preparation of Control Liposomes Labelled with a Fluorochrome Marker

In order to study whether or not liposomes are taken up by particular macrophage subsets in tissues, it may be helpful to study the distribution of control liposomes.

Control liposomes may be labelled, e.g., with the fluorochrome DiI, because they do not affect macrophages. As a result, the label will show the distribution pattern of the liposomes within tissues and their uptake by macrophages. We recommend not to use DiI-labelled clodronate liposomes for the following reason: clodronate liposomes will kill the macrophages. As a consequence, the DiI label will be redistributed as soon as the macrophages are dying and from that time on, the label does no longer represent the actual distribution of the liposomes. So, liposomes should either contain clodronate to eliminate macrophages or DiI to demonstrate the uptake of liposomes by macrophages. Combination may lead to misinterpretation.

Materials

- 2.5 mg/mL DiI solution in 100% ethanol
- Sterile PBS for injection, containing 8.2 g NaCl, 1.9 g Na_2HPO_4.$2H_2O$, 0.3 g NaH_2PO_4.$2H_2O$ at pH 7.4 per liter

Method

1. Add 10 μL DiI solution per milliliter of liposome suspension.
2. Shake liposome suspension thoroughly.
3. Incubate 10 minutes at RT (dark).
4. Centrifugate liposomes at 20,000 × g for 10 minutes.
5. Remove supernatant.
6. Add sterile PBS and resuspend.
7. Centrifugate liposomes at 20,000 × g for 10 minutes.
8. Add sterile PBS to original volume.
9. Store labelled liposomes in dark at 4°C.

Determination of Clodronate Content

Materials

- 10.0 mg/mL standard clodronate solution in distilled water, pH adjusted to 7.1 to 7.3 with NaOH
- 4 mM $CuSO_4$ solution
- PBS, containing 8.2 g NaCl, 1.9 g Na_2HPO_4.$2H_2O$, 0.3 g NaH_2PO_4.$2H_2O$ at pH 7.4 per liter
- 0.65% HNO_3 solution
- Distilled water
- Saline
- Phenol 90%
- Chloroform, analytical grade
- 16-mL glass tubes, caps with Teflon® (E.I. du Pont de Nemours & Company, Inc., Wilmington, Delaware, U.S.A.) inlay
- 10-mL polystyrene tubes
- Spectrophotometer

Method: Extraction of Clodronate from Liposomes

1. Dispense in separate glass tubes: 1 mL of the clodronate liposome suspension, 1 mL of standard clodronate solution, and 1 mL PBS.

2. Add 8 mL of phenol/chloroform (1:2) to each tube.
3. Vortex and shake the tubes extensively.
4. Hold the tubes at RT for at least 15 minutes.
5. Centrifuge (1100 × g) the tubes at 10°C for 10 minutes.
6. Hold the tubes at RT until clear separation of both phases (at least 10 minutes).
7. Transfer the aqueous (upper) phase to clean glass tubes using a Pasteur pipette.
8. Add 6 mL chloroform per tube: re-extract the solution by extensive vortexing.
9. Hold the tubes for at least five minutes at RT.
10. Centrifugate (1100 × g) the tubes at 10°C for 10 minutes.
11. Transfer the aqueous phase (without any chloroform) to 10 mL plastic tubes using a Pasteur pipette. These are the samples for determination of clodronate concentration.

Method: Determination of Clodronate Concentration

1. Prepare a standard curve using 0, 10, 20, 40, 50, 70, and 80 μL of the extracted standard clodronate solution added with saline to a total volume of 1 mL per tube.
2. Dilute the samples with saline to a total volume of 1 mL per tube until they are within range of the standard curve. A suspension of clodronate liposomes prepared according to the protocol above contains about 6 mg clodronate per 1 mL suspension.
3. Add 2.25 mL 4 mM $CuSO_4$ solution, 2.20 mL distilled water and 0.05 mL HNO_3 solution to each tube, containing 1 mL sample or standard.
4. Vortex all tubes vigorously.
5. Read the samples at 240 nm using a spectrophotometer and determine the clodronate concentration.

REFERENCES

1. Van Rooijen N, Sanders A. Elimination, blocking and activation of macrophages: three of a kind? J Leuk Biol 1997; 62:702.
2. Van Rooijen N. The liposome mediated macrophage "suicide" technique. J Immunol Meth 1989; 124:1.
3. Van Rooijen N, Sanders A. Liposome mediated depletion of macrophages: mechanism of action, preparation of liposomes and applications. J Immunol Meth 1994; 174:83.
4. Van Rooijen N, Sanders A. Kupffer cell depletion by liposome-delivered drugs: comparative activity of intracellular clodronate, propamidine and ethylenediaminetetraacetic acid (EDTA). Hepatology 1996; 23:1239.
5. Van Rooijen N, Sanders A, Van den Berg T. Apoptosis of macrophages induced by liposome-mediated intracellular delivery of drugs. J Immunol Meth 1996; 193:93.

314   van Rooijen and van Kesteren-Hendrikx

6. Fleisch H. Bisphosphonates in bone disease: from the laboratory to the patient, 1993.
7. Van Rooijen N, Van Kesteren-Hendrikx E. Clodronate liposomes: perspectives in research and therapeutics. J Liposome Res 2002; 12:81.
8. http://www.ClodronateLiposomes.org.
9. Van Rooijen N, Kors N, Van den Ende M, Dijkstra CD. Depletion and repopulation of macrophages in spleen and liver of rat after intravenous treatment with liposome encapsulated dichloromethylene diphosphonate. Cell Tis Res 1990; 260:215.
10. Van Rooijen N, Kors N, Kraal G. Macrophage subset repopulation in the spleen: differential kinetics after liposome-mediated elimination. J Leuk Biol 1989; 45:97.
11. Barbé E, Huitinga I, Dopp EA, Bauer J, Dijkstra CD. A novel bone marrow frozen section assay for studying hematopoietic interactions in situ: the role of stromal bone marrow macrophages in erythroblast binding. J Cell Sci 1996; 109:2937.
12. Alves-Rosa F, Stanganelli C, Cabrera J, Van Rooijen N, Palermo MS, Isturiz MA. Treatment with liposome-encapsulated clodronate as a new strategic approach in the management of immune thrombocytopenic purpura (ITP) in a mouse model. Blood 2000; 96:2834.
13. Jordan MB, Van Rooijen N, Izui S, Kappler J, Marrack, P. Liposomal clodronate as a novel agent for treating autoimmune hemolytic anemia in a mouse model. Blood 2003; 101:594.
14. Schumann J, Wolf D, Pahl A, et al. Importance of Kupffer cells for T cell-dependent liver injury in mice. Am J Pathol 2000; 157:1671.
15. Delemarre FGA, Kors N, Kraal G, Van Rooijen N. Repopulation of macrophages in popliteal lymph nodes of mice after liposome mediated depletion. J Leuk Biol 1990; 47:251.
16. Biewenga J, Van der Ende B, Krist LFG, Borst A, Ghufron M, Van Rooijen N. Macrophage depletion and repopulation in the rat after intraperitoneal administration of Cl$_2$MBP- liposomes: depletion kinetics and accelerated repopulation of peritoneal and omental macrophages by administration of Freund's adjuvant. Cell Tis Res 1995; 280:189.
17. Thepen T, Van Rooijen N, Kraal G. Alveolar macrophage elimination in vivo is associated with an increase in pulmonary immune responses in mice. J Exp Med 1989; 170:499.
18. Polfliet MMJ, Goede PH, Van Kesteren-Hendrikx EML, Van Rooijen N, Dijkstra CD, Van den Berg TK. A method for the selective depletion of perivascular and meningeal macrophages in the central nervous system. J Neuroimmunol 2001; 116:188.
19. Marin-Teva JL, Dusart I, Colin C, Gervais A, Van Rooijen N, Mallat M. Microglia promote the death of developing Purkinje cells. Neuron 2004; 41:535.
20. Van Lent PLEM, Van De Hoek A, Van Den Bersselaar L, et al. In vivo role of phagocytic synovial lining cells in onset of experimental arthritis. Am J Pathol 1993; 143:1226.
21. Barrera P, Blom A, Van Lent PLEM, et al. Synovial macrophage depletion with clodronate containing liposomes in rheumatoid arthritis. Arthritis Rheum 2000; 43:1951.

22. Bergh A, Damber JE, Van Rooijen N. Liposome-mediated macrophage depletion: an experimental approach to study the role of testicular macrophages. J Endocrinol 1993; 136:407.

23. Claassen I, Van Rooijen N, Claassen E. A new method for removal of mononuclear phagocytes from heterogenous cell populations "in vitro," using the liposome-mediated macrophage "suicide" technique. J Immunol Meth 1990; 134:153.

24. Sunderkotter C, Nikolic T, Dillon MJ, et al. Subpopulations of mouse blood monocytes differ in maturation stage and inflammatory response. J Immunol 2004; 172:4410.

25. Leenen PJM, Radosevic K, Voerman JSA, et al. Heterogeneity of mouse spleen dendritic cells: in vivo phagocytic activity, expression of macrophage markers and subpopulation turnover. J Immunol 1998; 160:2166.

26. Litzinger DC, Buiting AMJ, Van Rooijen N, Huang L. Effect of liposome size on the circulation time and intraorgan distribution of amphipathic polyethylene-glycol containing liposomes. Biochim Biophys Acta 1994; 1190:99.

27. Van Rooijen N, Ter Hart H, Kraal G, Kors N, Claassen E. Monoclonal antibody mediated targeting of enzymes: a comparative study using the mouse spleen as a model system. J Immunol Meth 1992; 151:149.

28. Huitinga I, Van Rooijen N, De Groot CJA, Uitdehaag BMJ, Dijkstra CD. Suppression of experimental allergic encephalomyelitis in Lewis rats after elimination of macrophages. J Exp Med 1990; 172:1025.

29. Tran EH, Hoekstra K, Van Rooijen N, Dijkstra CD, Owens T. Immune invasion of the central nervous system parenchyma and experimental allergic encephalomyelitis, but not leukocyte extravasation from blood, are prevented in macrophage-depleted mice. J Immunol 1998; 161:3767.

17

Lung Surfactants: Correlation Between Biophysical Characteristics, Composition, and Therapeutic Efficacy

Oleg Rosenberg, Andrey Seiliev, and Andrey Zhuikov

Department of Medical Biotechnology, Central Research Institute of Roentgenology and Radiology, Ministry of Health, St. Petersburg, Russia

INTRODUCTION

Lung surfactant is a lipoprotein complex covering the alveolar epithelial surface of the lungs (1). It was discovered about 50 years ago when the pathogenesis of respiratory failure, which some premature newborns suffered from immediately after birth, was being investigated. In 1959, Avery and Mead (2) first found out that bronchoalveolar lavage (BAL) fluid of newborns with the disease of hyaline membrane, which is now known as respiratory distress syndrome in infants (IRDS), lowered surface tension less than BAL of healthy newborns.

Lung surfactant is synthesized in type II pneumocytes, stored in the lamellar bodies (LBs), and secreted to the alveolar space (3). It reduces the surface tension at the air–water interface from 72 mN/m to 20 to 25 mN/m and makes alveolar ventilation and gas exchange possible preventing alveoli from collapsing, i.e., it ensures respiratory mechanics. Surfactant also prevents pulmonary edema formation and provides host defense properties in the lung.

Abnormalities of pulmonary surfactant system have been described in IRDS (2), acute lung injury (ALI), and acute respiratory distress syndrome (ARDS) (4–6), pneumonia (7–10), cystic fibrosis (11,12), idiopathic pulmonary fibrosis (13,14), atelectasis (15), radiation injury (16), asthma (17–23), chronic obstructive pulmonary diseases (COPD) (24), sarcoidosis (25), tuberculosis (26,27), and others (24). The surfactant system undergoes both qualitative and quantitative alterations. In ARDS, the main biochemical abnormalities comprise an 80% fall in the total phospholipids (PLs), decrease in comparative content of dipalmitoyl phosphatidylcholine (DPPC), phosphatidylglycerol (PG) and other lipid fractions, and loss of surfactant-associated proteins (5). Surfactant function is also inhibited by leaked plasma proteins, oxygen radicals, and proteases in the alveolar compartment.

In 1980, Fujiwara et al. (28) first demonstrated high therapeutic efficiency of PL extract from bovine lung with the addition of palmitic acid (PA) and DPPC in IRDS. Surfactant therapy of IRDS is considered to be one of the major advances in neonatology in our time. About 10 preparations of lung surfactant have been developed and applied for IRDS treatment. This success induced the attempts of application of exogenous surfactants in the treatment of ALI/ARDS and other lung diseases. However, clinical trials in ARDS have had rather conflicting results (29). Parallel with efficient usage of surfactants (30,31), some studies did not result in any improvement in either oxygenation or survival (32). Among the reasons for the failure can be different etiology of ARDS (33), late surfactant administration (31), wrong dose (33), mode of delivery (32,34,35), difference in the surfactants themselves, and mistakes in planning and conducting of clinical trials (36).

In this article, we have made an attempt to analyze the experience in the clinical application of exogenous lung surfactants and discuss some conditions of their usage whose observing or neglecting can lead to success or failure of the treatment. We have tried to answer the questions of what is an ideal formulation of pulmonary surfactant and what is the mode of its application for the treatment of different lung diseases.

BASIC BIOPHYSICAL AND BIOCHEMICAL PROPERTIES OF THE LUNG SURFACTANT SYSTEM

Composition of Lung Surfactant

The composition of the surfactant may vary with such factors as species, age, lung compartment, disease states, diet, method of isolation, and so on (37). Surfactant isolated from lung BAL of healthy mammals consists of about 90% lipids and 10% proteins. Ten percent to twenty percent of the lipids are neutral and the remaining 80% to 90% is PL. About 80% of PL is phosphatidylcholine (PC), about 50% to 60% of PC is DPPC, and about 10% of PL is PG. There are also small quantities of phosphatidylethanolamine (PE), phosphatidylserine (PS), phosphatidylinositol (PI), and sphingomyelin (SM) (1,37–39).

About a half of protein fraction of surfactant is composed of four surfactant-associated proteins: SP-A, SP-B, SP-C (40), and SP-D (41).

Whereas SP-B and SP-C are extremely hydrophobic low-molecular-weight proteins, SP-A and SP-D are hydrophilic high-molecular-weight proteins from the protein family of collectins. SP-A represents 4% of surfactant and SP-B and SP-C each make up less than 1% (37).

The lipid and protein components of the surfactant are assembled and packaged in type II cells as LB, which are then secreted into the airspace and form tubular myelin, the direct precursor to the surfactant film at the air–liquid interface. LB and tubular myelin are dense forms of alveolar surfactant. The less dense and smaller aggregates of surfactant are formed during respiratory motion. They are taken up by type II cells or by macrophages, which results in a consistent ratio between functionally active large surfactant aggregates and dysfunctional small aggregates in normal lung.

Functions of Lung Surfactant System

Initially, surfactant was thought to be a key player only in the biophysical behavior of the lung. It is known that during the cycle of inspiration and expiration, fast and repeated alteration of alveolar surface size and, correspondingly, the area of surfactant cover occur. The surface tension of water which covers glicocalex of alveolar cells is 72 mN/m. Surfactant adsorption on alveolar surface decreases the surface tension to 23 mN/m, which facilitates the work of breath and provides respiratory mechanics (42).

Experimental data in vitro (42) and in vivo (43–45) shows that the surface tension at compression (expiration) falls to about 0 mN/m at the water–air interface (42). However, both we and other investigators have been confused by the lack of physical sense in this finding (46). We think that the following statements can explain surface phenomenon in inspiration/expiration cycle more profoundly. The quantity of surface-active molecules in water phase of alveoli is much more than necessary for monolayer formation on the air–water interface. Therefore, the molecule adsorption on the surface is maximum, and the surface tension coincides with one on the PL–air interface and is about 25 mN/m (42,47,48). Furthermore, many experimental data show that the surfactant film on the air–water interface may consist (probably partly) of not one but three layers (42,49,50).

The high concentration of surfactant molecules on the interface means that when the surface area decreases, they come tightly to each other; and on reaching the tightest packing, repulsive force will result in exertion in the film, which will compensate the force compressing the surface. In rheology, it is named concatenation of viscosity and elasticity. The force that compresses the surface is surface tension on air–water interface in alveolar (25 mN/m after adsorption). At pressure reduction (expiration), this force tries to reduce the surface. Finally, elastic stress will balance the surface tension force, and the resulting "force" will be equal to zero. This is the resulting surface force, which is measured as surface tension. Surfactant surface tension cannot be less than 25 mN/m, (PL surface tension on air–water interface), while the resulting surface "force" can fall to zero. Because surfactant film is not solid, its molecules are squeezed out of the surface of the water phase. Surfactant bilayer located under the monolayer may prevent molecule squeezing out and increase the stability of the film.

When the surface area is the least at expiration, and surfactant film is in the condition of its maximum compression, the force of elastic tension is practically completely balanced by surface tension force and resulting "force" is equal to zero. Therefore, there are no reasons for the following reduction of alveolar surface and its collapse. The available data on surface forces in surfactant films on air–water interface can be explained by this concept.

Although stabilizing the lungs is undoubtedly the major physiological function of surfactant, there is evidence that surfactant system may also serve other functions: it affects the permeability of the alveolar–capillary barrier to soluble compounds (51) and contributes to innate and adaptive immunity of the lung. Surfactant proteins act as a first-line defense against invading microorganisms and viruses (51–53). Moreover, they possess binding capacity for aeroallergens, highlighting the possible role of the pulmonary surfactant system in allergic diseases such as asthma (54,55).

Every component of surfactant complex plays its own role in polyfunctional surfactant activities. The key element in all pulmonary surfactants, DPPC, is considered to be the most important component with respect to its biophysical function (56). Anionic PL, especially PG, are responsible for modulating the properties of surfactant interfacial films, improving their stability during compression, and facilitating the adsorption and refining of PL on the air–lipid interface. PG can stimulate uptake of liposomal PC by type II cells (57). PA interacts with DPPC and/or SP-B to increase the movement of surfactant from the subphase and to stabilize the surfactant complex at the air–water interface (58–61). Cholesterol may play an important role in the lateral phase organization of surfactant structures (62).

Of particular interest are the specific surfactant-associated proteins that control the normal lifecycle of endogenous surfactant. SP-B and SP-C are mainly important for the biophysical properties of surfactant. SP-A and SP-D contribute essentially to host defense, which is realized in two ways: interaction with potentially injurious agents and alteration of the behavior of immune cells (63). SP-A and SP-D bind various microorganisms (64,65), lipids, and other exogenous substances. They stimulate alveolar macrophages (AM) (5,65–68) and influence the behavior of mast cells, dendric cells, and lymphocytes (69). SP-A inhibits the maturation of dendric cells, whereas SP-D enhances the ability of the cells to take up and present antigen, thereby enhancing adaptive immunity. SP-D may reduce the number of apoptotic cells (70,71). Transgenic models (SP-A null mice and SP-D null mice) demonstrates the importance of these proteins in the setting of bacterial and virus pneumonia (72). SP-A and SP-D have differential roles in modulating the inflammatory response to noninfectious lung injury (73). The overall effect of SP-D might be anti-inflammatory, whereas SP-A can contribute to both pro- and anti-inflammatory activity.

SP-B and SP-C play an important role in lung mechanics. Genetic deactivation of the SP-B gene induces irreversible and lethal respiratory failure both at birth (74,75) and in adults (76) due to incapability to maintain an opened respiratory surface. However, the controversial role of SP-B in monolayer refining and formation of a DPPC enriched layer is being discussed. It is thought now that SP-B brings lateral stability to the DPPC-rich monolayer of PL by both electrostatic and hydrophobic interactions (77). The analysis of the structure of lipid films at the nanoscopic level suggests that SP-B and SP-C alter the structure of surfactant films to optimize film rheological behavior under the dynamic conditions imposed by the lungs (78,79). Besides SP-A, SP-B is necessary for the formation of tubular mielin from secreted LB material. SP-B plays a role in host defense of the lung together with SP-A (80–82). SP-C, the smallest pulmonary surfactant-associated polypeptide, can have several functions: it contributes to the formation and dynamics of surfactant films at the air–liquid interface (83,84), prevention of surfactant inactivation by serum proteins, modulation of surfactant PL turnover, and binding to bacterial lipopolysaccharides (LPS).

ABNORMALITIES OF LUNG SURFACTANT IN DIFFERENT PATHOLOGIES

Infant Respiratory Distress Syndrome

The surfactant deficiency in IRDS results in direct biophysical consequences, i.e., high abnormalities in the mechanical properties of the respiratory system (84). There is evidence that variation in the level of surfactant-associated proteins expression or genetic variation in their genes is associated with IRDS (85) and congenital pulmonary alveolar proteinosis (86,87).

Acute Respiratory Distress Syndrome

ARDS described in 1967 by Ashbaugh et al. (4) can develop after the action of both direct injurious factors such as pneumonia, aspirated toxic agents, gastric contents, and others (direct ARDS), and as a result from inflammatory processes due to numerous systematic disorders such as sepsis, multitrauma, multiple blood transfusions, and others (indirect ARDS). It is associated with biochemical and biophysical abnormalities in the surfactant. In ARDS, marked increase in alveolar surface tension is observed. It resulted from a lack of surface-active compounds, changes in PL, fatty acid, neutral lipid, and surfactant-associated proteins; loss of the surface-active large surfactant aggregate fraction; inhibition of surfactant functions by leaked plasma proteins, inflammatory mediators, oxygen radicals, and proteases in the alveolar compartment; incorporation of surfactant PL and proteins into polymerizing fibrin (4,6,39,88,89).

The studies of the PL composition of bronchoalveolar lavage fluid (BALF) samples from patients with ARDS discovered the overall reduction of PL content; significant change in the distribution of PL classes including a marked decrease in PG, increase in the portion of the minor components (PE, PS, PI, and SM), and reduction of PC; significant decrease (to about 80% of control values) of the portion of PA, and the increase of the portion of unsaturated fatty acids in PL; nearly twice reduction of DPPC (6,88–90). In ARDS, a significant decline of SP-A, SP-B, and SP-C but not of SP-D was demonstrated (6,88,90,91). SP-A and SP-B levels remained decreased at least within 14 days after ARDS beginning (91).

Surfactant disturbance also involves some abnormalities at the higher levels of its structural organization. In model lung injury and ARDS (6,92,93), an increase of smaller surfactant aggregates occurs. It is paralleled by a loss of SP-B and surface activity. The increase in air–blood barrier permeability in ARDS causes plasma protein leakage into alveolar space. Among them, albumin (94,95), hemoglobin (96), and particularly fibrinogen or fibrin monomers (95–99) have strong surfactant-inhibitory properties. The presence of SP-B and SP-C in physiologic quantities reduces the sensitivity of surfactant to fibrinogen inhibition (99,100). The process of fibrinogen polymerization in surfactant presence results in loss of surfactant PL from the soluble phase due to their binding to fibrin strands, which is accompanied by the complete loss of surface activity in these areas (101,102). The surface activity can be largely restored by adding fibrinolytic agents (103,104).

Other mechanisms leading to surfactant dysfunction include nitration of some surfactant-associated proteins (particularly SP-A), degradation of surfactant lipid components due to increased phospholipase activity, and direct oxidation of surfactant (92).

The abnormalities of lung surfactant in ARDS cause dramatic pathophysiologic changes: alteration in lung mechanics, alveolar instability, atelectasis, and the decrease of lung compliance, which results in impairment of gas exchange (105), and a decreased resistance to secondary lung infection. Although the exact contribution of individual surfactant component to the alveolar host defense system is not completely clear, the marked decrease in SP-A content (6,89–91) and the evidence of degradation of SP-A in vivo in the lungs of ARDS patients (106) suggest a loss of opsonizing capacity to pathogens (90,107).

Very few data are available on the influence of surfactant treatment on biochemical and biophysical parameters of surfactant in ARDS (35,108). BALs were performed three hours prior to, and 15 to 18 hours and 72 hours after, surfactant administration to the patients and healthy volunteers (35). Surfactant treatment resulted in a marked increase in the lavagable PL, but predominance of the alveolar surfactant-inhibitory proteins was still encountered. Essential or even complete normalization of the PL profile, large surfactant aggregates fraction, SP-B and SP-C (but not SP-A) content, and the fatty acid composition of the PC was noted. So, surfactant administration in severe ARDS causes restoration of surfactant properties.

Asthma

Accumulating data indicate that airway obstruction, which is thought to be caused by smooth muscle constriction, mucosal edema, and secretion of fluid into the airway lumen, may partly be due to a dysfunction of pulmonary surfactant (54,55,109,110). Surfactant obtained from BAL and sputum of patients with asthma has decreased surface activity and changes in composition (17). It has been shown in animal models of asthma that though the change in the amount of surfactant is little, it may be in a less functional form (111). Cheng et al. (112) demonstrated that, in a guinea-pig model of chronic asthma, the surfactant pool size and content of large surfactant aggregates was decreased.

Pneumonia

The surfactant in BAL fluid from patients with pneumonia has reduced PC and PG content, and alterations in fatty acid composition. These changes are qualitatively similar to those registered in patients with ARDS. The amount of SP-A is also decreased and the surfactant surface tension lowering function is disturbed, partly due to the alterations in lipid components (6). As found in other conditions, where hydrophilic surfactant protein content is diminished, host defense functions may be impaired.

Tuberculosis

In experimental tuberculosis model (26), it was shown that the neutral lipids increase in BAL, whereas the total PL decreases. The enhancement of the permeability of endothelia and alveolar cell membranes results in intracellular edema and liquid leakage into alveolar space. Metabolic processes in type II cells and, therefore, the synthesis and recycling of new surfactant are disturbed, resulting in its deficiency. The functions of AM are also impaired: incomplete phagocytosis results in *Mycobacterium tuberculosis* persisting in AM. Antituberculosis drugs usually stop inflammation development in tuberculosis animal model, but long application of these drugs, for example, the combination of isoniazid, rifampicin, and ethambutol, causes disturbances of biosynthetic processes in type II cells (26).

Surfactant abnormalities often result in very severe consequences, even death. So the attempts to stop this process by means of surfactant administration seem to be quite promising and a logical way for the treatment of these pathologies.

EXOGENOUS LUNG SURFACTANTS AND METHODS OF OBTAINING THEM

Available preparations of lung surfactant can be divided into two types: the preparations made of synthetic compounds and the preparations of natural origin (Table 1).

Synthetic Preparations of Lung Surfactant

The design of synthetic preparations is based on the studies of the functions of different surfactant components with following construction of the

Table 1 The Preparation of Lung Surfactant

Chemical name	Trade name	Source	Specific proteins
Synthetic surfactants			
Pumactant	ALEC		None
Colfosceril	Exosurf		None
KL4, sinapultide, lucinactant	Surfaxin		Synthetic peptide KL4
rSP-C, lusupultide	Venticute		Recombinant SP-C
Natural surfactants			
Nonmodified surfactants			
SF-RI1	Alveofact	Lavaged bovine lung	SP-B, SP-C
Surfactant-BL	Surfactant-BL	Minced bovine lung	SP-B, SP-C
Calfactant	Infasurf	Lavaged bovine lung	SP-B, SP-C
Modified surfactants			
Surfactant TA	Surfacten	Minced bovine lung	SP-B, SP-C
Beractant	Survanta	Minced bovine lung	SP-B, SP-C
Poractant alfa	Curosurf	Minced porcine lung	SP-B, SP-C
HL-10	Surfactant HL-10	Minced porcine lung	SP-B, SP-C
CLSE	BLES	Lavaged bovine lung	SP-B, SP-C
Human surfactant			
Amniotic fluid derived	Amniotic fluid derived	Amniotic fluid	SP-A, SP-B, SP-C
Surfactant-HL	Surfactant-HL	Amniotic fluid	SP-B, SP-C

Abbreviations: CLSE, calf lung surfactant extract; BLES, bovine lipid extract surfactant; SP, surfactant-associated proteins.

preparations from the substitutes that can be obtained easier, cheaper, or safer. The ability of surfactants to decrease surface tension and increase oxygen concentration in blood was thought to be its most important function. Four synthetic preparations are known: Exosurf, ALEC, Surfaxin, and Venticute.

Exosurf (Glaxo-Wellcome, Inc., Research Triangle Park, North Carolina, U.S.A.) is a protein-free preparation devised by J Clements. It is composed of 85% DPPC, 9% hexadecanol, and 6% tylaxopol, in the form of powder. DPPC serves biophysical functions of surfactant, whereas hexadecanol imitate the functions of surfactant proteins, PG, and other lipids to some degree (37). Hexadecanol facilitates secondary spreading and sorption of DPPC on liquid surface. Tylaxopol is a strong detergent, that contributes to DPPC dispersion. The preparation is delivered at a dose of 67.5 mg/kg body weight. Now, Exosurf marketing is very limited.

ALEC (Pumactant, Britannia Pharmaceutical, Redhill, Surray, U.K.) is a protein-free surfactant composed of DPPC and PG in weight ratio 7:3 (113). It was usually used as a suspension in physiological solution, in two to four doses, 100 mg in 1 to 1.2 mL each.

Surfaxin (KL4, Discovery Laboratories, Doylestown, Pennsylvania, U.S.A.) is a suspension in 0.9% NaCl containing DPPC and palmitoyl-oleoyl-PG in the ratio of 3:1, 15% of PA and 3% of synthetic SP-B-like peptide, Sinapultide. The latter is amphiphathic helix of repeated subunits of one lysine and four leucines (114). The manufacturing method is the following. First, the peptide in the mixture of chloroform/methanol (1:1) is added to the mixture of DPPC and PG (1:10), heated up to 43°C, and dried either in N_2 current or under vacuum. Dried sediment is then resuspended in water at 43°C, added NaCl up to 0.9%, and incubated during one hour. The mixture can be exposed to several cycles of freezing and thawing.

Venticute (Byk Gulden, Kinslum; Atlanta Pharma, Konstanz, Germany) contains 1.8% of rSP-C, 63% of DPPC, 28% of palmitoyl-oleoyl PG, 4.5% of PA, and 2.5% of $CaCl_2$ after suspension in 0.9% of NaCl. rSP-C is a sequence of 34 amino acids and differs from human SP-C by amino acid substitutes. Phenylalanine in four and five positions of amino acid sequence of native protein substitutes for cysteine, and isoleucine substitutes for methionine. These substitutes are made to intensify the interaction between rSP-C and PL, stabilize the film at the air–water interface, and finally prevent molecular aggregation (115).

Surfactant Preparations of Natural Origin

The preparations of natural origin can be divided into two subgroups: modified natural surfactants (Surfacten, Survanta, Curosurf, and Surfactant-HL-10) and nonmodified natural surfactants [Alveofact, Infasurf, bovine lipid extract surfactant (BLES), Surfactant-BL, Surfactant-HL, and human surfactant from amniotic fluid]. They are obtained from bovine and porcine lungs or from human amniotic fluid and contain surfactant-associated proteins and all classes of PL.

Modified Natural Surfactant Formulations

Surfacten (Surfactant TA, Tokyo Tanabe, Japan) is the first commercial preparation of lung surfactant developed by Fujiwara et al. in 1980 (28). To obtain Surfacten, the cow lungs are minced and extracted by organic solvents. Ballast proteins, neutral lipids, and nonlipid admixture are removed. Then the product is modified by adding DPPC, PA, and triglycerides. The final freeze-dried product contains 48% DPPC, 16% unsaturated PC, SM, 4% triglycerides, 8% fatty acids, 7% cholesterol, and about 1% SP-B and SP-C. Surfacten is administered as a sonicated emulsion at a dose of 100 mg/kg body weight, in concentration of 25 mg/mL of PL. Electron microscopy of pellets of Surfacten demonstrates heterogeneity of forms comprising lamellae, vesicles of different sizes, and amorphous substance resembling protein (116). Surfacten is marketed in Japan and Southeast Asia.

Survanta (Beractant, Abbot Ltd., Ross Laboratories, Columbus, Ohio, U.S.A.) is modified Surfacten. It is a natural bovine lung extract comprising PL, neutral lipids, fatty acids, and SP-B and SP-C with the adding of DPPC, PA, and tripalmitin for improving tension-lowering properties and standardizing the finished product. Unlike Surfacten, Survanta is produced as a frozen suspension. The preparation contains 25 mg/mL PL (including 11.0–15.5 mg/mL DPPC), 0.5–1.75 mg/mL triglycerides, 1.4–3.5 mg/mL free-fatty acids, no cholesterol, and less than 1% proteins. Electron microscopy shows that the preparation consists of about 55% crystals and 45% lamellar-vesicular forms. It is administered at a dose of 4 mL/kg (37).

Curosurf (Poractant alfa, Chiesi Farmaceutichi S.P.A., Parma, Italy) is a surfactant from porcine lungs. Its production consists of several stages: water–salt extraction of minced porcine lung, centrifugation, chloroform–methanol extraction, and liquid–gel column chromatography on Lipidex-5000. The fraction of polar lipids is resolved in chloroform and filtered consecutively through filters of 0.45 and 0.2 µm. After the removal of organic solvent, the sediment is suspended in 0.9% NaCl with sodium bicarbonate (pH 6.2) and sonicated at 50 W, 40 kHz. Curosurf contains about 99% polar lipids (30–35% DPPC) and about 1% SP-B and SP-C in the ratio of 1:2. Neutral lipids and cholesterol are removed (117), that is why it is considered to be modified natural surfactant. The finished product is 1.5 or 3 mL emulsion with PL concentration of 80 mg/mL. Ninety percent of Curosurf emulsion is the particles with the size less than 5 µm. It is used at a dose of 120 to 200 mg/kg (118).

Nonmodified Natural Lung Surfactants

The surfactants from human amniotic fluid—one developed in the United States (California) (119) and Surfactant-HL (Biosurf, Russia)

(120,121)—are the closest to the pulmonary surfactant in situ. The former surfactant contains surfactant PL and SP-A, SP-B and SP-C, whereas Surfactant-HL contains PL and SP-B and SP-C. Although these preparations were efficient in clinical trials, they are not produced because of the difficulty with obtaining raw material.

Alveofact (SF-RI 1, Thomae GmbH, Biberach/Riss, Germany) is a chloroform–methanol extract of BAL bovine lung. It comprises 88% PL, 4% cholesterol, 8% other lipids, and 1% SP-B and SP-C. It contains relatively higher amount of SP-B (37).

Infasurf (Calfactant, Forrest Labs, St. Louis, Missouri, U.S.A.) is a chloroform–methanol extract of neonatal calf lung lavage. It comprises 35 mg/mL PL, 55% to 70% of which is saturated PC, SP-B, and SP-C (122).

BLES (BLES Biochemicals, Inc., London; Ontario, Canada) is isolated by organic extraction of bovine lung lavage. It comprises 63% saturated PC, 32% other PL, 2% SP-B and SP-C, and no neutral lipids (37,123).

Surfactant-BL (Biosurf, St. Petersburg, Russia) is isolated from bovine lung. Its manufacturing consists of the following stages. Lung tissue is homogenized to pieces with the side sizes not more than 5 mm in the stream of 0.9% NaCl. Debris and cells are removed by centrifugation. Supernatant is frozen at $-20°C$ and thawed at $+4°C$ to increase the size of surfactant aggregates, which allows raising the output of intermediate material. The suspension is centrifuged, at $10,000 \times g$, for 30 minutes, at $+4°C$. The precipitate is resuspended in water and extracted by chloroform–methanol mixture (124). The lower phase of two-phase system is collected, organic solvents are removed by rotary evaporation, the dry residuum is resuspended in water, and lyophilized. The preparation comprises 75% to 80% PL, 5% to 6% neutral lipids, 9% to 11% free cholesterol and its ethers, 1.8% to 2.5% SP-B and SP-C, and 3% to 4% nonidentified components. It should be mentioned that it contains all classes of PL of natural surfactant: PC, 62% to 70% of all PL (66% of PC is DPPC); lysolecithine, 1.1%; SM, 9.7%; PE, 13%; PI + PS, 6.8%; PG + diphosphatidyl glycerine, 5%; and nonidentified lipids containing phosphorus, 1.6%. The group of neutral lipids (5–6%) comprises triglycerides (4.5–5.5% of PL), diglycerides (1%), and free-fatty acids (the quantity is not estimated). Electron microscopy of the emulsion of Surfactant-BL shows that the preparation consists of aggregates of 1.6 to 1.8 µm, which in their turn are formed by 0.2 to 0.5 µm vesicles and does not contain crystal structures. We think that the presence of the aggregates shows the nativity of the preparation because they derive from self-assembly (121). The preparation is permitted for newborns and adults. The dose of Surfactant-BL is 75 mg/kg body weight for newborns and 6 mg/kg every 12 hours for adults with ARDS. Two to three administrations are usually enough for the course. Surfactant-BL is marketed in Russia.

The presented data show that available commercial preparations of lung surfactants vary a lot in their composition and properties.

Methods of Obtaining Lung Surfactants

The properties of surfactant preparations very much depends on the approaches applied for their obtaining. Synthetic surfactants are produced by mixing PL, usually from soy (DPPC, palmitoyl-oleoyl PG), with long-chain spirits (hexadecanol) and emulsifiers (tylaxopol), and in some cases synthetic peptide KL-4 (Surfaxin) or rSP-C (Venticute). To obtain Surfaxin, a peptide is added to PL in the mixture of chloroform–methanol with following removal of organic solvents in the current of rare gases, repeated emulsification in NaCl, and heating at 43°C. Several cycles of freezing–thawing or sonication are used. Such techniques are widely used in liposome technology for better peptide building into PL membrane and producing more homogenous preparations. However, electron microscopy shows that the preparations with narrow spectrum of saturated PL have crystal structure in emulsion, which might cause poor interaction with alveolar epithelium.

The methods of obtaining modified and nonmodified natural surfactants differ from each other in several ways. First, different raw materials are used: BAL or minced lung. Alveofact and Infasurf are extracted from BAL, which results in less lung tissue components in preparations (37). For production of other surfactants, either water–salt extraction of minced lung with following precipitation by ultracentrifugation of crude unpurified surfactant with subsequent extraction by organic solvent mixture (Curosurf) or extraction of minced lung by the mixtures of the same solvents (Surfacten, Survanta, Surfactant-HL 10) is used. Sometimes, neutral lipids and cholesterol are removed from lipid extract by precipitation with acetone, in other cases by means of liquid–gel column chromatography on Lipidex-5000 (Curosurf).

The design of many surfactants used to be aimed at making the substance, which could only lower surface tension with maximum efficiency. That is why the components, which deteriorated this parameter (neutral lipids, cholesterol and its ethers, ballast proteins and SP-A and SP-D), are removed from finished products. This can also result in the loss of many native surfactant components (plasmalogen and other minor PL, nonidentified substances), which are very important for improving surface properties (125,126). To compensate the loss, DPPC, PA, and tripalmitin are added to Surfacten and Survanta. The content of surfactant minor lipids that contribute to biophysical properties of the preparations was studied in three commercial surfactants: Alveofact, Curosurf, and Survanta (126). Lipid compositions had strong differences. Survanta had the highest portion of unsaturated PL and the lowest portion of acid-containing PL. The highest plasmalogen and acid-containing PL concentrations were found in Curosurf. Different lipid compositions could explain some of the differences in surface viscosity. PL pattern and minor surfactant lipids are important for biophysical activity. The removed components may also be responsible for innate local immunity of lungs, host defense properties, and increase of mucociliary clearance.

Marked differences among surfactants were observed in vitro in the presence of possible surfactant inhibitors (127). Inactivation effect of fibrinogen, albumin, and hemoglobin was studied with various surfactants. Curosurf and Survanta were inhibited by all three proteins, whereas BLES and Alveofact demonstrated low sensitivity (128).

The characteristics of surfactant preparations listed above lead to very different results of their clinical usage. The next part of the article is devoted to the clinical results of surfactant replacement therapy.

THERAPEUTICAL EFFICACY OF DIFFERENT LUNG SURFACTANTS

Bonĉuk-Dayanikli et al. (37) described the requirements for ideal therapeutic surfactant, which include the attributes of any ideal preparation and characteristics specific for surfactants: mimic effect of pulmonary surfactant in vitro, nonimmunogenicity, ability to improve gas exchange, lung mechanics and functional residual capacity, resistance to inactivation, optimal distribution characteristics, known clearance mechanisms, and minimal toxicity. Furthermore, the preparation must possess such properties of lung surfactant in situ as host defense ability and innate immunity (53).

Infant Respiratory Distress Syndrome

Although none of the surfactants meets all these requirements, the efficient application of surfactant replacement therapy for IRDS was started in 1980 (28). Not all the newborns with IRDS respond positively to surfactant administration, which can be explained by different degree of prematurity and infection constituent. Considerable experience in IRDS treatment and some clinical studies showed that synthetic protein-free Exosurf is less efficient than Curosurf and Survanta (129,130). The wide application of surfactants for IRDS treatment allowed reducing mortality rate significantly. It has been shown that newborns treated with surfactant have much less respiratory problems later compared to the newborns without surfactant treatment. Now surfactants are being used more and more in other lung pathologies in newborns such as meconium aspiration, innate pneumonia, and so on.

Acute Lung Injury and Acute Respiratory Distress Syndrome

The pathophysiology of ALI/ARDS is much more complex, that is why the development of optimal treatment strategies is a challenge. ARDS is caused by secondary surfactant deficiency. The first attempt of surfactant application for ARDS treatment was made in 1987 (131). Since then, rather controversial data have been obtained (Table 2). In spite of the introduction of some modern techniques for ARDS treatment such as "safe" conventional mechanical ventilation (CMV), usage of the concept of "open lung" (132), and so on, the mortality rate due to ALI and ARDS is still very high, and according to consolidated data on 10 European countries it was 53% in 2003. So, the development of new approaches for ALI and ARDS treatment is well-justified.

Table 2 shows that the majority of clinical studies registered positive alterations in oxygenation and lung compliance, though significant reduction of mortality rate was achieved only with the application of natural surfactants (27,30,133–138). Randomized clinical trials (RCT) of different surfactants in accordance with evidence-based medicine (EBM) requirements

Table 2 Surfactant Application in Acute Lung Injury/Acute Respiratory Distress Syndrome

Surfactant trade name	Number of patients	Mode of administration and dose	Result	References
Exosurf	725	Aerosolized surfactant 5 mg/kg for five days	No effect	(32)
Survanta	59	50–100 mg/kg, via an endotracheal tube	Mortality reduction from 43–18.8%	(108)
Infasurf	21	2.8 g/m², via an endotracheal tube	Mortality reduction	(133)
Alveofact	27	200–500 mg/kg, via bronchoscope	Mortality rate 44% compared to calculated rate of 74%	(134)
Venticute	448	200–400 mg/kg up to four intratracheal instillations	No effect	(135)
Surfactant-HL-10	35	200 mg/kg intratracheal instillations	Mortality reduction	(29)
Surfaxin	22	50–60 mg/kg, via bronchoscope (lavage) 400 mL of emulsion	Significant decrease in mortality	(29)
Surfactant-BL	183	10–12 mg/kg, via bronchoscope	Reduction of mortality rate to 15% (direct lung injury), and 25% to 30% (indirect lung injury)	(27,136)

resulted in the negative third phases of RCT (29,63,135). The only exception is Surfaxin and Venticute, whose clinical trials are in process at the moment (29,63).

The contradictions of the results of the efficiency of surfactant therapy in ALI/ARDS, and negation of the prospects of surfactant application due to some unsuccessful attempts (29,63,135), have induced us to analyze possible reasons for failure. They can be the following:

- Late administration of surfactant preparations
- Incorrect therapeutic dose and methods of preparation administration
- The injustice of EBM principle usage in patients in critical conditions
- Great variety in surfactant compositions

Timing for Surfactant Administration

Surfactant therapy is usually started very late, within first 48 to 72 hours of CMV or even later (29,134,135,138). High efficiency of early surfactant administration compared to late administration was first demonstrated for the treatment of the children and adults with ALI and ARDS (31,136,139). Multicenter case-uncontrolled clinical trials of Surfactant-BL were carried out in 58 patients with ALI and ARDS who met the requirements of American-Europe consensus conference (AECC) of 1994 (140). The patients had oxygenation index [arterial oxygen tension (PaO_2)/inspiratory oxygen fraction (FiO_2) ratio] equal to 119.4 ± 5.7 mmHg, and lung injury score (LIS) 3.04 ± 0.25 before surfactant administration. The analysis of treatment results allows dividing the patients into two groups: those who responded to surfactant positively (81.03%) and those who did not respond to surfactant administration (18.87%). In the first group, 24 hours after administration, PaO_2/FiO_2 ratio increased by 78.4% and LIS decreased by 57.9%. After 6.4 ± 1.2 days, 70.7% of the patients of the first group were weaned from CMV, the mortality rate was 14.9%, whereas the mortality in the second group was 90%. The main difference in therapeutic modes was the period of time between the moment of PaO_2/FiO_2 ratio drop less than 200 mmHg and surfactant administration: it was 18.7 ± 2.72 hours in the first group and 31.9 ± 5.6 hours in the second group (31,136).

Searching Therapeutic Dose and Method of Surfactant Administration

The question of a dose seems to be very difficult. Some investigators (35,138) believe that high doses of the preparation are necessary for successful treatment of ARDS. This approach is based both on clinical experience and data about inhibiting effect of leaking plasma proteins. We think that the calculation of leaking inhibitors quantity was based on wrong assumptions (6,35). Protein quantity was measured in pooled BAL samples. The increase in permeability of alveolar–capillary barrier in ARDS does not mean that it is permeable completely: the leakage of the proteins has a definite speed. After the first lavage, the protein concentration in the following lavage sample will be very small, and protein concentration gradient between capillary and airspaces is created, which causes the following protein leakage. So the total protein measured in pooled lavage fractions does not reflect the true situation in alveolar space. To test this assumption, we assessed albumin, fibrinogen, and total protein content separately in five successive 40-mL portions of lavage fluids. The period of time between the lavages should be minimal, and usually it was 30 to 60 seconds. The sharp increases in protein content in BALF are turned out to be followed by deep falls with protein concentration equal to zero in some patients (Fig. 1). So we think that the protein content in airspaces was significantly overestimated

The fact that healthy adult has 3 to 15 mg/kg of surfactant also supports the application of the less dose of preparation. Nevertheless, the therapeutic dose for the majority of surfactants is very high, and in some cases reaches 200 to 800 mg/kg per course (30,134,135,138). Such high and variable doses can be explained only by the diversity of surfactants. Surfactant-BL is the only surfactant whose therapeutic dose, 6 to 12 mg/kg for ALI/ARDS (136,141), is close to the surfactant content in vivo (121).

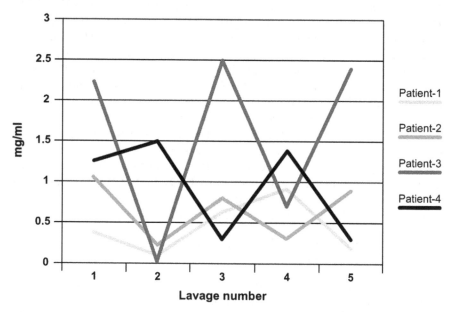

Figure 1 Protein content in separate bronchoalveolar lavage fluid fractions of patients with acute lung injury/acute respiratory distress syndrome. Total protein content separately in five successive 40-mL portions of lavage fluids was determined. The period of time between the lavages was 30 to 60 seconds. The sharp increases in protein content in bronchoalveolar lavage fluid are turned out to be followed by deep falls with protein concentration equal to zero in some patients.

Experimental data prove that only 4.5% of surfactant reaches alveolar surface at aerosol way of administration (32). Clinical trials demonstrated that aerosol way is less efficient in newborns (142) and inefficient in adults (32). Larger volumes are better for particle distribution among different parts of lungs, but at the same time the preparations hardly reach injured lung areas (143). Now the most efficient way of administration is considered to be endobronchial administration of preparation into every lung segment.

The Injustice of EBM Principle Usage in Patients in Critical Conditions

The correctness of observing the principles of EBM in patients in critical condition is questionable (36,144–146). Not long ago, mortality rate in ARDS reached 60% to 70%. Such conditions are considered to be fatal, and RCT of the preparations, whose efficiencies has been proved in experiments or has physiological ground, are not justified (146). The interpretation of the results of RCT is complicated very much by wide heterogeneity of the patients with ARDS who must not be enrolled in the same groups according to the etiology and severity of the disease. For example, additional analysis of the negative results of RCT of Venticute for ALI/ARDS treatment showed that it significantly reduced mortality in direct lung injury (29,63,135). Surfactant-BL was proved to be more efficient in direct lung injury compared to systemic lesion (29,147–150). Carrying out clinical studies of the surfactants in homogenous groups of patients gave more promising results and allowed recommending the treatment of patients of certain etiology of ARDS (147,148,150,151).

Another very important thing, which is not taken into account by EBM at planning RCT, is the number of patients treated in one particular intensive care unit (ICU). The desire to minimize the period of the third phase of RCT, which should involve a large number of patients, causes the distribution of the patients among many hospitals. For example, the third phase of RCT of Venticute (135) enrolled 448 patients treated in 109 hospitals. So, on average, four patients (two treated with surfactant and two control patients) were in each hospital, which makes data comparison incorrect. The procedure of surfactant treatment is quite complex. Differences in basic therapy and respiratory support methods as well as very small experience in surfactant application can affect the result.

Different Surfactant Preparations Have Different Therapeutic Effect

The efficiency of surfactant therapy depends on the composition of a chosen preparation (152). Higher efficiency of Survanta and Curosurf for IRDS treatment compared to Exosurf (153) and Exosurf inefficiency for ARDS treatment (32,154) prove that natural surfactants give better responses than protein-free synthetic surfactants. Phase II of clinical study of recombinant SP-C (Venticute) in patients with ARDS showed marked improvements in the oxygenation index, ventilator-free days, and the percentage of successfully weaned patients. However, mortality rate in this group was 29% compared to 33% in the control. Patients delivered up to 200 mg/kg of total PL in four doses (154).

The application of modified natural surfactants Curosurf (137) and Survanta (30,108) demonstrated gas exchange responses (30,108,137) and, in case of Survanta, a trend toward reduced mortality. BALF analysis revealed partially improved surfactant functions (108).

The most promising are the results of clinical application of natural nonmodified surfactants: Alveofact (35,138) and Surfactant-BL (27,31,136, 147,148,151). Uncontrolled multicenter study showed that bronchoscopic application of a high dose of Alveofact in patients with severe ARDS and septic shock is both feasible and safe, resulting in pronounced improvement in gas exchange and far-reaching, though incomplete, restoration of the severely changed biochemical and biophysical surfactant properties (35,138). A total of 15 patients survived the 28-day study period (mortality rate 44.4%, compared to a calculated risk of death for the given acute physiology, age, and chronic health evaluation (APACHE) II scores of 74.0%) (138). Another controlled, randomized, open-label study of the efficiency of Infasurf in 42 children with ARDS demonstrated a rapid improvement in oxygenation, reduced duration of CMV, and an earlier discharge from the pediatric ICU in the surfactant treatment group (133,136).

Multicenter uncontrolled clinical trials of Surfactant-BL have been carried out in the patients with ALI and ARDS of different etiology such as sepsis, multiple trauma, multiple transfusion, aspiration of gastric content, thromboembolism of lung artery, severe pneumonia, thermochemical burns of respiratory tracts, and postbypass lung injury (Table 3). Surfactant administration at a dose of 6 to 12 mg/kg per course reduced significantly the duration of CMV and 28-day mortality rate (from 60% to 23.2%). The mortality rate in the patients who responded to surfactant administration was 15%. Seven patients with severe burns of respiratory tracks treated by Surfactant-BL survived compared to 1 survivor of 15 patients in the control group (150).

Several ways of improving surfactants are under study (152). The investigators have been developing some substitutes for natural surfactant components: first, either synthetic or recombinant surfactant proteins or their analogues to generate proteins that are free of animal contaminants; second, PL analogues that may improve surface activity of surfactant and be resistant to phospholipase and, third, the substances to prevent surfactant inactivation, for example, such nonionic polymers as dextran or polyethylene glycol (155).

We think that only surfactant preparations with complex and similar to native surfactant in situ composition and structure can have high therapeutic effect. These preparations can not only improve biophysical

Table 3 Surfactant-BL Application in Homogenous Groups of Patients with ALI Acute Respiratory Distress Syndrome

ALI/acute respiratory distress syndrome etiology	Type of lung injury	Number of patients	28-day survival
Aspiration of gastric content	Direct	18	17 (94%)
Severe pneumonia	Direct	26	22 (85%)
Respiratory tract burns	Direct	11	10 (91%)
Complication after pulmonectomy in tuberculosis, ALI	Indirect	26	24 (92%)
Sepsis	Indirect	28	17 (61%)
Massive hemotransfusion	Indirect	16	10 (68%)
Postbypass lung injury	Indirect	36	25 (69%)
Severe multiple trauma	Indirect	22	15 (68%)

Abbreviation: ALI, acute lung injury.

parameters of injured lung but also serve as a substrate and stimulator for own endogenous surfactant synthesis (156), involve uninjured lung parenchyma areas in respiration, contribute to lung parenchyma immunity, lung defense system, and removal of toxic compounds from alveolar space with sputum.

Surfactant Therapy of Other Lung Disease

The finding of abnormalities of pulmonary surfactant system in practically all lung pathologies encourages the attempts of surfactant treatment of others than IRDS and ARDS lung disease. The experience in this field is not very wide. Surfactant preparations have been used for the treatment of pneumonia (27,157), atelectasis (158), asthma (159–161), and tuberculosis (27,162–164).

Limited experience with selective bronchial instillation of surfactant in a patient with pneumonia has suggested the possibility of benefit (157). Surfactant-BL was used in more than 60 children (from 9 months to 14 years old) with acute bronchopneumonia complicated by stable atelectasis. The treatment resulted in significant reduction of the number of fibrobronchoscopies, an increase in complete and partial atelectasis solvability (158).

Clinical use of surfactant in asthma is currently under investigation. A study in which 12 asthmatic children received aerosolized bovine surfactant indicated that there were no changes in lung functions (159). In another clinical investigation, 11 adult asthmatic patients with stable airway obstruction were given aerosolized surfactant six hours after an asthma attack (160). All patients showed an improvement in pulmonary function. The investigation of the effect of a porcine natural surfactant on inflammatory changes in patients with mild asthma following segmental allergen challenge

(165) showed that allergen-induced inflammatory response was increased by surfactant pretreatment compared to placebo. It is unknown whether this pulmonary action is restricted to one specific preparation or true for various formulations. The researchers conclude that surfactant treatment of patients with asthma may require specifically designed preparations (23,54,110).

There is limited information on the value of surfactant treatment of patients with COPD. In a single study of the effect of surfactant PL in COPD, patients with chronic bronchitis who received aerosolized PL three times daily for two weeks had a modest dose-related improvement in mucociliary transport and airflow compared to that in patients who received saline (166). The Discovery Labs has been developing an aerosolized surfactant solid form to treat hospitalized COPD patients. Work with aerosolized surfactant has demonstrated improved pulmonary function in such patients.

The surfactant application in lung tuberculosis is quite efficient (63,162–164). Surfactant-BL was used in complex therapy for 52 patients with lung tuberculosis (162–164). All the patients discharged bacteria with sputum and had multidrug resistance. Small doses of surfactant (25 mg a day, 0.3 mg/kg) were administered according to a specially designed scheme during two months together with four to five antituberculosis preparations. Six to eight weeks after the beginning of the treatment, 85.7% of patients demonstrated conversion of sputum to negative (vs. 65% in the control group); two to four months later, 94% of patients had infiltrate resolutions (vs. 67% in the control group), and 83% of patients had reduction or close of cavities (vs. 47% in the control group).

CONCLUSION

Although there are still a lot of questions regarding feasibility, efficiency, and methods of surfactant therapy for the diseases others than IRDS, the future of surfactant preparations seems to be quite promising. The application of surfactant preparations in patients with direct lung injury is more efficient than in the patients with indirect lung injury. Surfactant application can be fearlessly recommended for the patients with aspiration of gastric content, severe burns of respiratory tracts, severe pneumonia, lung contusion, and others. In any case, the analysis of the efficiency of surfactant therapy should be carried out in homogenous groups of patients as the definition of ARDS is too broad and includes the variety of patients with different and extremely complex pathophysiologies. Some patients may be more responsive to exogenous surfactant than others.

The therapeutic efficiency of surfactant formulations varies a lot. It is necessary to emphasize that the closer preparation composition and structure are to the characteristics of the surfactant in situ, the better results it has. The preparation must be administered as early after the onset of ALI/ARDS as possible, preferably within the first 24 hours. Later administration often leads to failure. The dose varies essentially depending on the chosen surfactant preparation. The application of surfactants for some subacute pulmonary diseases and tuberculosis requires working out the treatment regimens different from ones in ALI/ARDS.

REFERENCES

1. King RJ, Clements JA. Surface active materials from dog lung: composition and physiological correlations. Am J Physiol 1972; 223:715.
2. Avery ME, Mead J. Surface properties in relation to atelectasis and hyaline membrane disease. Am J Dis Child 1959; 97:517–523.
3. Gross NK, Barnes E, Narine KR. Recycling of surfactant in black and beige mice: pool sizes and kinetics. J Appl Physiol 1988; 64:1027.
4. Ashbaugh DG, et al. Acute respiratory distress in adults. Lancet 1967; 2:319.
5. Lewis JF, Jobe AH. Surfactant and the adult respiratory distress syndrome. Am Rev Respir Dis 1993; 147:218–233.
6. Schmidt R, et al. Alteration of fatty acid profiles in different pulmonary surfactant phospholipids in acute respiratory distress syndrome and pneumonia. Am J Respir Crit Care Med 2001; 163:95.
7. Gunther A, et al. Surfactant alterations in severe pneumonia, ARDS and cardiogenic lung edema. Am J Respir Crit Care Med 1996; 153:176.
8. Baughman RP, et al. Decreased surfactant protein A in patients with bacterial pneumonia. Am Rev Respir Dis 1987; 147:653.
9. Baughman RP, et al. Changes in fatty acids in phospholipids of the bronchoalveolar fluid in bacterial pneumonia and in adult respiratory distress syndrome. Clin Chem 1984; 30:521.
10. Lewis J, Veldhuizen R. The role of surfactant in pneumonia and sepsis. Appl Cardiopulm Pathophysiol 2004; 13(1):55 (abstracts).
11. Griese M, Birrer P, Demirsoy A. Pulmonary surfactant in cystic fibrosis. Eur Respir J 1997; 10:1983.
12. von Bredow C, Birrer P, Griese M, Surfactant protein A and other bronchoalveolar lavage fluid proteins are altered in cystic fibrosis. Eur Respir J 2001; 17:716.
13. McCormack FX, et al. Idiopathic pulmonary fibrosis. Abnormalities in the bronchoalveolar lavage content of surfactant protein A. Am Rev Respir Dis 1991; 144:160.
14. Robinson PC, et al. Idiopathic pulmonary fibrosis. Abnormalities in the bronchoalveolar lavage fluid phospholipids. Am Rev Respir Dis 1988; 137:585.
15. Oyarzun MJ, Stevens P, Clemens JA. Effect of lung collapse on alveolar surfactant in rabbits subjected to unilateral pneumothorax. Exp Lung Res 1989; 15:90.
16. Hallman M, et al. Changes in surfactant in bronchoalveolar lavage fluid after hemithorax irradiation in patients with mesothelioma. Am Rev Respir Dis 1990; 141:998.
17. Hohlfeld J, et al. Dysfunction of pulmonary surfactant in asthmatics after segmental allergen challenge. Am J Respir Crit Care Med 1999; 159:1803.
18. Wright SM, et al. Altered airway surfactant phospholipids composition and reduced lung function in asthma. J Appl Physiol 2000; 89:283.

19. Cheng G, et al. Increased levels of surfactant protein A and D in bronchoalveolar lavage fluids in patients with bronchial asthma. Eur Respir J 2000; 16:831.
20. Wang JY, et al. Allergen–induced bronchial inflammation is associated with decreased levels of surfactant proteins A and D in a murine model of asthma. Clin Exp Allergy 2001; 31:652.
21. Haczku A, et al. The late asthmatic response is linked with increased surface tension and reduced surfactant protein B in mice. Am J Physiol Lung Cell Mol Physiol 2002; 283:L755.
22. Ackerman SJ, et al. Hydrolysis of surfactant phospholipids catalyzed by phospholipase A2 and eosinophil lysophospholipases causes surfactant dysfunction: a mechanism for small airway closure in asthma. Chest 2003; 123:355S.
23. Hohlfeld JM, et al. Eosiniphil cationic protein (ECP) alters pulmonary structure and function in asthma. J Allergy Clin Immunol 2004; 113:96.
24. Devendra G, Spragg RG. Lung surfactant in subacute pulmonary disease. Respir Res 2002; 3:19–30.
25. Günther A, et al. Surfactant abnormalities in idiopathic pulmonary fibrosis, hypersensitivity pneumonitis and sarcoidosis. Eur Respir J 1999; 14:565.
26. Erokhin VV. Morphofunctional condition of lung cells at tuberculosis inflammation. In: Erokhin VV, Romanova LK, eds. Cell Biology of Normal and Pathological Lungs. Moscow: Medicine, 2000:422 (Russian).
27. Rosenberg O, et al. Surfactant therapy for acute and chronic lung diseases. Appl Cardiopulm Pathophysiol 2004; 13(1):78.
28. Fujiwara T, et al. Artificial surfactant therapy in hyaline membrane disease. Lancet 1980; 1:55.
29. Spragg RG. Current status of surfactant treatment of ARDS/ALI. Appl Cardiopulm Pathophysiol 2004; 13(1):88.
30. Gregory TJ, et al. Survanta supplementation in patients with acute respiratory distress syndrome (ARDS). Am J Respir Crit Care Med 1994; 49:A125.
31. Rosenberg OA, et al. When to start surfactant therapy (ST-therapy) of acute lung injury? Eur Respir J 2001; 18(suppl 38):P153, 7s (11th ERS Annual Congress, Berlin).
32. Anzueto A, et al. Aerosolized surfactant in adults with septic-induced acute respiratory distress syndrome. N Engl J Med 1996; 334:1417.
33. Anzueto A. Exogenous surfactant in acute respiratory distress syndrome: more is better. Eur Respir J 2002; 19:787.
34. MacIntire NR, et al. Efficiency of the delivery of aerosolized artificial surfactant in intubated patients with the adult respiratory distress syndrome. Am J Respir Crit Care Med 1994; 149:A125.
35. Günter A, et al. Bronchoscopic administration of bovine natural surfactant in ARDS and septic shock: impact of biophysical and biochemical surfactant properties. Eur Respir J 2002; 19:797.
36. Marini JJ. Advances in the understanding of acute respiratory distress syndrome: summarizing a decade of progress. Curr Opin Crit Care 2004; 10(4):265.
37. Bonĉuk-Dayanikli P, Taeusch WH. Essential and nonessential constituents of exogenous surfactants. In: Robertson B, Taeusch HW, eds. Surfactant Therapy for Lung Disease, Lung Biology in Health and Disease. Vol. 84. New York: Marcel Dekker, Inc., 1995:217.

38. Sanders RL. The composition of pulmonary surfactant. In: Farrell PM, ed. Lung Development: Biological and Clinical Perspectives. Vol. 1. New York: Academic Press, 1982:193.
39. Günther A, et al. Surfactant alteration and replacement in acute respiratory distress syndrome. Respir Res 2001; 2:353.
40. Possmayer F. A proposed nomenclature for pulmonary surfactant-associated proteins. Am Rev Respir Dis 1988; 138:990.
41. Persson A, et al. Purification and biochemical characterization of CP4 (SP-D), a collagenous surfactant-associated protein. Biochemistry 1989; 28:6361.
42. Schurch S, Bachofen H. Biophysical aspects in the design of a therapeutic surfactant. In: Robertson B, Taeusch HW, eds. Surfactant Therapy for Lung Disease, Lung Biology in Health and Disease. 84. New York: Marcel Dekker, Inc., 1995:3.
43. Goerke J, Clements JA, Alveolar surface tension and lung surfactant. In: Handbook of Physiology. The Respiratory System. Mechanics of Breathing. Vol. 3. Bethesda, Am Phisiol Soc, 1986:Sect 3, Part I, 247.
44. Mead J, Collier C. Relation of volume history of lungs to respiratory mechanics in anesthetized dogs. J Appl Physiol 1959; 14:669.
45. Horie T, Hildebrandt J. Dynamic compliance, limit cycles, and static equilibria of excised cat lung. J Appl Physiol 1971; 31:23.
46. Bangham AD. Artificial lung expanding compound (ALEC^{tm}). In: Lasic, Papahajopoulos, eds. Medical Applications of Liposomes. Elsevier Science, 1998:445.
47. Schürch S, Goerke J, Clements JA. Direct determination of volume- and time-dependence of alveolar surface tension in excised lungs. Proc Natl Acad Sci USA 1978; 75:3417.
48. Bachofen H, et al. Relations among alveolar surface tension, surface area, volume and recoil pressure. J Appl Physiol 1987; 62:1878.
49. Bachofen H, et al. Experimental hydrostatic pulmonary edema in rabbit lungs. Morphology Am Rev Respir Dis 1993; 147:989.
50. Bachofen H, Schürch S, Weibel IR. Experimental hydrostatic pulmonary edema in rabbit lungs. Barrier lesions. Am Rev Respir Dis 1993; 147:997.
51. Wollmer P, Jonson B, Lachmann B. Evaluation of lung permeability. In: Robertson B, Taeusch HW, eds. Surfactant Therapy for Lung Disease, Lung Biology in Health and Disease. 84. New York: Marcel Dekker, Inc., 1995:199.
52. LaForce FM, Kelley WJ, Huber JL. Inactivation of staphylococci by alveolar macrophages with preliminary observations on the importance of alveolar lining material. Am Rev Respir Dis 1973; 108:783.
53. van Iwaarden FJ, van Golde LMJ. Pulmonary surfactant and lung defense. In: Robertson B, Taeusch HW, eds. Surfactant Therapy for Lung Disease, Lung Biology in Health and Disease. 84. New York: Marcel Dekker, Inc., 1995:75.
54. Hohlfeld J, Fabel H, Hamm H. The role of pulmonary surfactant in obstructive airway disease. Eur Respir J 1997; 10:482.
55. Hohlfeld JM. The role of pulmonary surfactant in asthma. Major components of surfactant (functions). Respir Res 2002; 3:4.
56. Goerke J. Pulmonary surfactant: functions and molecular composition. Biochim Biophys Acta 1998; 1408:79.

57. Bates SR, Ibach PB, Fisher AB. Phospholipids co-isolated with rat surfactant protein C account for the apparent protein-enhanced uptake of liposomes into lung granular pneumocytes. Exp Lung Res 1989; 15:695.
58. Longo ML, et al. A function of lung surfactant protein SP-B. Science 1993; 261:453.
59. Cocksgutt A, Absolom D, Possmayer F. The role of palmitic acid in pulmonary surfactant: enhancement of surface activity and prevention of inhibition by blood proteins. Biochim Biophys Acta 1991; 1085:248.
60. Holm BA, et al. Biophysical inhibition of synthetic lung surfactants. Chem Phys Lipids 1990; 52:243.
61. Tanaka Y, et al. Development of synthetic lung surfactants. J Lipid Res 1986; 27:475.
62. Orgeig S, Daniels CB. The roles of cholesterol in pulmonary surfactant: insights from comparative and evolutionary studies. Comp Biochem Physiol A Mol Integr Physiol 2001; 129:75.
63. Floros J, et al. Pulmonary surfactant-update on function, molecular biology and clinical implications. Current Respir Med Rev 2005; 1:77.
64. Crouch E, Wright JR. Surfactant proteins A and D and pulmonary host defense. Ann Rev Physiol 2001; 63:521.
65. Phelps DS. Surfactant regulation of host defense function in the lung: a question of balance. Pediatr Pathol Mol Med 2001; 20:269.
66. Hawgood S. The hydrophilic surfactant protein SP-A: molecular biology, structure, and function. In: Robertson B, Van Golde LMG, Batenburg JJ, eds. Pulmonary Surfactant: From Molecular Biology to Clinical Practice. Amsterdam: Elsevier, 1992:33.
67. Erpenbeck VJ, et al. Surfactant protein D regulates phagocytosis of grass pollen-derived starch granules by alveolar macrophages. Appl Cardiopulm Pathophysiol 2004; 13:31.
68. Wright JR. Immunomodulatory functions of surfactant. Physiol Rev 1997; 77:931.
69. Wright JR. Surfactant: a pulmonary link between innate and adaptive immunity. Appl Cardiopulm Pathophysiol 2004; 13:104.
70. Clark H, et al. Surfactant protein reduces alveolar macrophage apoptosis in vivo. J Immunol 2002; 169:2892.
71. Clark HW, et al. A recombinant fragment of human surfactant protein D reduces alveolar macrophage apoptosis and pro-inflammatory cytokones in mice developing pulmonary emphysema. Ann NY Acad Sci 2003; 1010:113.
72. Lewis J, Veldhuizen R. The role of surfactant in the pathophysiology of acute lung injury. Appl Cardiopulm Pathophysiol 2004; 13:53.
73. Kaplan JH, et al. Differential response of collectin deficient mouse models to bleomycin induced lung injury. Appl Cardiopulm Pathophysiol 2004; 13:51.
74. Nogee LM, et al. A mutation in the surfactant protein B gene responsible for fatal neonatal respiratory disease in multiple kindreds. J Clin Invest 1994; 93:1860.
75. Clark JC, et al. Targeted disruption of the surfactant protein B gene disrupts surfactant homeostasis, causing respiratory failure in newborn mice. Proc Natl Acad Sci USA 1995; 92(17):7794.

76. Melton KR, et al. SP-B deficiency causes respiratory failure in adult mice. Am J Physiol 2003; 285:L543.
77. Cochrane CG. A critical examination of the role of SP-B in alveolar expansion. Appl Cardiopulm Pathophysiol 2004; 13:27.
78. Cruz A, et al. Analysis by scanning force microscopy of the micro- and nano-structure of Langmuir-Blodgett pulmonary surfactant films. Appl Cardiopulm Pathophysiol 2004; 13:30.
79. Cruz A, et al. Effect of pulmonary surfactant protein SP-B on the micro- and nanostructure of phospholipids films. Biophys J 2004; 86:308.
80. Miles PR, et al. Pulmonary surfactant inhibits LPS-induced nitric oxide production by alveolar macrophages. Am J Physiol 1999; 276:L186.
81. van Iwaarden JF, et al. Alveolar macrophages, surfactant lipids, and surfactant protein B regulate the induction of immune responses via the airways. Am J Respir Cell Mol Biol 2001; 24:452.
82. Ikegami M. SP-B protects lung from inflammation. Appl Cardiopulm Pathophysiol 2004; 13:45.
83. Weaver TE, Conkright JJ. Function of surfactant proteins B and C. Annu Rev Physiol 2001; 63:555.
84. Na Nkorn P, et al. The influence of truncated surfactant-protein C (SP-C17) on the surface properties of lipid-monolayers at the air-water interface. Appl Cardiopulm Pathophysiol 2004; 13:62.
85. Cotton RB, Olsson T. Lung mechanics in respiratory distress syndrome. In: Robertson B, Taeusch HW, eds. Surfactant Therapy for Lung Disease, Lung Biology in Health and Disease. Vol. 84. New York: Marcel Dekker, Inc., 1995:121.
86. Floros J, et al. Dinucleotide repeats in the human surfactant protein B gene and respiratory distress syndrome. Biochem J 1995; 305:583.
87. Nogee LM, et al. Deficiency of pulmonary surfactant protein B in alveolar proteinosis. N Engl J Med. 1993; 328:406.
88. Hallman M, et al. Evidence of lung surfactant abnormality in respiratory failure. Study of bronchoalveolar lavage phospholipids, surface activity, phospholipase activity, and plasma myoinositol. J Clin Invest 1982; 70:673.
89. Gregory TJ, et al. Surfactant chemical composition and biophysical activity in acute respiratory distress syndrome. J Clin Invest 1991; 88:1976.
90. Pison U, et al. Surfactant abnormalities in patients with respiratory failure after multiple trauma. Am Rev Respir Dis 1989; 140:1033.
91. Green KE, et al. Serial changes in surfactant-associated proteins in lung and serum before and after onset of ARDS. Am J Respir Crit Care Med 1999; 160:1843.
92. Lewis JF, et al. Altered alveolar surfactant is an early marker of acute lung injury in septic adult sheep. Am J Respir Crit Care Med 1994; 150:123.
93. Veldhuizen R, et al. Pulmonary surfactant subfractions in patients with the acute respiratory distress syndrome. Am J Respir Crit Care Med 1995; 152:1867.
94. Seeger W, Stohr G, Wolf HR. Alteration of surfactant function due to protein leakage: special interaction with fibrin monomer. J Appl Physiol 1985; 58:326.

95. Cockshutt AM, Weitz J, Possmayer F. Pulmonary surfactant-associated protein A enhances the surface activity of lipid extract surfactant and reverses inhibition by blood proteins in vitro. Biochemistry 1990; 29:8424.
96. Holm BA, Notter RH. Effects of hemoglobin and cell membrane lipids on pulmonary surfactant activity. J Appl Physiol 1987; 63:1434.
97. Fuchimukai T, Fujiwara T, Takahashi A. Artificial pulmonary surfactant inhibition by proteins. J Appl Physiol 1987; 62:429.
98. Seeger W, et al. Surface properties and sensitivity to protein-inhibition of a recombinant apoprotein C-based phospholipids mixture in vitro-comparison to natural surfactant. Biochim Biophys Acta 1991; 1081:45.
99. Seeger W, Gunther A, Thede C. Differential sensitivity to fibrinogen inhibition of SP-C- vs. SP-B-based surfactants. Am J Physiol 1992; 262:L286.
100. Venkitaraman AR, et al. Biophysical inhibition of synthetic phosphilipid-lung surfactant apoprotein admixtures by plasma proteins. Chem Phys Lipids 1991; 57:49.
101. Seeger W, et al. Lung surfactant phospholipids associate with polymerizing fibrin: loss of surface activity. Am J Respir Cell Mol Biol 1993; 9:213.
102. Gunther A, et al. Surfactant incorporation markedly alters mechanical properties of a fibrin clot. Am J Respir Cell Mol Biol 1995; 13:712.
103. Gunther A, Markart P, Kalinowski M, et al. Cleavage of surfactant-incorporated fibrin by different fibrinolytic agents. Kinetics of lysis and rescue of surface activity. Am J Respir Cell Mol Biol 1999; 21:738.
104. Schermuly RT, et al. Conebulization of surfactant and urokinase restores gas exchange in perfused lungs with alveolar fibrin formation. Am J Physiol Lung Cell Mol Physiol 2001; 280:L792.
105. Petty TL, et al. Abnormalities in lung elastic properties and surfactant function in adult respiratory distress syndrome. Chest 1979; 75:571.
106. Baker CS, et al. Damage to surfactant-specific protein in acute respiratory distress syndrome. Lancet 1999; 353:1232.
107. Pison U, et al. Altered pulmonary surfactant in uncomplicated and septicemia-complicated courses of acute respiratory failure. J Trauma 1990; 30:19.
108. Gregory TJ, et al. Bovine surfactant therapy for patients with acute respiratory distress syndrome. Am J Respir Crit Care Med 1997; 155:1309.
109. Kurashima K, et al. Surface activity of sputum from acute asthmatic patients. Am J Respir Crit Care Med I 1997; 155:1254.
110. Hohlfeld JM. Potential role of surfactant in asthma. Appl Cardiopulm Pathophysiol 2004; 13:44.
111. Liu M, Wang L, Enhorning G. Surfactant dysfunction develops when the immunized guinea-pig is challenged with ovalbumin aerosol. Clin Exp Allergy 1995; 25:1053.
112. Cheng G, et al. Compositional and functional changes of pulmonary surfactant in a guinea-pig model of chronic asthma. Respir Med 2001; 95:180.
113. Morley CJ, et al. Dry artificial lung surfactant and its effect on very premature babies. Lancet 1981; 1:64.
114. Revak SD, et al. The use of synthetic peptides in the formation of biophysically and biologically active pulmonary surfactants. Pediatr Res 1991; 29:460.

115. Häfner D, Germann P-G, Hauschke D. Effects of rSP-C surfactant on oxygenation and histology in a rat-lung-lavage model of acute lung injury. Am J Respir Crit Care Med 1998; 158:270.
116. Taeusch HW, et al. Characterization of bovine surfactant for infants with respiratory distress syndrome. Pediatrics 1986; 77:572.
117. Curstedt T, et al. Artificial surfactant based on different hydrophobic low-molecular-weight proteins. In: Lachmann B, ed. Surfactant Replacement Therapy in Neonatal and Adult Respiratory Distress Syndrome. Berlin: Springer-Verlag, 1988:332.
118. Wiseman LR, Bryson HM. Porcine-derived lung surfactant: a review of the therapeutic efficacy and clinical tolerability of a natural surfactant preparation (Curosurf) in neonatal respiratory distress syndrome. Drugs 1994; 48:386.
119. Hallman M, et al. Isolation of human surfactant from amniotic fluid and pilot study of its efficacy in respiratory distress syndrome. Pediatrics 1983; 71:473.
120. Shalamov VIu, et al. Efficacy of domestic surfactant from human amniotic fluid - surfactant-HL in complex intensive care therapy of newborns with RDS. Ross Vest Perin Ped 1999; 44:29 (Russian).
121. Rosenberg OA, et al. Pharmacological properties and therapeutic efficacy of the domestic preparations of lung surfactants. Biull Eksp Biol Med 1998; 126:455 (Russian).
122. Notter RH, et al. Lung surfactant replacement in premature lambs with extracted lipids from bovine lung lavage: effects of dose, dispersion technique and gestational age. Pediatr Res 1985; 19:569.
123. Yu S, et al. Bovine pulmonary surfactant: chemical composition and physical properties. Lipids 1983; 18:522.
124. Bligh EG, Dayer WJ. A rapid method of total lipid extraction and purification. Can J Biochem Physiol 1959; 37:911.
125. Bernhard W, et al. Commercial versus native surfactants surface activity, molecular components, and the effect of calcium. Am J Respir Crit Care Med 2000; 162:1524.
126. Rudiger M, et al. Naturally derived commercial surfactants differ in composition of surfactant lipids and in surface viscosity. Am J Physiol Lung Cell Mol Physiol 2005; 288:L379.
127. Gunther A, Seeger W. Resistance to surfactant inactivation. In: Robertson B, Taeusch HW, eds. Surfactant Therapy for Lung Disease, Lung Biology in Health and Disease. Vol. 84. New York: Marcel Dekker, Inc., 1995:269.
128. Seeger W, et al. Surfactant inhibition by plasma proteins: differential sensitivity of various surfactant preparations. Eur Respir J 1993; 6:971.
129. Halliday HL. Natural vs synthetic surfactant in neonatal respiratory distress syndrome. Drugs 1996; 51:226.
130. Arnold C, et al. A multicenter, randomized trial comparing synthetic surfactant with modified bovine surfactant extract in the treatment of neonatal respiratory distress syndrome. Pediatrics 1996; 97:1.
131. Lachmann B. Surfactant replacement in acute respiratory failure: animal studies and first clinical trials. In: Lachmann B, ed. Surfactant Replacement Therapy. New York: Springer-Verlag, 1987:212.

132. Lachmann B. Open up the lung and keep the lung open. Intensive Care Med 1992; 13:319.
133. Willson DF, et al. Instillation of calf lung surfactant extract (Calfactant) is beneficial in pediatric acute hypoxemic respiratory failure. Crit Care Med 1999; 27:188.
134. Walmrath D, et al. Bronchoscopic surfactant administration in patients with severe adult respiratory distress syndrome and sepsis. Am J Respir Crit Care Med 1996; 154:57.
135. Spragg RG, et al. Effect of recombinant surfactant protein C–based surfactant on the acute respiratory distress syndrome. N Engl J Med 2004; 351:884.
136. Bautin A, et al. Multicentre clinical trial of surfactant-BL for treatment of adult respiratory distress syndrome. Clin Stud Med 2002; 2:8 (Russian).
137. Spragg RG, et al. Acute effects of a single dose of porcine surfactant on patients with the adult respiratory distress syndrome. Chest 1994; 105:195.
138. Walmrath D, et al. Bronchoscopic administration of bovine natural surfactant in ARDS and septic shock: impact on gas exchange and haemodynamics. Eur Respir J 2002; 19:805.
139. Tsibul'kin EK. Our experience in the use of a Russian preparation of pulmonary surfactant in the treatment of acute respiratory distress syndrome and severe pneumonia in children. Anesteziol Reanimatol 1999; 2:61 (Russian).
140. Bernard GR, et al. The American-European consensus conference on ARDS: definitions, mechanisms, relevant outcomes, and clinical trial coordination. Am J Respir Crit Care Med 1994; 149:818.
141. Rosenberg O, et al. Surfactant therapy of all dose-effect relationship. Am J Respir and Crit Care Med 2002; 165:8.
142. Rusanov SIu, et al. Comparative efficacy of endotracheal aerosol and microstream introduction of Surfactant-BL in complex intensive care therapy of newborns with severe RDS. Ros Vest Perinat Ped 2003; 1:26 (Russian).
143. Spragg RG. Acute respiratory distress syndrome. Surfactant replacement therapy. Clin Chest Med 2000; 21:531.
144. Liolios A. Evidence-based medicine in intensive care unit. A debate 23rd International Symposium of Intensive Care and Emergency Medicine, Program and abstracts, Brussels, Belgium, Mar 18–21, 2003.
145. Marini JJ. Evidence-based medicine in intensive care unit. A debate 23rd International Symposium of Intensive Care and Emergency Medicine, Program and abstracts, Brussels, Belgium, Mar 18–21, 2003.
146. Greenhalgh T. How to Read a Paper: The Basics of Evidence Based Medicine. 2nd ed. London: BMJ Books, 2001.
147. Osovskikh V, Seiliev A, Rosenberg O. Exogenous surfactant: a potent drug for gastric content aspiration? Eur Respir J 2002; 20:336.
148. Osovskikh V, Seiliev A, Rosenberg O. ARDSp and ARDSexp: different responses to surfactant administration. Eur Respir J 2003; 22:551.
149. Satoh D, et al. Effect of surfactant on respiratory failure associated with thoracic aneurysm surgery. Crit Care Med 1998; 10:1660.
150. Tarasenko M, et al. Surfactant therapy–the real chance to survive for patients with severe inhalation injury. Eur Respir J 2004; 24:4127.

151. Bautin AE, et al. Natural surfactant preparation in complex therapy of acute respiratory distress syndrome in adults following open-heart surgeries. Anesteziol Reanimatol 2003; 55 (Russian).

152. Taeusch HW, Karen LU, Ramierez-Schrempp D. Improving pulmonary surfactants. Acta Pharmacol Sin 2002; 11:15.

153. Holliday LH, Speer CP. Strategies for surfactant therapy in established neonatal respiratory distress syndrome. In: Robertson B, Taeusch HW, eds. Surfactant Therapy for Lung Disease, Lung Biology in Health and Disease. Vol. 84. New York: Marcel Dekker, Inc., 1995:443.

154. Walmrath D, et al. Treatment of ARDS with a recombinant SP-C (rSP-C) based synthetic surfactant. Am J Respir Crit Care Med 2000; 161:A379.

155. Lu K, et al. Polyethylene glycol/surfactant mixtures improve lung function after HCl and endotoxin lung injuries. Am J Respir Crit Care Med 2001; 164:1531.

156. Bunt JEH. Surfactant therapy stimulates endogenous surfactant synthesis in premature infants. Crit Care Med 2000; 28:3383.

157. Mikawa K, et al. Selective intrabronchial instillation of surfactant in a patient with pneumonia: a preliminary report. Eur Respir J 1993; 6:1563.

158. Rosenberg OA, et al. Liposomic forms of natural lung surfactants for drug delivery to lung replacement therapy of different lung diseases. Proc. Liposome Advanced Progress in Drug Vaccine Delivery, London, Dec 13–17, 1999:44.

159. Oetomo SB, et al. Surfactant nebulization does not alter airflow obstruction and bronchial responsiveness to histamine in asthmatic children. Am J Respir Crit Care Med 1996; 153:1148.

160. Kurashima K. A pilot study of surfactant inhalation in the treatment of asthmatic attack. Arerugi 1991; 40:160.

161. Dulkys Y, et al. Natural porcine surfactant augments airway inflammation after allergen challenge in patients with asthma. Am J Respir Crit Care Med 2004; 169:578.

162. Rosenberg O, et al. The application of the liposome form of lung surfactant (Surfactant-BL) for complex treatment of multi-drug resistant tuberculosis. 6th International Conference "Liposome Advances", London, Dec 15–19, 2003:61.

163. Erokhin VV, et al. Peculiarities of macrophagal composition of bronchial washings in destructive TB patients after using surfactant-BL (S-BL). Proc. 13th ERS Annual Congress, Vienna. Eur Respir J 2003; 22:340.

164. Lovacheva OV, et al. Use of surfactant-BL (S-BL) in complex treatment of pulmonary TB patients. Proc. 13th ERS Annual Congress, Vienna. Eur Respir J 2003; 22:521.

165. Erpenbeck VJ, et al. Natural porcine surfactant augments airway inflammation after allergen challenge in patients with asthma. Am J Respir Crit Care Med 2004; 169:78.

166. Anzueto A, et al. Effects of aerosolized surfactant in patients with stable chronic bronchitis: a prospective randomized controlled trial. JAMA 1997; 278:1426.

18

Liposomes in Cancer Immunotherapy

Michel Adamina, Reto Schumacher, and Giulio C. Spagnoli

Institute for Surgical Research and Hospital Management, University of Basel, Basel, Switzerland

INTRODUCTION

Malignant diseases remain a leading cause of mortality and disability worldwide despite impressive advances in early diagnosis and treatment.

Immunotherapy is one of the many experimental treatments that have been proposed over four decades (1), melanoma and bladder tumors being among the first cancers targeted. Besides a steeply rising incidence [in 2000, roughly 60,000 new diagnosed melanomas and 16,000 deaths in Europe (2)], an important particularity of melanoma is its immunogenicity, early recognized and challenged with various clinical immunization protocols.

The characterization of the first melanoma-associated antigen in 1991 has opened the era of antigen-specific active immunotherapy (3). In the past decade, many clinical centers developed immunization strategies for treatment of metastatic melanoma (4). Simultaneously, refinements of molecular biology allowed for the identification of tumor-associated antigens (TAA) across a wide panel of tumors (5) and opened the scope of cancer immunotherapy to many cancers. Immunotherapy is designed as a systemic treatment, able to eradicate not only the primary tumor but also any disseminated metastases that may reside in the body's organs and tissues.

CANCER IMMUNOTHERAPY

Cancer immunotherapies fight against billions of tumor cells in a dynamic process that involves downregulation of tumor major histocompatibility complex (MHC) and/or TAA, selection of resistant tumor clones, and other mechanisms of escape as well as local/systemic immunodepression (6). Moreover, TAA are frequently also expressed in nontransformed cells. Thus, the immune system is likely to have developed some degree of tolerance toward TAA. Furthermore, nearly all preventive vaccines developed for infectious diseases are effective in that they induce an antibody response, unfit to target TAA mostly expressed intracellularly, which rely on the induction of antigen-specific cytotoxic T lymphocytes (CTL). Finally, alum, the adjuvant commonly included in commercial vaccines, is unable to support CTL generation (7).

Peculiar to antigens recognized by CTL is that they are produced inside the cells, then physiologically degraded in peptidic fragments (= epitopes) and presented on cell surfaces within MHC class I molecules' grooves. Thus, mimicking physiological pathways would suggest immunization based on the use of virus recombinant for TAA capable of infecting antigen-presenting cells (APC) (8–10). However, low expression of recombinant antigens may result from replication inactivation of the viruses, recommended to improve the safety of these reagents, and virus specific immune responses may further limit their efficacy over time. Alternatively, APC might be exogenously loaded with synthetic TAA, binding the small percentage of human leukocyte antigen (HLA) class I molecules that are present on cell surfaces in empty state.

Accordingly, epitopes have been injected in the absence (11) or presence (12–14) of adjuvants or cytokines, or loaded ex vivo on APC before injection. However, the exogenous loading of MHC molecules by soluble epitopes is quite inefficient. Moreover, peptides frequently represent poor immunogens possibly because of their fast hydrolysis by plasma or cell-associated peptidases (15,16). To circumvent this difficulty, peptide analogues resistant to enzymatic digestion have been designed for a number of epitopes (17,18) and carrier formulations like liposomes (19) have been developed, which may create a depot effect and be preferentially endocytosed by APC.

Liposomes in Cancer Immunotherapy

Different types of liposomes have long been used as vaccine components (20–22). Interestingly, antigens taken up within particles were shown to be presented more effectively by dendritic cells than antigens taken up from solution (23). Furthermore, liposomes containing peptides derived from MUC1 human TAA and monophosphoryl lipid A as adjuvant were shown to be able to induce specific immune responses in vitro (24).

Small liposomes (≤100 nm) containing gel-state phospholipids [such as distearyl phosphatidyl choline (DSPC)] and cholesterol markedly extend their stability in vivo (25) and display a high absorption in lymphnodes (26) (up to 70%) upon subcutaneous injection. However, sterically stabilized liposomes (SSL) (27,28) are characterized by polyethylene glycol (PEG) coating, yet display a much more prolonged bioavailability and decreased clearance from the reticuloendothelial system, as compared to conventional liposomes. Noteworthy, SSL containing a model antigenic protein have been shown to induce both class I and class II restricted immune responses in mice (29) upon presentation by professional APC.

Long immunization courses are known to be required before specific responsiveness to TAA becomes evident and eventually translates in clinical responses (30,31). The particulate form of the encapsulated TAA may allow for a better uptake and presentation by professional APC, and the stabilization conferred by the PEG coating allow for a sustained release of the TAA in biological systems.

We constructed SSL containing immunodominant HLA-A2.1 restricted CTL epitopes derived from Mart-1 melanoma TAA (32,33) and we showed that antigenic peptides encapsulated in SSL display a significantly higher capacity to resist plasma enzymatic hydrolysis, to stimulate T-cell proliferation, and to induce specific CTL responses than their soluble (S) counterparts in vitro.

MATERIALS AND METHODS

Liposome Production and Characterization (19)

SSL were generated upon solubilization in chloroform of cholesterol, DSPC, and distearyl phosphatidyl ethanolamine (DSPE)-PEG^{2000}. The molar ratio used was 1:0.65 phospholipids to cholesterol. The lipids were then dried down in a rotatory evaporator. Upon rehydration and mechanical dispersion into a solution of Mart-1 peptides in phosphate buffered saline (PBS), multilamellar liposomes were formed. SSL batches were then submitted to five cycles of freeze–thaw. Sizing down of SSL to 100 nm small unilamellar vesicles was achieved through extrusion. Separation of SSL encapsulated and soluble peptide was obtained by gel chromatography and extensive dialysis. Finally, SSL suspensions were sterile filtered. SSL preparations were stored at 4°C in the dark to prevent lipid oxidation. Size of liposomes in individual SSL batches was tested by photon correlation spectroscopy. Unilamellarity of the liposomes was verified by electron microscopy. Quantification of SSL encapsulated peptides was performed by fluorescamine assay. SSL preparations were negative (<10 pg/mL) when tested for endotoxin contamination.

Immunological Assays

NA-8 melanoma cells not expressing the TAA under investigation served as targets of specific CTL in ^{51}Cr release assays, upon pulsing with S or

SSL encapsulated Mart-1$_{27-35}$ (M27-35) or Mart-1$_{26-35}$ (M26-35) epitopes. These reagents were also used to stimulate CTL proliferation, measured as 3H-Thymidine incorporation, in the presence of immature dendritic cells (iDC) as APC. Induction of TAA-specific CTL was attempted in healthy donors' peripheral blood mononuclear cells (PBMC) and in patient-derived tumor-infiltrating lymphocytes (TIL), and monitored by ^{51}Cr release assays and tetramer staining. For a detailed description of "Materials and Methods" (19).

RESULTS

Immunorecognition of Mart-1 Epitopes by Specific CTL

The capacity of SSL-based antigenic preparations to sensitize target cells to the killing by CTL recognizing both M26-35 and M27-35, was tested (Fig. 1). NA-8 melanoma cells were preincubated in the presence of TAA or control peptides in soluble form or included in SSL. Maximal killing by specific CTL was detected upon incubation with 0.4 μg/mL of soluble M27-35, but significantly lower (0.016–0.001 μg/mL) concentrations of M26-35. Encapsulation into SSL did not modulate the targeting capacity of M26-35. In contrast, inclusion of M27-35 in SSL resulted in a fivefold improvement of their sensitizing capacity, with maximal killing observed at 0.08 μg/mL TAA concentration.

Antigenic Peptide Degradation in the Presence of Human Plasma

We then asked whether encapsulation into SSL could protect peptides from plasma degradation. As previously reported (17), S M27-35 was rapidly degraded (80% loss of activity) within 45 minutes of incubation in the presence of human plasma. SSL M27-35 effectively protected the TAA, with a loss of activity limited to 30%. On the other hand, S M26-35 proved more resistant to hydrolysis with no loss of activity being detectable upon 45 minutes incubation in the presence of plasma (Fig. 2).

Induction of Specific CTL Proliferation by SSL

The efficacy of immunotherapy is critically dependent on a high expansion of specific CTL, whose proliferation not only requires adequate costimulation but also an increased density of MHC/peptide complexes on APC, as compared to mere targeting of cytotoxic activity (34,35).

We comparatively analyzed the capacity of S or SSL encapsulated M26-35 to induce antigen-specific proliferative responses in CTL clones. HLA matched allogenic iDC served as APC in the presence of TAA. At saturating

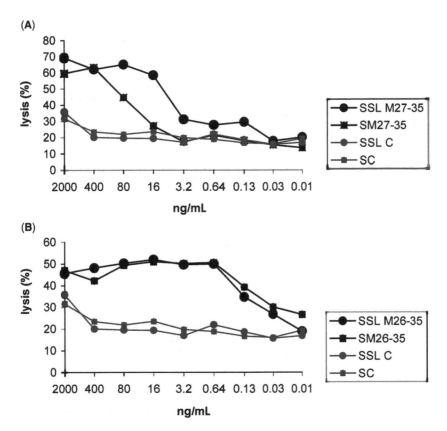

Figure 1 Immunorecognition of Mart-1 epitopes by specific cytotoxic T lymphocytes (CTL). NA-8 target cells were ^{51}Cr labeled and incubated in the presence of graded amounts of M27-35 (**A**) or M26-35 (**B**), or control peptides in solution (*S*) or encapsulated into liposomes sterically stabilized liposomes (SSL). Cells were then washed and cultured in the presence of a specific CTL clone at 5:1 effector target ratio. Supernatants were then harvested and ^{51}Cr release was measured by a gamma counter. Data are reported as mean percentage-specific target cell lysis from triplicate wells. Standard deviations, never exceeding 10% of the reported values, were omitted. *Source*: From Ref. 19.

concentrations, CTL proliferation was comparably induced by either antigen formulations. However, at lower antigen concentrations ($\leq$200–25 ng/mL), a significantly higher proliferation was detected upon stimulation by SSL M26-35, as compared to S M26-35 (Fig. 3).

   Notably, greater than 80% inhibition of CTL proliferation was observed after prolonged (six hours) preincubation of S M26-35 with iDC, as compared to 50% for SSL M26-35. The decreased stimulatory capacity of

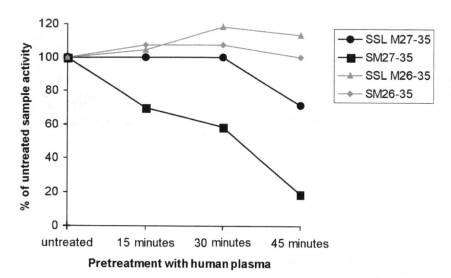

Figure 2 Epitope degradation in the presence of human plasma. M27-35 (1 μg/mL) or M26-35 (8 ng/mL) peptides in S or sterically stabilized liposomes were incubated for the indicated times in the presence of undiluted human plasma. The different preparations were then used to pulse ^{51}Cr labeled NA-8 cells for two hours. Upon coculture in the presence of a specific cytotoxic T lymphocytes clone at 5:1 effector target ratio in triplicate samples, supernatants were harvested and ^{51}Cr release was measured. Data are reported as percentage of the cytotoxic activity detected in the presence of the corresponding untreated M27-35 or M26-35 preparations. Standard deviations, never exceeding 10% of the reported values were omitted. *Source*: From Ref. 19.

either TAA upon prolonged incubation with APC was partially circumvented by increasing antigen concentrations (Fig. 4).

SSL Induce Specific CTL in PBMC from Healthy Donors and from Melanoma-Infiltrating Lymphocytes

We compared the capacity of S M26-35 and SSL M26-35 formulations to induce specific CTL by stimulating PBMC from healthy donors. Following repeated (2,3) stimulations, specific cytotoxic activity could be generated. In a number of donors CTL were only detectable in cultures stimulated with SSL M26-35 and no specific cytotoxicity was observed upon stimulation with S M26-35 (Fig. 5).

Regarding TIL from melanoma patients, SSL and SM26-35 were equally effective at 20 μg/mL TAA concentration, with induced CTL precursor frequencies (CTLp) on limiting dilution analysis (9) of about $20/10^6$ (Fig. 6, *panel A*). However, at lower (physiological) antigen concentration, while SSLM26-35 peptide retained its full immunogenic capacity,

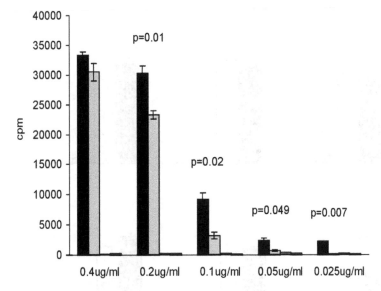

Figure 3 Induction of specific cytotoxic T lymphocytes (CTL) proliferation by Mart-1 tumor-associated antigens. Cells from a Mart-1 specific CTL clone were cocultured with irradiated, human leukocyte antigen (HLA) matched, immature dendritic cells and M26-35 or control peptides in solution (*shaded and white columns,* respectively), or included into sterically stabilized liposomes (*black and striped columns,* respectively). Proliferation of triplicate wells was assessed by 3H-Thymidine on day 3 of culture.

SM26-35 peptide was virtually unable to induce specific responsiveness. Comparable results were also observed on tetramer staining with specific reagents (Fig. 6, *panel B*).

DISCUSSION

Cancer immunotherapy requires highly immunogenic reagents (30,36,37). A number of clinical protocols currently rely on the in vivo administration of soluble TAA (12,30,36) or the ex vivo pulsing of APC with TAA prior to reinfusion (37,38). However, synthetic TAA are subject to degradation by soluble or cell membrane–associated peptidases (15–17) and may represent poor immunogens. In this context, because persistence of the antigen (39) is emerging as a critical factor for the induction of specific immune responses, the respective bioavailabilities of different vaccine formulations assume crucial relevance.

Liposomes have long been used to carry drugs, proteins, peptides, or DNA (21,22,40), and induction of CTL by liposome-carried antigens has

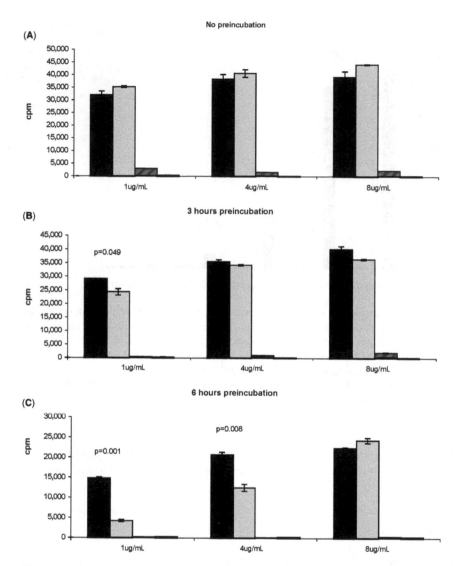

Figure 4 Antigen persistence on immature dendritic cells (iDC). iDC were cultured for three or six hours (**B** and **C**, respectively) in the presence of M26-35 or control peptides in solution (*shaded and white columns*, respectively) or encapsulated into sterically stabilized liposomes (*black* respectively) at the indicated concentrations. A specific cytotoxic T lymphocyte (CTL) clone was then added. Data reported (**A**) refers to wells where iDC, CTL, and immunogenic materials were simultaneously added without preincubation steps. 3H-Thymidine incorporation of triplicate wells was measured on day 3. *Source*: From Ref. 19.

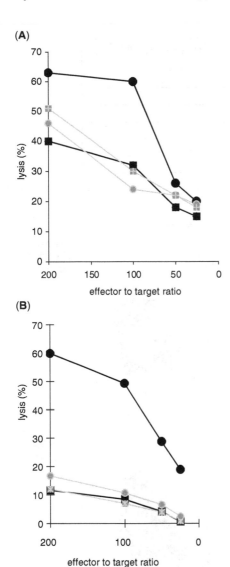

(A)

(B)

Figure 5 Induction of Mart-1 specific cytotoxic T lymphocytes by tumor-associated antigens encapsulated into sterically stabilized liposomes (SSL). Healthy donor PBMC were cultured for one week in the presence of M 26-35 or control peptides at 10 μg/mL in solution (*gray symbols*) or encapsulated into SSL (*black symbols*). rIL-2 (20 U/mL) was added and cultures were restimulated on day 10 and weekly thereafter. Cytotoxicity was tested against target cells pulsed with S M26-35 (●) or control peptide (■). Data from two donors (**A** and **B**) are reported. *Source*: From Ref. 19.

(A)

	CTLp frequency per 10⁶
SM26-35 20μg/mL	24
SM26-35 2μg/mL	1
SM26-35 0,2μg/mL	2
SSL M26-35 20μg/mL	22
SSL M26-35 2μg/mL	25
SSL M26-35 0,2μg/mL	1

(B)

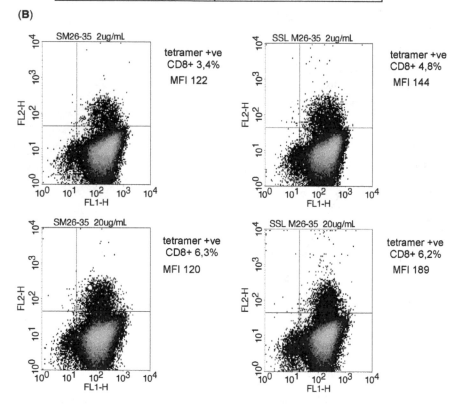

Figure 6 Induction of tumor-associated antigens–specific CTL in melanoma derived TIL. (**A**) Melanoma-derived TIL were stimulated in a limiting dilution setting with decreasing amounts of SM26-35 or SSLM26-35. CTL precursor frequencies are expressed as number of specific effectors per 10^6 CTL. (**B**) Melanoma-derived TIL were stimulated in bulk cultures in the presence of the different antigen formulations and concentrations, and stained with FITC-labeled anti-CD8 and PE labeled M27-35 tetramers. Percentages of tetramer-positive CD8+ cells and mean fluorescence intensities are shown. *Abbreviations*: CTL, cytotoxic T lymphocytes; TIL, tumor-infiltrating lymphocytes; SSL, sterically stabilized liposomes; FITC, fluorescein isothiocyanate. *Source*: From Ref. 19.

been demonstrated in animal models (29,41), but there is a paucity of data on humans (24,42).

We have developed an innovative vaccine formulation for clinical tumor immunotherapy by using epitopes from the Mart-1 melanoma TAA (32,33) and liposome technology. Considering the peculiarity of tumor-specific immune response (43), we have focused on SSL liposomes endowed with increased bioavailability, albeit showing reduced cellular uptake, possibly related to steric hindrance due to PEG coating (28). However, effectiveness of immunotherapy critically depends on the expansion of specific T-cells. CTL proliferation requires more stringent molecular interactions than mere CTL-mediated target killing (34,44). Remarkably, TAA encapsulation into SSL provides a significant advantage over its soluble counterpart in CTL immunorecognition and proliferation assays, particularly when physiologically amounts of TAA were used, or prolonged incubations in the presence of plasma and APC were applied. Noteworthy, these data are consistent with a recent report (41) indicating that in vivo presentation of exogenous antigen to CTL is markedly inefficient due to the rapid turnover on APC. Accordingly, specific CTL stimulation can only be achieved in the presence of high epitope concentrations. In this context, encapsulation of TAA into SSL is likely to provide enhanced immunogenicity.

Most importantly, M26-35 peptides encapsulated into SSL efficiently induce specific CTL in PBMC from healthy donors, following few restimulation courses. In our experience, CTL induction can be induced by soluble peptides as well (17,33), but usually requires longer culture times (more than four restimulations). Stimulation assays using melanoma TIL as responder cells further support an enhanced immunogenicity of SSL M26-35, as detected by CTLp frequency analysis and tetramer staining.

CONCLUSION

Our data indicate that SSL-containing antigenic peptides could represent an effective alternative to soluble synthetic epitopes in cancer immunotherapy, both for direct in vivo administration and for ex vivo pulsing. Their lack of intrinsic immunogenicity, favoring repeated use, low cost, and easy manufacture under good medical pratice (GMP) conditions suggest broad clinical applicability.

REFERENCES

1. Woodruff MF, Nolan B. Preliminary observations on treatment of advanced cancer by injection of allogeneic spleen cells. Lancet 1963; 13:426.
2. de Vries E, Coebergh JW. Cutaneous malignant melanoma in Europe. Eur J Cancer 2004; 40:2355.

3. Van der Bruggen P, et al. A gene encoding an antigen recognized by cytolytic T lymphocytes on a human melanoma. Science 1991; 254:1643.

4. Adamina M, Oertli D. Antigen specific active immunotherapy: lessons from the first decade. Swiss Med Wkly. In press.

5. Bolli M, Kocher T, Adamina M, et al. Tissue microarray evaluation of Melanoma antigen E (MAGE) tumor-associated antigen expression: potential indications for specific immunotherapy and prognostic relevance in squamous cell lung carcinoma. Ann Surg 2002; 236:785.

6. Dunn GP, et al. Cancer immunoediting: from immunosurveillance to tumour escape. Nat Immunol 2002; 3:991.

7. O'Hagan DT, et al. Recent advances in the discovery and delivery of vaccine adjuvants. Nat Rev Drug Discov 2003; 2:727.

8. Rosenberg SA, et al. Immunizing patients with metastatic melanoma using recombinant adenoviruses encoding MART-1 or gp100 melanoma antigens. J Natl Cancer Inst 1998; 90:1894.

9. Oertli D, et al. Rapid induction of specific cytotoxic T lymphocytes against melanoma associated antigens by a recombinant vaccinia virus vector expressing multiple immunodominant epitopes and costimulatory molecules in vivo. Hum Gene Ther 2002; 13:569.

10. Zajac P, et al. Phase I/II clinical trial of a nonreplicative vaccinia virus expressing multiple HLA-A0201-restricted tumor-associated epitopes and costimulatory molecules in metastatic melanoma patients. Hum Gene Ther 2003; 14:1497.

11. Parmiani G, et al. Cancer immunotherapy with peptide based vaccines: What have we achieved? Where are we going? J Nat Cancer Inst 2002; 94:805.

12. Rosenberg SA, et al. Immunologic and therapeutic evaluation of a synthetic peptide vaccine for the treatment of patients with metastatic melanoma. Nat Med 1998; 4:321.

13. Peterson AC, et al. Immunization with Melan-A peptide-pulsed peripheral blood mononuclear cells plus recombinant human interleukin-12 induces clinical activity and T-cell responses in advanced melanoma. J Clin Oncol 2003; 21:2342.

14. Weber J, et al. Granulocyte-macrophage colony-stimulating factor added to a multipeptide vaccine for resected Stage II melanoma. Cancer 2003; 97:186.

15. Cavazza A, et al. Hydrolysis of the tumor-associated antigen epitope gp100(280–288) by membrane-associated and soluble enzymes expressed by immature and mature dendritic cells. Clin Immunol 2004; 111.

16. Albo F, et al. Degradation of the tumor antigen epitope gp100(280–288) by fibroblast-associated enzymes abolishes specific immunorecognition. Biochim Biophys Acta 2004; 1671.

17. Blanchet JS, et al. A new generation of Melan-A/MART-1 peptides that fulfill both increased immunogenicity and high resistance to biodegradation: implications for molecular anti-melanoma immunotherapy. J Immunol 2001; 167:5852.

18. Smith JW, et al. Adjuvant immunization of HLA-A2-positive melanoma patients with modified gp100 peptide induces peptide-specific CD8+ T-cell responses. J Clin Oncol 2003; 21:1562.

19. Adamina M, et al. Encapsulation into sterically stabilized liposomes improves the immunogenicity of melanoma associated Mart-1/Melan-A epitope. Br J Cancer 2004; 90:263.

20. Gluck R. Liposomal presentation of antigens for human vaccines. Pharm Biotechnol 1995; 6:325.
21. Alving CR, et al. Liposomes as carriers of peptide antigens: induction of antibodies and cytotoxic T lymphocytes to conjugated and unconjugated peptides. Immunol Rev 1995; 145:5.
22. Gregoriadis G, et al. Vaccine entrapment in liposomes. Methods 1999; 19:156.
23. Shen Z, et al. Cloned dendritic cells can present exogenous antigens on both MHC class I and class II molecules. J Immunol 1997; 158:2723.
24. Agrawal B, et al. Rapid induction of primary CD4+ and CD8+ T cell responses against cancer-associated MUC-1 peptide epitopes. Int Immunol 1998; 157: 2089.
25. Senior J, Crawley JC, Gregoriadis G. Tissue distribution of liposomes exhibiting long half-lives in the circulation after intravenous injection. Biochim Biophys Acta 1985; 839:1.
26. Oussoren C, Storm G. Liposomes to target the lymphatics by subcutaneous administration. Adv Drug Deliv Rev 2001; 50:143.
27. Woodle MC, Lasic DD. Sterically stabilized liposomes. Biochim Biophys Acta 1992; 1113:171.
28. Allen TM. Long-circulating (sterically stabilized) liposomes for targeted drug delivery. Trends Pharmacol Sci 1994; 15:215.
29. Ignatius R, et al. Presentation of proteins encapsulated in sterically stabilized liposomes by dendritic cells initiates CD8(+) T-cell responses in vivo. Blood 2000; 96:3505.
30. Marchand M, et al. Tumor regressions observed in patients with metastatic melanoma treated with an antigenic peptide encoded by gene MAGE-3 and presented by HLA-A1. Int J Cancer 1999; 80:219.
31. Coulie P, van der Bruggen P. T-cell responses of vaccinated cancer patients. Curr Opin Immunol 2003; 15:131.
32. Kawakami Y, et al. Identification of the immunodominant peptides of the MART-1 human melanoma antigen recognized by the majority of HLA-A2-restricted tumor infiltrating lymphocytes. J Exp Med 1994; 180:347.
33. Valmori D, et al. Enhanced generation of specific tumor reactive CTL in vitro by selected Melan-A/MART-1 immunodominant peptide analogues. J Immunol 1998; 160:1750.
34. Gervois N, et al. Suboptimal activation of melanoma infiltrating lymphocytes (TIL) due to low avidity of TCR/MHC-tumor peptide interactions. J Exp Med 1996; 183:2403.
35. Lanzavecchia A, Sallusto F. Regulation of T cell immunity by dendritic cells. Cell 2001; 106:263.
36. Jaeger E, et al. Generation of cytotoxic T-cell responses with synthetic melanoma-associated peptides in vivo: implications for tumor vaccines with melanoma-associated antigens. Int J Cancer 1996; 66:162.
37. Thurner B, et al. Vaccination with mage-3A1 peptide-pulsed mature, monocyte-derived dendritic cells expands specific cytotoxic T cells and induces regression of some metastases in advanced stage IV melanoma. J Exp Med 1999; 190:1669.
38. Nestle FO, et al. Vaccination of melanoma patients with peptide- or tumor lysate-pulsed dendritic cells. Nat Med 1998; 4:328.

39. Zinkernagel RM, Hengartner H. Regulation of the immune response by antigen. Science 2001; 293:251.
40. Lasic DD. Novel applications of liposomes. Trends Biotechnol 1998; 16:307.
41. Ludewig B, et al. Rapid peptide turnover and inefficient presentation of exogenous antigen critically limit the activation of self-reactive CTL by dendritic cells. J Immunol 2001; 166:3678.
42. Neidhart J, et al. Immunization of colorectal cancer patients with recombinant baculovirus-derived KSA (Ep-CAM) formulated with monophosphoryl lipid A in liposomal emulsion, with and without granulocyte-macrophage colony-stimulating factor. Vaccine 2004; 22:773.
43. Zinkernagel RM. Immunity against solid tumors? Int J Cancer 2001; 93:1.
44. Schild H, et al. Limit of T cell tolerance to self proteins by peptide presentation. Science 1990; 247:1587.

19

Preparation of Oil-in-Water Emulsions Stabilized by Liposomes for Vaccine Delivery: Expected Immune Responses[†]

Gary R. Matyas and Carl R. Alving

Division of Retrovirology, Department of Vaccine Production and Delivery, U.S. Military HIV Research Program, Walter Reed Army Institute of Research, Silver Spring, Maryland, U.S.A.

INTRODUCTION

Emulsions have been used extensively in immunology and vaccine research for decades. The best-known emulsion is Freund's adjuvant, which is a water-in-oil emulsion (1). Complete Freund's, containing mycobacteria, is universally regarded as too toxic for human use, but incomplete Freund's (lacking mycobacteria) has been used as a potent adjuvant formulation for a diversity of vaccines in more than a million people (2–4). Freund's incomplete adjuvant is currently being developed for certain human and veterinary vaccines (5). Other vaccines using emulsion technology have been developed and are either licensed or in clinical trials (6–9). An oil-in-water emulsion (MF59) has been developed and manufactured by Novartis, Emeryville, California, U.S.A. It has been studied extensively in animals and is used in a licensed influenza vaccine in Europe (10–12). Prior to the 2004–2005 influenza season, more that 11 million doses had been

---

[†] Disclaimer: The information contained herein reflects the views of the authors and does not represent those of the Department of the Army or the Department of Defense.

administered and the vaccine was proven to be safe. Clinical trials have also recently been conducted with MF59 adjuvanted antigens for cytomegalovirus (13) and HIV (14). Glaxo SmithKline, Brentford, Middlesex, U.K. has developed an oil-in-water emulsion (AS02) that has been used in clinical trials for a malaria vaccine (15–17), melanoma tumor expressing MAG3 protein (18), and hepatitis B (19). This emulsion contains monophosphoryl lipid A and QS-21 as adjuvants.

We have developed an oil-in-water emulsion using light mineral oil, and liposomes which has been tested as a therapeutic vaccine platform in phase I and II clinical trials in colo-rectal and prostate cancer patients (2,20–23). These liposomes, called Walter Reed liposomes, have been shown to be potent adjuvants by themselves (24–27). During emulsification process, some of the liposomes are broken to stabilize the emulsion (2,28,29). This liposomal emulsion has been shown to induce both antibody and cellular immunity (2,20–28,30). In this chapter, we describe the manufacture of Walter Reed liposomes (29–32), their subsequent emulsification with light mineral oil, and factors affecting emulsion stability (29,30).

MANUFACTURE OF LIPOSOMES

Glassware

The glassware used in the manufacture of liposomes should be thoroughly cleaned and free of residual detergents. Phosphate-based detergents should be avoided. Disposable glass pipets also should be avoided, because the blue graduations are soluble in chloroform and will contaminate the liposomes. The preparation of liposomes utilizes rotary evaporation and lyophilization, which concentrates contaminants leading to impure and potentially leaky liposome formulations. There are several cleaning methods appropriate for the glassware that can be used for liposomes manufacture. (i) *Chromic-sulfuric acid*—this method has been utilized for many years and thoroughly cleans the glassware. Extensive rinsing with deionized water is required to ensure complete removal of the acid. Chromic-sulfuric acid is extremely caustic and represents safety concerns for personnel when used. In addition, spent chromic-sulfuric acid must be disposed as hazardous wastes. (ii) *RBS-35* from Pierce Biotechnology (Rockford, Illinois, U.S.A.) is a sodium hydroxide–based cleaning solution that contains phosphate, which can easily be rinsed away. We have used this method for several years, but concerns about disposal of phosphate-based cleaning agents in wastewater disposal system have caused us to switch to alternative agents. Pierce Biotechnology now has a phosphate-free formulation called RBS-pF. However, we have not tested this reagent. (iii) *Crystal Simple Green* from Chagar Corporation (Hamden, Connecticut, U.S.A.) is a biodegradable, noncaustic cleaning agent. It is diluted 1:10 and the glassware are soaked in it at least for 24 hours. The glassware are removed and extensively washed with deionized water prior to use. We have even used Crystal Simple Green as a cleaning agent for glass pipets. (iv) *CIP 100*® formulated alkaline cleaner (Steris Mentor, Ohio, U.S.A.) is a phosphate-free, potassium hydroxide–based proprietary cleaning agent that we have used extensively for the washing of glassware prior to Current Good Manufacturing Practices (cGMP) manufacturing of liposomes for human clinical trials.

In order to ensure that the glassware are free of endotoxin, the washed glassware are depyrogenated. The openings of the glassware are covered with aluminum foil and the glassware is then placed in an oven set at 220°C for at least three hours. We typically bake the glassware overnight. The glassware are allowed to cool to room temperature prior to use. Those pieces of glassware that need to remain sterile are only unwrapped in a biological safety cabinet.

Preparation of Chloroform and Lipid Stocks

Liposomes are made by first mixing the lipids dissolved in chloroform. Chloroform is highly unstable, undergoing free radical degradation and is consequently sensitive to light, heat, and oxygen. It is essential that the chloroform be highly purified, stabilized by the addition of ethanol at 0.75% or higher concentrations, and used within 90 days of manufacture. Either chloroform is purchased that meets these criteria directly from Honeywell Burdick & Jackson (Muskegon, Michigan, U.S.A.) or distilled before use. For distillation, the chloroform is placed in a 3-L round bottom distillation flask that is placed in an electric heating mantel. A water-cooled distillation column is attached to the flask. Chloroform boils at 61°C to 62°C. The first 100–200 mL is discarded to ensure that the chloroform is pure. Immediately after distillation absolute ethanol is added to a final concentration of 0.75%. The chloroform is stored in a brown bottle after distillation.

Highly purified lipids are essential to obtaining liposomes that do not leak antigen. Oxidized or unsaturated fatty acids in lipids adversely affect the uniform stacking of the lipids in the membrane allowing for leakage of liposome contents. We use synthetic, unsaturated phospholipids 1,2-dimyristoyl-*sn*-glycerol-3-phosphocholine (DMPC), and 1,2-dimyristoyl-*sn*-glycero-3-phospho-*rac*-(1-glycerol) (DMPG) together with cholesterol from Avanti Polar Lipids, Inc. (Alabaster, Alabama, U.S.A.). We have found these lipids to be highly purified and have excellent lot-to-lot reproducibility. The adjuvant used in the liposomes is lipid A purified from the lipopolysaccaride from *Salmonella minnesota* R595. It is essentially in the monophosphate form and can be obtained from Avanti Polar Lipids Inc., List Biological Laboratories (Campbell, California, U.S.A.) or Corixa Corporation (Seattle, Washington, U.S.A.).

The phospholipids are dissolved in chloroform at the concentrations indicated in Table 1. The lipids are placed in a volumetric flask using a

Table 1 Preparation of Lipid Stock Solutions

Lipid	Molecular weight	Stock concentration	Mass required	Volume (mL)	Solvent
DMPC	667.94	180 mM	30.5 g	250	Chloroform
DMPG	688.86	20 mM	3.32 g	250	Chloroform
Cholesterol	386.66	150 mM	14.5 g	250	Chloroform
Lipid A	1955.81	1 mg/mL	10 mg	10	Chloroform: methanol (9:1, v/v)

Abbreviations: DMPC, dimyristoyl-*sn*-glycerol-3-phosphocholine; DMPG, dimyristoyl-*sn*-glycero-3-phospho-*rac*-(1-glycerol).

wide-bore glass funnel. Chloroform is added to the funnel, rinsing any residual lipids into the flask. DMPC and cholesterol are readily soluble at the concentrations used. Chloroform is added just below the neck and the flask is shaken to facilitate the dissolution of the lipids. Chloroform is then added up to the calibration line. DMPG is just soluble in 100% chloroform at the 20 mM concentration used. The flask can be warmed in a 37°C water bath, to facilitate dissolution. In addition, a stir bar can be placed in the flask to stir and dissolve the DMPG. Up to 10% methanol can be added to the DMPG to aid in dissolving, but methanol tends to lead to foaming during rotary evaporation, and therefore its use is not recommended.

Lipid A stocks are made at 1 mg/mL in chloroform:methanol mixture in the ratio 9:1 (v/v). To minimize lipid A contamination of the laboratory, lipid A is not weighed. The manufacturer's weight is used directly and the chloroform:methanol is added to the vial of lipid A. The solution is then transferred to a graduated cylinder with a ground glass stopper. Chloroform:methanol is added to obtain a 1 mg/mL solution. The small amount of methanol used in lipid A solution does not represent a major problem with foaming during rotary evaporation.

The ground glass stoppers on the volumetric flasks containing the lipid solutions are taped down to ensure that they do not work loose during storage. The lipid solutions are stored at −20°C for up to 90 days or until the chloroform has expired. The solutions are warmed to room temperature prior to use. The DMPG solution should be warmed in the 37°C water bath to aid in redissolving the DMPG, but is allowed to return to room temperature before use.

Mixing of Lipids, Rotary Evaporation, and Application of High Vacuum

Walter Reed liposomes are formulated with a lipid molar ratio of 9:7.5:1 for DMPC:cholesterol:DMPG. The phospholipids concentration of the

Table 2 Lipids Needed to Make 20 mL of a 150 mM Phospholipid Phosphate Liposomes with a Lipid A Dose of 20 µg/50 µL Injection Dose

Lipid	Stock solution	Amount of lipid needed	Volume (mL)
DMPC	180 mM	2700 µmol	15
Cholesterol	150 mM	2250 µmol	15
DMPG	20 mM	300 µmol	15
Lipid A	1 mg/mL	8 mg	8

Abbreviations: DMPC, dimyristoyl-*sn*-glycerol-3-phosphocholine; DMPG, dimyristoyl-*sn*-glycero-3-phospho-*rac*-(1-glycerol).

liposomes can range from 25 to 250 mM. For the preparation of liposomal emulsions, phospholipid concentrations need to be 100 mM or higher. The lipid solutions described in the previous section are formulated so that equal volumes are used for each of the solutions—DMPC, cholesterol, and DMPG. Lipid A is added to give the correct dose required for the immunization volume. Table 2 contains an example in which 20 mL of 150 mM phospholipid phosphate liposomes with a lipid A dose of 20 µg/50 µL injection dose are to be made.

The lipids are transferred to pear-shaped or round bottom flask. The flask size used is approximately 10 times the volume of lipids used. Pear-shaped flasks are used for volumes up to 250 mL because the pear shape gives a more uniform dried lipid layer. For flask sizes greater than 250 mL, round bottom flasks are used due to lack of availability of larger pear-shaped flasks. For example, in Table 2, 500 mL round bottom flask would be used to add 53 dl of lipid solution. The lipid solutions are measured with glass pipets or graduated cylinders and transferred to the flask. If lipid antigens such as glycolipids, prostaglandins, or sterols are to be used, they are dissolved in chloroform or the appropriate solvents and are added to the round bottom flask.

The flask is placed on a rotary evaporator (Büchi, Model EL131, Brinkman Instruments Inc., Westbury, New York, U.S.A.) fitted with a ground-glass solvent trap, which was depyrogenated before use. A circulating water bath is set at 4°C and used to cool the condensing coils of the evaporator. The water bath of the evaporator is set a 40°C. An oil-free vacuum pump (Büchi, Model B-171 Westbury, New York U.S.A.) is connected to the evaporator. The vacuum is set to 200 mbar. The flask is lowered approximately one-third of the way into the water bath and rotated at 80 rpm. The solvent will rapidly evaporate and collect in the collection reservoir. The flask contents will become viscous and the volume of the solvent stream collecting in the solvent will stop flowing. The vacuum is then increased 100 mbar and flask rotation is increased to 150 rpm. Evaporation is allowed to continue until the solvent is removed and a dry lipid layer coats the flask. The evaporation process should be monitored for foaming. Foaming will carry the lipids out of the flask up into the rotary evaporator effectively decreasing the lipids added to the flask and will contaminate the evaporator. The solvent in the sample flask will bubble and boil, but if foam develops, the stopcock should be immediately turned to release some of the vacuum. The flask should be raised out of the water bath or the vacuum pressure reduced and the stopcock closed, thereby allowing the evaporation to continue.

After the solvent has evaporated, the flask should be removed from the evaporator and immediately covered with a piece of autoclaved, hardened ashless filter paper (No. 541, Whatman International Ltd., Maidstone, U.K.). The filter paper is secured in place with the help of a rubber band. The flask is then placed in a desiccator, which has a silicon-sealing ring (Dryseal, Wheaton Scientific Products Inc., Millville, New Jersey, U.S.A.). Vacuum grease should be avoided. The desiccator is attached to the oil-free vacuum pump, which is set at as high a vacuum pressure as possible (11 mbar). This ensures that the vacuum pump will constantly run, as it is unable to obtain this pressure. This application of high vacuum removes any residual solvents left over from the rotary evaporation process. High vacuum should be applied for a minimum of one hour for small amounts of liposomes (1–5 mL) to 12 to 24 hours for large amounts of liposomes (>100 mL). The lipids used in the example should be under vacuum for at least three hours, because it is 20 mL of 150 mM liposomes.

Lyophilization

After removing the dried lipids from the desiccator, the flask is placed in a biological safety cabinet. Antigen solution can be added directly to the dried lipids for incorporation into liposomes. However, we have found that suspension of the dried lipids in water and subsequent lyophilization prior to the addition of antigen increases the antigen encapsulation efficiency and reduces the variability of encapsulation of antigen (32,33). The filter paper is removed and water for injection is added to the flask. The final volume of the liposome solution should be 50 mM to ensure easy manipulation. Because the lipids take up a considerable volume, approximately, 50% of the final volume should be added as water to the dried lipids. The flask is stoppered with a depyrogenated ground-glass stopper and is then shaken until all of the lipids are removed from the flask. The final volume is then adjusted with water to the appropriate volume. For example, where 20 mL of 150 mM liposomes are being made, the hydrated liposomes should be 60 mL. The hydrated liposomes are transferred to depyrogenated glass vaccine vials. The vials are typically not filled more than half full. The size and number of vials used is not important. However, smaller vaccine vials have a smaller surface area and require longer lyophilization time. In addition, the authors have found that the use of multiple vials for lyophilization and subsequent encapsulation with antigen and pooling of the liposomes allows for a more reproducible encapsulation efficiency of antigen (32,33). After the hydrated liposomes are transferred to the vials, the flask is rinsed twice with water to ensure that the lipids are removed. Each rinse is divided evenly between all of the vials. The vials are stoppered with autoclaved three-prong butyl lyophilization stoppers using sterile forceps. For example, the hydrated lipids should be dispersed between 20-mL vials.

The vials containing the liposomes are frozen by placing them in a −80°C freezer or directly in the lyophilizer (Advantage model with computer control and a lyophilization chamber with stoppering capacity, The VirTis Company, Gardiner, New York, U.S.A.) at −50°C to −60°C. The freezing time depends on the size of the vials and the volume placed in the vials, but typically is four to six hours. It is imperative that the vials be completely frozen. Application of vacuum to the vial containing the liquid pulls the liquid out of the vials and subsequently destroys the samples.

The vials are transferred to the lyophilizer and stoppers are pulled up to the notch in the stopper. This allows the slit in the stopper to be above the rim of the vial exposing the frozen hydrated liposomes to the vacuum. The operator should wear sterile gloves and use sterile forceps during this process. The lyophilizer is programmed as indicated in Table 3. In order to ensure that the vials are frozen prior to the application of vacuum and did not partially thaw when the stoppers were loosened, an additional freezing time of two hours is applied prior to the start of the vacuum (Step 1). The authors have used this lyophilization program extensively and have obtained consistently good results, but they have not done extensive experiments to determine if this is the optimal lyophilization program. The time required for Step 6 depends on the vial size, sample volume, and convenience for the operator. Dried samples have a fluffy appearance, while partially dried samples have a fluffy top and sides with a solid core making it difficult to distinguish between the two states while viewing the samples through the Plexiglas door of the lyophilizer. The samples are very stable under the conditions used for lyophilization. The authors typically allow the lyophilization process to continue until the

Table 3 Program for Lyophilization of Hydrated Lipids

Step	Temperature (°C)	Time (hr)	Vacuum pressure (mbar)
1 (refreeze)	−60 to −50	2	0
2	−40	2	100
3	−20	10	100
4	0	2	100
5	10	6	100
6 (hold)	10	Until removed	100

next morning. For example, they lyophilize for at least 48 hours, but would most likely continue the lyophilization until the next morning (60 to 64 hours total). The vials are stoppered by the stoppering device within the lyophilizer, while the vacuum is still applied, thus, allowing the samples to be stored under vacuum. After stoppering, the vials are removed from the lyophilizer and examined closely to ensure that the samples have completely lyophilized. If a vial contains a frozen chunk or an area of wet sample, the lyophilization is not complete. If this should occur, the samples are transferred to a biological safety cabinet and allowed to be completely thawed. In order to continue the lyophilization process, the sample must be entirely hydrated. It may be necessary to remove the stoppers and add additional water-for-injection to the vials to completely hydrate the sample. However, sufficient water should be added only to completely hydrate the sample. It is not necessary to add water to the full starting volume. New sterile stoppers should be placed back on the samples and samples are frozen and lyophilized as described above.

Completely dry samples can be used immediately for encapsulation of antigen or can be stored at less than or equal to −20°C. Tear-away aluminum crimp seals (Wheaton Scientific Products Inc. Millville, New Jersey, U.S.A.) should be placed on the vials prior to storage. We have stored the lyophilized lipids under these conditions for six years without any detectable degradation of the lipids. Because there is no oxygen present to oxidize the lipids, no aqueous media to catalyze hydrolysis of the lipids, and no microorganisms to degrade the lipids, the lipids are expected to be stable for extensive periods of time, perhaps decades.

Encapsulation of Antigen

The vials of lyophilized lipids are sprayed with 70% ethanol or isopropanol and moved to the biological safety cabinet. If they were stored frozen, they are allowed to come to room temperature before use. In addition, the alcohol is allowed to evaporate before use. The antigen is dissolved or diluted in the appropriate buffer and sterilized by filtration through a 0.2-μm low-binding syringe filter (Millex GV, Millipore Corp., Billerica, Massachusetts, U.S.A.). The antigen is added to the vials so that the total phospholipid concentration of the vial is approximately 200 mM. Prior to the addition of antigen, the vacuum should be released by puncturing the stopper with a needle attached to a sterile syringe that has the barrel removed. The antigen can be added to the vial with a needle and syringe by directly puncturing the stopper and injecting the desired volume. Alternatively, the aluminum seal and stopper can be removed with sterile forceps and the antigen added with a pipeting device. For example, 0.75 mL of antigen would be added to each vial. The vial is restoppered with sterile butyl stoppers and seals. The vials are shaken to wet the lipids and then are stored at 4°C. We routinely incubate the liposomes for 2.5 days. We have not rigorously studied the time required for optimal encapsulation of antigen, but at least an overnight incubation is required.

Encapsulation of antigen is a complex process involving a number of factors including the antigen itself. Table 4 summarizes the encapsulation efficiency of several different antigens. True encapsulation represents the antigen trapped within the aqueous volume of the liposomes. Antigen charge, hydrophobicity, and the availability of lipid-binding sites are several properties, which may affect encapsulation efficiency. For example, the 100% encapsulation efficiency observed for the ricin A subunit (Table 4) clearly indicates that there are probably numerous factors accounting for this high association of the ricin A subunit with liposomes (37).

The buffer and pH used for encapsulation may also affect encapsulation efficiency. Although any buffer can be used for the encapsulation of antigen, the choice of buffers should be made carefully for a number of reasons: (i) pH less than 6.0 should be avoided as acid catalyzes the hydrolysis of the fatty acid from the sn-2 position of the phospholipids leading to the

Table 4 Percent Encapsulation of Different Antigens in Liposomes

Antigen	Encapsulation (%)
R32NS1 (recombinant protein from *Plasmodium falciparum* circumsporozoite protein fused to NS-1) (24)	21
NKPKDELDYENDIEKKICKMEKCS (synthetic peptide from positions 367–390 of *P. falciparum* circumsporozoite protein) (34)	33
Recombinant gag protein derived from HIV-1 IIIB[a]	34
Kallikrein (porcine)	40
KSA (recombinant human colo-rectal cancer antigen manufactured in *Escherischia coli*) (35)	40
Prostate-specific antigen	50
KSA (recombinant human colo-rectal cancer antigen manufactured in baculovirus)	52
Recombinant envelope glycoprotein, oligomeric gp140 of HIV-1 IIIB	52
Conalbumin	55
Ovalbumin	60
Anthrax protective antigen	60
CGP18 (a synthetic peptide RIQRGPGRAFVTIGK derived from the tip of the V3 loop of the envelope glycoprotein, gp120, of HIV-1 IIIB with CG added to the amino terminus) (36)	62
Bovine serum albumin	77
Ricin A subunit (37)	100

[a]Personal communication with Mangala Rao, Department of Vaccine Production and Delivery, WRAIR.

formation of lysophospholipids (38). The authors have observed that liposomes formulated at pH 3.5 have 30% lysophospholipid after 1.5 year of storage at 4°C, while liposomes at pH 7.2 have only 7.2% lysophospholipids after 5.5 years of storage at 4°C. (ii) pH greater than 8.5 should also be avoided as it will induce saponification of the phospholipids (38). (iii) If the liposomes are going to be quantified by assay of phosphate (see below), phosphate buffers should be avoided. (iv) Physiological concentrations of salts should only be used to avoid differences in ionic strength, which may rupture the liposomes when they are injected. We routinely use 50 mM Tris-HCl-150 mM sodium chloride, pH 7.4, 0.9% saline (0.15 M sodium chloride), or in the case when phosphate is not going to be assayed, Dulbecco's phosphate-buffered saline (PBS) without calcium or magnesium. In practice, a small encapsulation experiment is done prior to the formulation of the vaccine to guide us with the encapsulation efficiency of the antigen. This experiment would produce 2–3 mL of final liposomes and we assume for this small volume experiment that the antigen encapsulation efficiency is 50%. Even if the encapsulation efficiency is 10% or 90%, the measurement of the antigen would still be in the linear range of the protein assay (see next section).

After incubation to encapsulate the antigen, the vials are removed from the refrigerator, placed in a biological safety cabinet and sprayed with 70% ethanol or isopropanol. After evaporation of the alcohol and warming the vials to room temperature, the seals and stoppers are removed from the vials with sterile forceps. The next step depends on whether the unencapsulated antigen is to be removed from the liposomes or not. If the unencapsulated antigen is not to be removed, the contents of the vials are removed and pooled in one vial or tube. The vials can be rinsed with sterile buffer and added to the liposome pool. The volume of the pool is measured and adjusted to the final volume. For example, the final volume would be 20 mL.

If the unencapsulated antigen is to be removed from the liposomes, the liposomes are transferred to autoclaved 50-mL polycarbonate screw-capped centrifuge tubes (Nalgene Company, Rochester, New York, U.S.A.). The tubes and caps should be autoclaved separately. If the caps are left on the tubes, the caps melt during autoclaving. For example, two vials of liposomes would be placed in one tube. The vials would be rinsed and the rinses added to the tubes. The tubes are then filled with cold buffer, capped, shaken, and centrifuged at 30,000 × g for 30 minutes at 4°C. A Sorvall RC-5B centrifuge (DuPont Company, Wilmington, Delaware, U.S.A.) is used with a SA600 rotor set at 15,000 rpm. Following centrifugation, the tubes are removed and transferred to the biological safety cabinets. The supernatant is aspirated off the pellet and discarded. Care should be taken during this process. The pellets are loose and easily resuspended during movement of the tube and aspiration. Approximately 25 mL of buffer is added to each tube and the tube is capped and shaken to resuspend the pellet. After resuspension of the pellet, the tube is filled with buffer, capped, and shaken.

The centrifugation process is repeated and the supernatant removed as described above. The pellets are then resuspended in a small volume and pooled. A small rinse of each tube is performed and transferred to the pool. The final volume is measured and adjusted to the desired volume.

The pooled liposomes are then vialed in depyrogenated vials, stoppered, and aluminum seals are crimped in place. The authors routinely use 2-mL vaccine vials with butyl stoppers (both from Wheaton Scientific Products, Inc. Milleville, New Jersey, U.S.A.). The liposomes are stored at 4°C, not frozen. Freezing will cause ice crystals to form in the membrane, causing leakage of the antigen.

The question as to whether the liposomes need to be washed to remove unencapsulated antigen has no clear cut answer. Unencapsulated antigen does not enhance the immune response generated by immunization with liposomes. Free antigen mixed with liposomes lacking encapsulated antigen does not induce higher immune responses than immunization with antigen alone (39,40). Similarly, the question as to whether the unencapsulated antigen induce tolerance to, or supress the response to, the encapsulated antigen, also has no clear cut answer. Studies have shown that immunization with free antigen alone did not induce antibody responses to antigen until weeks or months after immunization (41,42). This suggested that free antigen induced suppression that was subsequently lost with time. Baker et al. (43,44) have demonstrated the induction of T suppressor cells in animals immunized with capsular polysaccharide antigen and that lipid A can prevent the suppression from occurring (45,46). Thus, there is a possibility that free antigen in unwashed liposomes may induce suppression, but it has not been directly demonstrated. However, the liposomes do contain lipid A, and this would be expected to prevent suppression. We have immunized mice with liposome-encapsulated prostate specific antigen (PSA) that was washed to remove unencapsulated antigen, and have also immunized with the same formulation to which free PSA was added back. No differences were detected between the two groups for both antibody and cytotoxic T lymphocytes (CTL) responses (unpublished data). We have also used unwashed liposomes in the clinical trials for prostate and colo-rectal cancer and observed potent immune responses (2,20–27). Thus, with our limited experience, we have not observed immunosuppression by immunization with unwashed liposomes. However, the induction of suppression by free antigen in unwashed liposomes is at least a theoretical possibility.

Quantification and Characterization of Liposomal Encapsulated Antigen

If the liposomes are unwashed, an aliquot should be washed as described above to remove the unencapsulated antigen. Liposomes (50–250 μL) are pipeted into 13 × 100 mm disposable glass test tubes. 0.5 mL of chloroform is added to each tube and the tubes are vigorously vortexed. The chloroform is evaporated from the tubes using a Speed Vac (SC100, Savant Industries Inc., Farmingdale, New York, U.S.A.). The dried samples are solubilized by adding 200 µL of 15% sodium deoxycholate (Calbiochem-Novabiochem Corp., San Diego, California, U.S.A.) per tube and vigorously vortexed until the lipid pellet is no longer visible. We have observed that solubilization is aided by vortexing each sample for approximately a minute and allowing it to sit for approximately five minutes before continuing vortexing (31). Highly purified sodium deoxycholate is essential for the solubilization. Sodium deoxycholate from other sources may be less pure and, consequently, may need to be recrystallized from acetone prior to use. Protein is quantified using the assay described by Lowry et al. for low protein amounts (47). The assay should include a standard curve of purified antigen because purified proteins can have varying reactions when compared to bovine serum albumin run as a standard. Before reading the absorbance of the sample tubes, they should be centrifuged at $1875 \times g$ (Sorvall RT6000, 3000 rpm in an H-1000 rotor Duport Company, Wilmington, Delaware, U.S.A.) at room temperature for 10 minutes. The supernatant is removed and the absorbance measured. Empty liposomes can be used as control samples. The encapsulation efficiency is calculated by dividing the total antigen encapsulated divided by the amount of antigen added or if unwashed liposomes are used by the amount of antigen measured in the unwashed liposomes. Encapsulated antigen can also be quantified by high performance liquid chromotography (HPLC) (48,49) or amino acid analysis.

Encapsulated antigen can be monitored for degradation by sodium dodecyl sulfate-polyacrylamide gel electrophoresis (SDS-PAGE) (50). The lipids in the liposomes cause smearing of the gel and must be removed prior to running the samples on the gel. This can be achieved by using a modified method of chloroform:methanol extraction and phase separation described by Wessel and Flügge (51). Liposomes (100 µL) are placed in a 1.5-mL screw cap microfuge tube (Bio-Rad Laboratories, Hercules, California, U.S.A.). 0.4 mL of methanol is added and the tube is capped, vortexed, and centrifuged in an Eppendorf microfuge (Brinkman Instruments Inc., Westbury, New York, U.S.A.) for one minute at $8800 \times g$. Two-tenth milliliter of chloroform is added and the tube is capped, vortexed, and centrifuged as described above. Three-tenth milliliter of water is added and the tube is capped, vortexed, and centrifuged as described above. The upper phase is carefully removed and discarded. Three-tenth milliliter of methanol is added to the tube and it is capped, vortexed, and centrifuged for two minutes. The supernatant is removed and discarded. The tube is inverted on an absorbent paper and drained for at minimum of two minutes. After blotting, the residual solvent is evaporated from the tube under a stream of nitrogen gas. 15 µL each of water and sample solubilizer are added to the tube and the sample is processed for SDS-PAGE (50) and subsequent Western blot analysis (52) if desired.

Characterization of Liposomal Lipids

Phospholipids

Total phospholipids are quantified by measuring phosphate (53). In addition to measuring the phospholipids, the assay also measures the phosphate in lipid A or in the buffer if PBS or other phosphate buffers are used. The lipid A represents an insignificant amount of phosphate compared to that of the phospholipids. In the example described in this chapter, the lipid A ($\sim$4 µmol) is 0.13% of the phospholipids (3000 µmol). The phosphate assay is linear from 0.05 to 0.3 µmol phosphate. The liposomes must be diluted 1:10 in water and 10 to 30 µL of this dilution should be assayed by placing it in a glass tube (150 × 15 mm) (Kimble/Kontes Glass Inc., Vineland, New Jersey, U.S.A.). Thus, 10 µL of the liposomes used in the example would be 0.15 µmol. Potassium phosphate (1 mM in water) is used for a phosphate standard curve, and 0.7 mL of a 1:1 (v/v) mixture of 60% perchloric acid:sulfuric acid (conc.) is added to each tube. The tube is heated over an open flame until the solution turns yellow. For the liposome samples, the solution will first turn brown then clear and then turn yellow. The hydrolysis should be done only in a perchloric acid fume hood. The hood is stainless steel and equipped with water to wash the vents, ducts, and the interior of the hood. The water wash down prevents the accumulation of potentially explosive perchloric acid residue. After cooling, 0.5 mL of water and 4 mL of 1% ammonium molybdate is added to each tube and the tubes are mixed with a vortex mixer. Two-tenth milliliter of Fiske-Subbarow reducer (Sigma-Aldrich Co., St. Louis, Missouri, U.S.A.) (0.79 g/5 mL water) is added and the tube is vortexed. A glass marble is placed on the tube and the tube is boiled for 10 minutes. After cooling, the solution is transferred to a cuvette and the absorbance determined at 820 nm.

The integrity of the phospholipids can be assessed by thin layer chromatography (TLC) (54). In our experience, the major degradation products are fatty acid and lysophospholipids. The latter can be easily detected by TLC. The liposomes are diluted in chloroform to 3 µmol/mL. This would be a 1:50 dilution of the liposomes used in the example. The LK6 silica gel 60 TLC plates (Whatman International Ltd., Maidstone, U.K.) are first cleaned by running in acetone. After air-drying, the plates are stored under vacuum until use. 25 µL of diluted liposomes (75 nmol) is spotted on the plate in a 1-cm long area with a Hamilton syringe (Hamilton Company, Reno Nevada, U.S.A.). DMPC, DMPG, 1-myristoyl-*sn*-glycerol-3-phospho-choline (lysoPC) (Avanti Polar Lipids), and 1-myristoyl-*sn*-glycero-3-phospho-*rac*-(1-glycerol) (lysoPG) (Avanti Polar Lipids) are dissolved in chloroform at 2 mg/mL and 25 µL of each is spotted on the plate as standards. The plate is developed in filter-paper (Whatman No. 1)–lined TLC tanks in 188 mL of chloroform:methanol:water (65:25:4, v/v). After drying, phospholipids are visualized with molybdenum blue spray (Altech Associates Inc.,

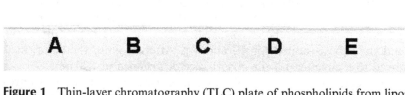

Figure 1 Thin-layer chromatography (TLC) plate of phospholipids from liposomes stored for 5.5 years at 4°C. Liposomes were extracted with chloroform: methanol and phospholipids were spotted on TLC plates developed in chloroform: methanol: water (65 : 25 : 4, v/v). Phospholipids were detected by spraying with molybdenum blue. Lane **A**, DMPC standard; **B**, DMFE standard; **C**, liposome sample; **D**, LysoPC; **E**, LysoPG. *Abbreviation*: DMPC, 1,2-dimyristoyl-*sn*-glycerol-3-phosphocholine; DMPE, dimyristoyl-*sn*-glycero-3-phospho-*rac*-(1-glycerol).

Deerfield, Illinois, U.S.A.). Phospholipids appear blue on a white background. The R_f values for DMPC, DMPG, lysoPC, and lysoPG are 0.35, 0.39, 0.15, and 0.23, respectively. A sample TLC plate is shown in Figure 1.

Cholesterol

Cholesterol is measured by the reaction with iron chloride (55). The assay is linear from 100 to 500 µg of cholesterol. Liposomes are assayed directly by placing 5–15 μL in a 16 × 125 mm glass screw cap test tubes with a Teflon-lined cap (Corning Inc., Corning, New York, U.S.A.). The cholesterol standard is prepared at 1 mg/mL in glacial acetic acid. 100, 200, 300, 400, and 500 μL of cholesterol are used as standards. Acetic acid is added to each tube to a final volume of 3.0 mL. Water (0.1 mL) is added to each tube. Two milliliter of 0.1% ferric chloride in acetic acid is added to each tube. The tubes are capped and vortexed. A light brown color develops which changes to purple in approximately one minute. After cooling to room temperature, the solutions are transferred to cuvettes and the absorbance read at 560 nm.

Cholesterol integrity is monitored by TLC (54). The principle oxidation product of cholesterol is 25-hydroxycholesterol. The liposomes are diluted in chloroform to obtain approximately 4 μmol/mL and 25 μl is spotted on an acetone washed LK6 silica gel 60 TLC plate. Cholesterol and 25-hydroxycholesterol (Sigma-Aldrich Co., St. Louis, Missouri, U.S.A.) (50 μg each) are spotted as standards. The plate is developed in filter paper lined TLC tank in 200 mL of benzene:ethyl acetate (3:2, v/v). After drying, the plate is sprayed with 50% sulfuric acid and baked in an oven at 120°C. The R_f for cholesterol and 25-hydroxycholesteol is 0.52 and 0.33, respectively. A sample plate is shown is Figure 2.

Lipid A

We do not routinely directly monitor the lipid A in the liposomes. TLC systems have been developed for lipid A (31,56), but the large quantity of phospholipids and cholesterol in the liposomes overload the TLC plates making lipid A characterization impossible by TLC. We rely on animal immunogenicity as a measure of the lipid A potency. HPLC systems have been developed for lipid A quantification and characterization (57–59) and it may be possible to adapt these methods for the analysis of lipid A in liposomes.

LIPOSOMAL OIL-IN-WATER EMULSIONS

Manufacturing of Liposomal Emulsions

Liposomal emulsions are made routinely just prior to injection (28–30). Light mineral oil (Sigma-Aldrich Chemical Co., or Spectrum Chemical Manufacturing Corp., Gardena, California, U.S.A.) is sterilized by filtration through a 0.2-μm polyethersulfone filter (Nalgene Nunc International, Rochester, New York, U.S.A.) and filled into 2-mL depyrogenated glass vaccine vials. The oil is stored at room temperature in the dark. Two 3-mL luer-lock syringes are placed on the two female connectors of a three-way stopcock (Kimble/Kontes Glass Company; cat. no. 420163-4503) (Fig. 3). A 21-gauge needle is placed on the male connector of the stopcock. A vial containing liposomes is placed on the needle and the liposomes are drawn into a single syringe. The light mineral oil is placed on the needle, the

Figure 2 Thin-layer chromatography (TLC) plate of cholesterol from liposomes stored for 5.5 years at 4°C. Liposomes were extracted with chloroform:methanol and lipids were spotted on TLC plates developed in benzene:ethylacetate (3:2, v/v). Lipids were detected by spraying sulfuric acid and charring. Lane **A**, Cholesterol; **B**, Liposomes; **C**, 25-hydroxycholesterol.

stopcock is turned and the light mineral oil is drawn up into the other syringe. No air spaces should be introduced into the syringes during this process. The stopcock is turned to allow passage between the two syringes and the needle is removed. The syringes are pushed alternatively to emulsify the samples. The pass rate should be 2 passes/sec for five minutes. The emulsion is pushed into one syringe. The stopcock is removed and needle is placed on the syringe for injection.

During the emulsification process, some of the liposomes are broken and the hydrophobic tails of the lipids are inserted into the oil droplet with

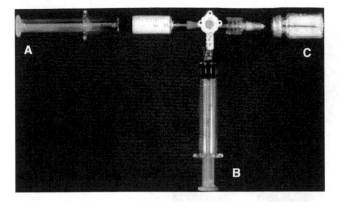

Figure 3 Photograph of the setup for the emulsification of liposomes with light mineral oil. Syringe (**A**) contained liposomes and syringe (**B**) contained the light mineral oil. A needle attached to the male connector of the stopcock was inserted into a vial of liposomes (**C**).

the hydrophilic head groups on the surface interacting with the aqueous solution; thus stabilizing the emulsion. A side effect of the emulsion process is that the antigen is released from the liposomes that are broken to stabilize the emulsion. However, a large number of liposomes remain intact (Fig. 4). Once injected, the emulsion serves as a depot for the slow release of liposomes.

Measurement of Emulsion Stability

In order to measure emulsion stability, the emulsion is placed in a 2-mL vaccine vial or a test tube. The vial is stoppered and the height of the emulsion is measured with a ruler. The emulsions are incubated at 4°C, 25°C, and 37°C. Four degrees is chosen to determine the time the emulsion can be stored before it breaks. Twenty-five degrees is chosen to determine the time the emulsion can be left at room temperature prior to injection. This is important for unstable emulsions. Thirty-seven degrees is chosen to approximate the stability of the emulsion after injection. At various intervals, the emulsions are removed from storage and the height of the total sample and separated phases are measured. There are two types of separations that can occur. The first is a water and oil separation with the clear layer of mineral oil floating on top of the aqueous liposome layer (Fig. 5A). The percentage of oil separated is calculated by the following formula:

$$100 \times \frac{\text{Height of oil (upper phase)/Total height of emulsion sample}}{F}$$

where F stands for the initial fraction of oil in the emulsion (percentage of oil added divided by 100). For example, if 0.1 mL of oil is emulsified with

(A)

(B)

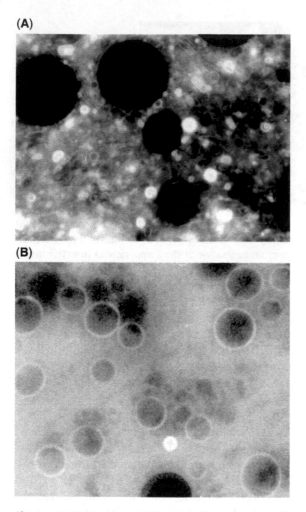

Figure 4 Photomicrographs of a liposomal emulsion showing liposomes containing trace amounts of N-NBD-PE (*bright areas*) and mineral oil droplets (*dark areas*). The oil droplets are outlined by a fluorescent ring, indicating that phospholipids from broken liposomes are coating the surface of the droplet to stabilize the emulsion. Panel (**A**) is a low magnification image. Panel (**B**) is a higher magnification image.

0.9 mL of liposomes, then the oil is 10% and, consequently, $F = 0.1$. The other type of separation that can occur is the separation of a lower cream layer from the emulsion (Fig. 5B and C). This is intermediate in the degradation process of the emulsion and eventually separates into an upper oil layer and lower aqueous liposome layer (30,60). The percentage separation for the

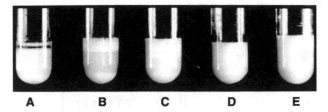

Figure 5 Physical appearance of liposomal emulsions after storage. (**A**) An emulsion has separated into a layer of light mineral oil on top and liposomes on the bottom. (**B**) and (**C**)-Emulsion have separated into layers of cream in the middle and liposomes on the top. (**D**) and (**E**)-Stable liposome emulsion that is not broken.

cream area is calculated by the formula:

$$100 \times \frac{\text{Height of the lower layer}}{\text{Total height}}$$

Purposeful Changes that Affect Emulsion Stability

There are two primary factors that can be manipulated to effect emulsion stability, liposome concentration, and percentage of oil in the emulsion. Emulsions made with 25-mM liposomal phospholipids are very unstable, breaking within a few hours. By increasing the phospholipid concentration of the liposomes, emulsions can be made to break in a few hours or a few days to being stable for long periods of times (Fig. 6). The volume of the liposomes can also be changed. Similarly, the stability of emulsion can be varied by changing the percentage of light mineral oil. Emulsions with low percentages of oil are unstable breaking within two days, while emulsions with high percentage of oil (42.5%) are stable for long periods of time (Fig. 7). In addition, a combination of these factors can be varied to change the stability of the emulsion. The emulsion that we used in clinical trials with prostate cancer patients contained 1 mL of PSA liposomes and 0.1 mL of light mineral oil (2,20). This emulsion started to break after eight hours at 4°C or room temperature. By changing the emulsion to 150 mM phospholipid liposomes (final concentration in the emulsion) and 40% light mineral oil, we were able to obtain an emulsion that was stable for three years at both 4°C and room temperature.

Other Factors that Affect Emulsion Stability

There are several other factors that, if not controlled rigorously, can affect emulsion stability. These were investigated in detail with the emulsion used for clinical trials in prostate cancer patients, i.e., 1 mL of 100 mM phospholipid

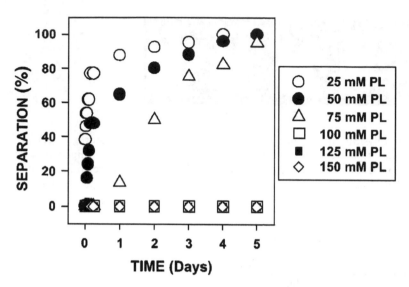

Figure 6 Effect of total liposomal phospholipid concentration on the stability of liposomal oil-in-water emulsions. Liposomes containing various amounts of phospholipids were emulsified with 40% light mineral oil. The amount of separation was measured as a function of time.

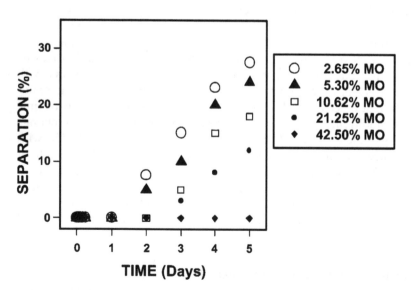

Figure 7 Effect of different amounts of light mineral oil on emulsion stability. Liposomes containing 100 mM phospholipids were emulsified with the percentage of oil indicated. The amount of separation was measured as a function of time.

liposomes containing 100 μg PSA and 200 μg of lipid A emulsified with 0.1 mL of light mineral oil as per the process described above. As stated above, the emulsion starts to break after eight hours of incubation at room temperature. The factors investigated include:

1. *Syringe*—Changing from 3-mL glass syringes (Hamilton Company) to B-D Glaspak syringes (glass barrels with plastic plungers containing rubber ends; Becton Dickinson, Franklin Lakes, New Jersey, U.S.A. cat. no. 5291) or B-D 3-mL plastic syringes had no effect on emulsion stability. The Glaspak syringe does not contain a luer-lock connector and the syringe can easily come off of the stopcock, especially at high oil concentrations, during the emulsion process.

2. *Bore size of the stopcock*—If the stopcock bore size was reduced from 0.1–0.05 cm, the emulsion did not break even at 10 hours. However, it was broken within 24 hours after emulsification. The use of the 0.05-cm bore stopcock increased the shear force during emulsification, thus increasing emulsion stability. However, the decrease in bore size made it difficult to push the solution through the stopcock when making the emulsion. When a 0.22-cm bore stopcock was used, the resulting emulsion broke in three hours. The large bore size caused a reduced shear force during emulsification, thus decreasing emulsion stability.

3. *Pass rate during emulsification*—Decreasing the pass rate from 2 passes/sec to 0.5 or 1 pass/sec dramatically reduced the stability of the emulsion. The emulsion broke at 0.25 and 0.5 hour, respectively, for the 0.5 and 1 pass/sec emulsion rate. Increasing the pass rate beyond 2 passes/sec is not physically possible when making the emulsion by hand.

4. *Duration of the emulsion process*—When the time of emulsification was decreased from five minutes to 30 seconds only minimal emulsification was obtained. There was a detectable phase separation 15 minutes after completion of the process. At the first and second minute, the emulsion broke at 0.5 and third hour, respectively. In contrast, there was no difference observed in the emulsion stability between the third and fifth minute of emulsification.

5. *Protein encapsulated in the liposome*—Liposomal emulsions made with liposomes lacking protein antigen have greater stability. An emulsion made with 10% light mineral oil and 100 mM phospholipid liposomes lacking protein antigen was stable for at least 24 hours, but started to break by 48 hours. The same emulsion made with PSA-encapsulated liposomes broke by eight hours. In addition, there are differences in emulsion stability based on the antigen encapsulated in the liposomes used to make the emulsion.

As described above, an emulsion containing PSA encapsulated in 150 mM phospholipid liposomes emulsified with 40% light mineral oil was stable for three years. In contrast, when the same emulsion was made with liposome encapsulated anthrax protective antigen, the emulsion broke after one month of incubation at 4°C. Interestingly, the formation of a cream layer during emulsion breakdown occurred only when antigen was present (Fig. 5B and C). Creaming does not occur with all antigens and predominately appears when high liposomal phospholipids concentrations are emulsified with higher percentages of light mineral oil.

Expected Immune Responses from Immunization with Liposomal Emulsions

We have used liposomal emulsions in studies with both mice and in humans. PSA, KSA, kallikrein, anthrax protective antigen, and HIV gp140 have been used as antigen in mice (28,30). High titers of antibodies and potent cellular immunity were induced in mice. The antibody responses tended to be reduced below those obtained with liposomes alone. Cellular immune responses were similar to those obtained with liposomes alone. In contrast, immunization of prostate cancer patients with liposomes containing PSA and lipid A induced minimal to no antibody or cellular immune responses (2,20). This was quite surprising because immunization with Walter Reed liposomes induced very strong immune responses in clinical trials for malaria and HIV vaccine (24,25,61). Vaccine research in prostate cancer patient represented a very challenging context for research. The patients were at the final stage, having failed all other conventional therapies, immunosuppressed, and had high levels of circulating PSA. In phase I clinical trials, immunization of the prostate cancer patients with liposome-encapsulated PSA formulated in an oil-in-water emulsion induced immune responses, either high titer antibody or cellular immune responses, or both, in 100% of the patients immunized. Thus, it appears that liposomal emulsions are potent adjuvant formulations for humans, but are not particularly effective (i.e., not better than liposomes by themselves) in mice. The ability of the liposomal emulsions to induce strong immune responses in other animals remains to be demonstrated. In summary, we have developed and described the manufacture of an adjuvant system using a liposomal emulsion that induces potent immune responses in humans.

REFERENCES

1. Freund J, Casals J, Page-Hosmer E. Sensitization and antibody formation after injection of tubercle bacilli and paraffin oil. Proc Soc Exp Biol Med 1937; 37:509.

2. Alving CR. Design and selection of vaccine adjuvants: Animal models and human trials. Vaccine 2002; 20:S56.
3. Davenport FM. Seventeen years' experience with mineral oil adjuvant influenza virus vaccines. Ann Allergy 1968; 26:288.
4. Stuart-Harris CH. Adjuvant influenza vaccines. Bull WHO 1969; 41:617.
5. Chang JCC, Diveley JP, Savary JR, Jensen FC. Adjuvant activity of incomplete Freund's adjuvant. Adv Drug Deliv Rev 1998; 32:173.
6. Kenney RT, Edelman R. Survey of human-use adjuvants. Expert Rev Vaccines 2003; 2:167.
7. Degen GJ, Jansen T, Schijns VEJC. Vaccine adjuvant technology: From mechanistic concepts to practical applications. Expert Rev Vaccines 2003; 2:327.
8. Aucouturier J, Dupuis L, Ganne V. Adjuvants designed fro veterinary and human vaccines. Vaccine 2001; 19:2666.
9. Sigh M, Srivastava I. Advances in vaccine adjuvants for infectious diseases. Curr HIV Res 2003; 1:309.
10. Podda A. The adjuvanted influenza vaccines with novel adjuvants: experience with the MF59-adjuvanted vaccine, Vaccine 2001; 19:2673.
11. Frey S, Poland G, Percell S, Podda A. Comparison of the safety, tolerability, and immunogenicity of a MF59-adjuvanted influenza vaccine and a non-adjuvanted influenza vaccine in non-elderly adults. Vaccine 2003; 21:4234.
12. Podda A, Del Giudice G. MF59-adjuvanted vaccines: Increased immunogenicity with an optimal safety profile. Expert Rev Vaccines 2003; 2:197.
13. Marshall BC, Adler SP. Avidity maturation following immunization with two human cytomegalovirus (CMV) vaccines: A live attenuated vaccine (towne) and a recombinant glycoprotein vaccine (gB/MF59). Viral Immunol 2003; 16:491.
14. O'Hagan DT, Signh M, Kazzaz J, et al. Synergistic adjuvant activity of immunostimulatory DNA and oil/water emulsions for immunization with p55 gag antigen. Vaccine 2002; 20:3389.
15. Alonso PL, Sacarlal J, Aponte JJ, et al. Efficacy of the RTS,S/AS02A vaccine against Plasmodium falciparum infection and disease in young African children: randomised controlled trial. Lancet 2004; 364:1411.
16. Garcon N, Heppner DG, Cohen J. Development of RTS,S/AS02: a purified subunit-based malaria vaccine candidate formulated with a novel adjuvant. Expert Rev Vaccines 2003; 2:231.
17. Heppner DG Jr., Kester KE, Ockenhouse, et al. Towards an RTS,S-based, multi-stage, multi-antigen vaccine against falciparum malaria: progress at the Walter Reed Army Institute of Research. Vaccine 2005; 23:2243.
18. Vantomme V, Dantinne C, Amrani N, et al. Immunologic analysis of a phase I/II study of vaccination of MAGE-3 protein combined with the AS02B adjuvant in patients with MAGE-3 positive tumors. J Immunother 2004; 27:124.
19. Vanderpapelière P, Rehermann B, Koutsoukos M, et al. Potent enhancement of cellular and humoral immune responses against recombinant hepatitis B antigens using AS02A adjuvant in healthy adults. Vaccine 2005; 23:2591.
20. Harris DT, Matyas GR, Gemella LG, et al. Immunologic approaches to the treatment of prostate cancer. Seminars Oncol 1999; 4:439.

21. Meidenbauer N, Harris DT, Spitler LE, Whiteside TL. Generation of PSA-reactive effector cells after vaccination with a PSA-based vaccine in patients with prostate cancer. The Prostate 2000; 43:88.

22. Meidenbauer N, Goodling W, Spitler L, Harris D, Whiteside TL. Recovery of ζ-chain expression and changes in spontaneous IL-10 production after PSA-based vaccines in patients with prostate cancer. British J Cancer 2002; 86:168.

23. Neidhart J, Allen KO, Barlow DL, et al. Immunization of colorectal cancer patients with recombinant baculovirus-derived KSA (Ep-CAM) formulated with monophosphoryl lipid A in liposomal emulsion, with and without granulocyte-macrophage colony-stimulating factor. Vaccine 2004; 22:773.

24. Fries LF, Gordon DM, Richards RL, et al. Liposomal malaria vaccine in humans: a safe and potent adjuvant strategy. Proc Natl Acad Sci USA 1992; 89:358.

25. Heppner DG, Gordon DM, Gross M, et al. Safety, immunogenicity, and efficacy of Plasmodium falciparum repeatless circumsporozoite protein vaccine encapsulated in liposomes. J Infect Dis 1996; 174:361.

26. Richards RL, Rao M, Wassef NM, Glenn GM, Rothwell SW, Alving CR. Liposomes containing lipid A serve as an adjuvant for induction of antibody cytotoxic T-cell responses against RTS,S malaria antigen. Infect Immun 1998; 66:2859.

27. Richards RL, Swartz GM, Schultz C, et al. Immunogenicity of liposomal malaria sporozoite antigen in monkeys: adjuvant effects of aluminum hydroxide and non-pyrogenic liposomal lipid A. Vaccine 1989; 7:506.

28. Muderhwa JM, Matyas GR, Spitler LE, Alving CR. Oil-in-water liposomal emulsions: characterization and potential use in vaccine delivery. J Pharm Sci 1999; 88:1332.

29. Matyas GR, Muderhwa JM, Alving CR. Oil-in water liposomal emulsions for vaccine delivery. Methods Enzymol 2003; 373:34.

30. Richards RL, Rao M, Vancott TC, Matyas GR, Birx DL, Alving CR. Liposomes-stabilized oil-in-water emulsions as adjuvants: increased emulsion stability promotes induction of cytotoxic T lymphocytes against an HIV envelope antigen. Immunol Cell Biol 2004; 82:531.

31. Alving CR, Shichijo S, Mattsby-Baltzer I, Richards RL, Wassef NM. Preparation and use of liposomes in immunological studies. In: Gregoriadis G, ed. Liposome Technology. Vol. III. 2nd ed. Interactions of Liposomes with the Biological Milieu. Boca Raton: CRC Press, 1993:317.

32. Wassef NM, Alving CR, Richards RL. Liposomes as carriers for vaccines. Immunomethods 1994; 4:217.

33. Alving CR, Ownes RR, Wassef NM. Process for making liposome preparation, United States Patent 6,007,838, issued Dec. 28, 1999.

34. White K, Krzych U, Gordon DM, et al. Induction of cytolytic and antibody responses using Plasmodium falciparum repeatless circumsporozoite protein encapsulated in liposomes. Vaccine 1993; 11:1341.

35. Herlyn M, Steplewski Z, Herlyn D, Koprowski H. Colorectal carcinoma-specific antigen: detection by means of monoclonal antibodies. Proc Nat Acad Sci U.S.A. 1979; 76:1438.

36. White WI, Cassatt DR, Madsen J, et al. Antibody and cytotoxic T-lymphocyte responses to a single liposome-associated peptide antigen. Vaccine 1995; 13:1111.

37. Matyas GR, Alving CR. Protective prophylactic immunity against intranasal ricin challenge induced by liposomal ricin A subunit. Vaccine Res 1996; 5:163.
38. Grit M, Crommelin DJA. Chemical stability of liposomes: implications for their physical stability. Chem Phys Lipids 1993; 64:3.
39. Thérien HM, Shahum E. Importance of physical association between antigen and liposomes in liposomes adjuvanticity. Immunol Lett 1989; 22:253.
40. Alving CR. Liposomes as carriers of antigens and adjuvants. J Immunol Meth 1991; 140:1.
41. Alving CR, Richards RL, Hayre MD, Hochmeyer WT, Witz RA. Liposomes as carriers of vaccines: development of a liposomal malaria vaccine. In: Gregoriadis G, Allison AC, Poste G, eds. Immunological Adjuvants and Vaccines. New York: Plenum Publishing Corp., 1989:123.
42. Richards RL, Hayre MD, Hockmeyer WT, Alving CR. Liposomes, lipid A, and aluminum hydroxide enhance the immune response to a synthetic malaria sporozoite antigen. Infect Immun 1988; 56:682.
43. Baker PJ. Regulation of magnitude of antibody response to bacterial polysaccharide antigens by thymus-derived lymphocytes. Infect Immun 1990; 58:3465.
44. Baker PJ, Fauntleroy MB, Stashak PW, Hiernaux Jr, Cantrell JL, Rudbach, JA. Adjuvant effects of trehalsoe dimycolate on the antibody response to type III pneumococcal polysaccharide. Infect Immun 1989; 57:912.
45. Baker PJ, Hraba T, Taylor CE, et al. Structural features that influence the ability of lipid A and its analogs to abolish expression of suppressor T cell activity. Infect Immun 1992; 60:2694.
46. Baker PJ, Hraba T, Taylor CE, et al. Molecular structures that influence the immunomodulatory properties of the lipid A and inner core region oligosaccharides of bacterial lipopolysaccharides. Infect Immun 1994; 62:2257.
47. Lowry OH, Rosenbrough NJ, Farr AL. Protein measurement with the folin phenol reagent. J Biol Chem 1951; 193:265.
48. Koppenhage FJ, Storm G, Underberg WJ. Development of a routine analysis method for liposome encapsulated recombinant interleukin-2. J Chromatogr B Biomed Sci Appl 1998; 716:285.
49. Lutsiak ME, kwon GS, Samuel, J. Analysis of peptide and lipopeptide content in liposomes. J Pharm Pharm Sci 2002; 5:279.
50. Laemmli UK. Cleavage of structural proteins during the assembly of the head of baceriophage T4. Nature 1970; 227:680.
51. Wessel D, Flügge UI. A method for the quantitative recovery of protein in dilute solution in the presence of detergents and lipids. Anal Biochem 1984; 138:141.
52. Towbin H, Staehelin T, Gordon J. Electrophoretic transfer of proteins from polyacrylamide gels to nitrocellulose sheets: procedure and some applications. Proc Natl Acad Sci U.S.A. 1979; 76:4350.
53. Gerlach E, Deuticke B. Eine einfache methode zur mikrobestimmung von phosphate in der papierchromatographie. Biochem Zeischrift 1963; 337:477.
54. Muderhwa JM, Wassef NM, Spitler LE, Alving CR. Effects of aluminum adjuvant compounds, tweens, and spans on the stability of liposome permeability. Vaccine Res 1996; 5:1.

55. Zlatkis A, Zak B, Boyle AJ. A new method for the direct determination of serum cholesterol. J Lab Clin Med 1953; 41:486.
56. Mattsby-Baltzer I, Alving CR. Lipid A fractions analysed by a technique involving thin-layer chromatography and enzyme-linked immunosorbent assay. Eur J Biochem 1984; 138:333.
57. Hagen SR, Thompson JD, Snyder DS, Myers KR. Analysis of a monophosphoryl lipid A immunostimulant preparation from *Salmonella minnesota* R595 by high-performance liquid chromatography. J Chrom 1997; 767:53.
58. Parleasak A, Bode C. Lipopolysaccharide determination by reversed-phase high-performance liquid chromatography after fluorescence labeling. J Chrom 1995; 711:277.
59. Kiang J, Wang LX, Tang M, Szu SC, Lee YC. Simultaneous determination of glucosamine and glucosamine 4-phosphate in lipid A with high-performance anion-exchange chromatography (HPAEC). Carbohyd Res 1998; 312:73.
60. Woodard LF, Jasman RI. Stable oil-in-water emulsions: preparation and use as vaccine vehicles for lipophilic adjuvants. Vaccine 1997; 15:1773.
61. McElrath MJ. Selection of potent immunological adjuvants for vaccine construction. Semin Cancer Biol 375; 1995:6.

20

A Biotechnological Approach to Formulate Vaccines Within Liposomes: Immunization Studies

Maria Helena Bueno da Costa

*Laboratório de Microesferas e Lipossomas—Centro de Biotecnologia,
Instituto Butantan, São Paulo, Brazil*

INTRODUCTION

The immunogenicity enhancements induced by particulate antigen vehicle such as liposome are unsurprising, because natural pathogens are also particulates and the immune system has evolved to deal with these. The biotechnological challenges to produce liposome vaccines are the stability of both, the antigen and liposome. Vaccine stability is an essential prerequisite to the successful development and dissemination of inexpensive, effective vaccine formulation. In many instances in the past, potential vaccine candidates have failed due to substantial losses at the production or downstream processing stages. There are reports from the World Health Organization (WHO) where labile entities have been successfully produced at the commercial level, only to be inactivated by inadequacies in the handling procedures during transportation and distribution. Such occurrences have contributed to the failure of many vaccination campaigns. Consequently, it is useful to take into consideration all of the process steps which is, starting from antigen production down to its stability, including encapsulation within stable or stabilized liposomes.

As examples of antigen downstream process, bacterial toxins such as those from tetanus and diphtheria must be detoxified to be used as vaccines.

During the detoxification process, there are the possibilities of formation of intra- and intermolecular bonds. Fortunately, in these cases, the formed antigens, named tetanus toxoid and diphtheria toxoid (Dtxd), became detoxified but retains their immunogenicity (1–3).

The primary concern in choosing a process for the encapsulation of protein antigenic vaccine within liposomes is the potential molecular instability imposed by general method conditions as disruptions in the quaternary structures because of oxidative conditions as sonication to form small liposomes, pH alterations, denaturation induced by heating, necessary to liposomes formed with lipids with phase temperature above 25°C (4). The antigens to be encapsulated must be resistant to these superimposed process conditions. The recombinant heat shock protein of 18 kDa from *Mycobacterium leprae* [18 kDa heat shock protein (18 hsp)] have been used in our laboratory as an alternative source of T-epitope on the development of second-generation vaccines, mainly for those produced with low immunogenicity character. Large-scale pure protein can be produced by fermentation at a low cost (5,6). Its biologically active conformation—expressed as immunological integrity in vitro—is preserved during its entrapment within vehicles such as poly(lactide-co-glycolide acid) (PLGA)-microspheres, liposomes, and proteic supramolecular aggregates (7,8). The 18 hsp resists a large range of temperatures (−20°C, 80°C). In contrast, it is known that freezing and dehydration-rehydration destabilizes the lipid bilayer of liposomes. It is known that the disaccharide trehalose stabilizes membranes by substitution of hydration water around the hydrophilic head of phospholipids (9). This phenomenon provides a diminution of the transition temperature gel-glass of the phospholipids avoiding disintegration, leakage, and fusion process (10). However, the high cost of trehalose for large-scale vaccine purposes practically impeaches its use as liposome cryoprotectant.

The objective of this chapter is to describe vaccine formulations encapsulated within liposomes (in the presence or absence of co-adjuvant) taking into consideration antigen/vehicle stabilities and choosing low costs process. The point is not to propose a new method to prepare liposomes. The central idea is to give new dresses for old vaccines by using classical and well-established liposome preparation method.

MATERIALS AND METHODS

Materials

The instruments used were Jobin-Yvon-SPEX CD6 dicrograph Instruments 1996, F2000 spectrofluorimeter (Hitachi), and Titerteck Multiscan MCC/340. The Shimadzu high-performance liquid chromatography (HPLC), Model LC-10AVP, a QC-PAK GFC 300 column (15 × 7.8 mm) was purchased from Shimadzu Co., Japan (Scanning electronic microscope JOEL, Model JSM 5200). Antivenom serum and horse serum albumin were gifted by

Divisão de Produção do Instituto Butantan (IBu). Trehalose was purchased from Sigma. Soy phosphatidylcholine (SPC) was purchased from Lukas Mayer (Hamburg, Germany), cholesterol (Chol), D,L-α-tocopherol (D,L-α-Toc), and mannitol from Fluka; the ODN1668, TCCATGACGTTCCTGATGCT was from Microsynth (Balgach, Switzerland). The gels Sepharose S1000, Q-Sepharose, and Sepharose 4B were from Pharmacia (Brazil). The monoclonal antibody L5 (specific for T epitope) was a gift from Dr A.C. Moreira Filho (Instituto de Ciências Biológicas, Universidade de São Paulo, SP, Brazil). Genetically selected mice of the H (high responder) line were maintained at the Laboratório de Imunogenética, Instituto Butantan (IBu), Balb-C females were from Biotério Central (IBu), and horses were from Fazenda São Joaquim (IBu). All other reagents were of analytical grade. Part of the experiments was executed at Laboratório de Sistemas Biomiméticos (Instituto de Química, Universidade de São Paulo, SP).

Methods

Coadjuvants and Antigens Preparations

Recombinant Soluble 18 kDa Heat Shock Protein, Hydrophobic Modified 18 Hsp and ODN1668: The co-adjuvant and T-epitope source—18 hsp, was cloned as exported protein in *Saccharomyces cerevisae*. The production of large-scale pure protein is easily obtained as described by our laboratory (5,6). The hydrophobic modified18 hsp (18 hsp-m), was obtained by esterifying the 18 hsp with palmitic acid as described by our laboratory (8). The poly-CG ODN1668 was obtained commercially.

PSC—The Polysaccharide from *Neisseria meningitides*, Sero-Group C: The PSC was prepared by a procedure suitable for scale-up in good manufacturer production (GMP) conditions as described by our laboratory (11).

Diphtheria Toxoid: The Dtxd was gifted by Seção de formulação-IBu. The Dtxd was subsequently purified as described by our laboratory (12). Briefly, the Dtxd was adsorbed on a Q-Sepharose column previously equilibrated with 500 mM Tris, pH 9.0. After washing the column with 500 mM Tris, pH 9.0, followed by a second wash with 100 mM NaCl in 500 mM Tris, pH 9.0, the Dtxd was eluted with 300 mM NaCl in 500 mM Tris, pH 9.0. The toxoid elutes with 95% purity (12).

Native and Modified Snake Venoms: The native pooled venoms from five different species of *Bothropic* (*B. alternatus*, *B. jararaca*, *B. jararacussu*, *B. moojeni*, and *B. neuwiedi*) and the venom of *Crotalus durissus terrificus* were gifted by Seção de Venenos-IBu. To chemically modify the snake venoms: the pool of *Bothropic* or *Crotalus durissus terrificus* venoms (5 mg/mL) were solubilized in manitol phosphate buffer (MPB)-pBB/Ethylene diamine tetra acetic acid (EDTA) buffer (20 mM phosphate, 295 mM mannitol, 5 mM EDTA, 3 mM 4-bromophenacyl bromide, pH 7.2). The solutions were incubated by two hours at 37°C with a gentle agitation.

Preparation of Liposomes

The lipids SPC (200 mg/mL); Chol (25 mg/mL); D,L-α-Toc (1.2 mg/mL) pBB (3 mg/mL) in CHCl$_3$: Methanol (2:1) were prepared as stock solutions and stored at −100°C. The buffer stock [generally phosphate buffered saline (PBS)] solutions were prepared in the presence of cryoprotectants. The cryoprotectants (295 mM mannitol or 400 mM trehalose final concentrations) were added, when necessary, during the preparation of the work solution (diluted buffers).

Encapsulation of 18 Hsp, Poly-CG, PSC or Diphtheria Toxoid: For these antigens, the lipid films were 22:5:0.18 molar ratios (SPC:Chol: α-Toc, respectively). The dried lipid film were solubilized with the buffer containing the antigen of interest and agitated in a vortex. The multilamellar vesicles were submitted to five cycles of freezing–thawing, followed by sequential filter extrusions. The extrusions in each filter (0.8, 0.4, and 0.2 μm) were repeated three to five times. All of these formulations were done in the presence or absence of 400 mM trehalose or 295 mM mannitol. Unencapsulated antigens or co-adjuvants—18 hsp, ODN1668 or Dtxd—were removed by applying the formulation on gel filtration columns (Sepharose S1000, Q-Sepharose, or Sepharose 4B). Unencapsulated PSC was removed by ultracentrifugation.

External Association of the 18 Hsp-m with Liposomes: The hydrophobic protein was externally associated spontaneously with empty liposomes previously prepared as described by our laboratory (8). The liposomes were composed by 22:5:1:0.18 molar ratios [SPC:Chol:Phosphatidic acid (PA): α-Toc, respectively]. All of these formulations were done in the absence or in the presence of 400 mM trehalose in PBS, pH 7.2.

Encapsulation of Snake Venoms: The lipid films were 22:5:0.18:1.0 molar ratios of SPC:Chol:α-Toc, pBB (patent pending), respectively. The bothropic-modified (Bm, 5 mg/mL) or crotalic (Cm, 5 mg/mL)-modified venoms in MPB-pBB/EDTA buffer (20 mM phosphate, 295 mM mannitol, 5 mM EDTA, 3 mM 4-bromophenacyl bromide, pH 7.2) were added to dried lipid film. The sequence of manipulations was similar to that of the other antigens. All of these formulations were done in the absence or in the presence of 295 mM mannitol in the described buffer. Unencapsulated Bm or Cm was removed by applying the formulation on gel filtration column (Sephacryl S1000).

Antigens Characterizations

Stabilities: In the presence or absence of trehalose of the 18 hsp and 18 hsp-m or Dtxd were followed by enzyme-linked immunosorbent assay (ELISA) and by spectroscopic studies as circular dichroism (CD) and intrinsic fluorescence.

The Dtxd Conformational Changes: The variations in distinct pHs and their relationships with different encapsulation ratios were observed by CD and fluorescence. Tryptophan fluorescence quenches (in these different pHs) were observed in the presence of acrylamide.

Stabilities of Native or Modified Snake: Venoms were observed by CD and ELISA.

Liposome Characterizations

Encapsulation Efficiencies: Encapsulation efficiencies were calculated after formulations elution from gel filtration columns. The lipid assay was done by Pi measurements. The proteic antigens were measured by HPLC monitored by absorbance at 230 nm (18 hsp); at 269 nm (Dtxd) or 280 nm (snake venoms); intrinsic fluorescence (18 hsp) and/or by ELISA (13–15). The PSC was measured by as sialic acid or by HPLC (monitored by refraction index or by absorbance at 212 nm) as developed by our laboratory (Bueno da Costa et al. A rapid and low cost method to quantify polysaccharide by HPLC. Submitted for publication).

Liposome Size, Homogeneity, and Stability: These parameters were determined by laser light scattering (Submicron Particle Sizer Model 370, Nicomp, Santa Barbara, U.S.A.); by gel filtration columns and freeze fractures (Electronic microscopy). The lyophilized samples were reconstituted either in water or in 9 g/L NaCl with 1 to 2 mL as final volumes. After solvent addition, the samples were vortexed and kept at room temperature for 60 to 120 minutes before the first measurement was made. When indicated, membrane stabilities were observed by freeze fracture.

The 18 hsp leakout was measured in vivo. Mice were injected with 18 hsp encapsulated within liposomes. These liposomes were previously prepared in the presence or in the absence of trehalose and freeze-dried. After reconstitution, the mice were injected with the formulation. The production of specific immunoglobulin (Ig)G and IgM were measured (8).

Immunization Protocols

Mice Immunizations: Genetically selected of the H line mice or Balb-C females were immunized intraperitoneally or subcutaneously with liposome formulation. Usually, the injections were composed of antigen (1, 2, 5, 10, or 50 µg antigen/mL) encapsulated within liposomes. A second injection with free antigen was usually given 30 to 90 days after the first immunization. The controls were injections of soluble antigen in PBS. At regular intervals, the blood was collected at the retro orbital plexus to measure specific antibody productions.

Horse Immunizations: *With diphtheria toxoid.* The horses were immunized subcutaneously twice. First immunization (t_0): 0.5 mg of Dtxd within liposomes. Second injection: 25.0 mg of Dtxd within liposomes, 13 weeks later (t_{13}). The animals were bleeding at t_0, t_7, t_{13}, t_{16}, t_{17}, t_{21} and t_{30}. The antibody productions were measured by ELISA or serum neutralization in vitro.

With snake venoms. The horses were immunized subcutaneously twice. First immunization (t_0): 2.5 mg of modified venoms (bothropic or crotalic) within liposomes. Second injection: 12.5 mg of modified venom (bothropic or crotalic) within liposomes, 13 weeks later (t_{13}). The animals were bleeding at t_0, t_7, t_{13}, t_{16}, t_{17}, t_{21}, and t_{30}. The antibody productions were measured by ELISA or serum neutralization in vitro.

Enzyme-Linked Immunosorbent Assays

ELISA for 18 Hsp Assay: Samples were added to the ELISA plates and, after two hours at 37°C, they were blocked with 10% skimmed milk. After 30 minutes, the L5 (a specific T-epitope monoclonal antibody) was added to the wells. The conjugate was added 30 minutes later, and, after a further 30 minutes, the substrate was added. After 15 minutes at RT, the reaction was stopped with H_2SO_4. The absorbance was automatically read at 450 nm in a Titertek Multiskan MCC/340.

For biological experiments, the immunized mice sera replaced the monoclonal antibody and the assay was performed as described. Antibody titers are the reciprocal serum dilution factor giving an absorbance value of 20% of the saturation value (8).

ELISA for PSC Assays: Plates were coated with a solution of 5 µg/mL poly-lysil-lysine followed by the addition of 20 µg/mL PSC. After 30 minutes, the antibody anti-PSC was added to the wells. The conjugate was added 30 minutes later, and, after a further 30 minutes, the substrate was added. After 15 minutes at RT, the reaction was stopped with H_2SO_4. The absorbance was automatically read at 450 nm in a Titertek Multiskan MCC/340.

For biological experiments, the immunized mice sera replaced the standard anti-PSC antibody and the assay was performed as described. Antibody titers are the reciprocal serum dilution factor giving an absorbance value of 20% of the saturation value (15).

ELISA for Diphtheria Toxoid Assays: Samples were added to the ELISA plates and after two hours at 37°C, they were blocked with 10% skimmed milk. After 30 minutes, the standard anti-Dtxd–specific IgG (developed in horses) was added to the wells. The conjugate was added 30 minutes later, and, after a further 30 minutes, the substrate was added. After 15 minutes at room temperature, the reaction was stopped with H_2SO_4. The absorbance was automatically read at 450 nm in a Titertek Multiskan MCC/340.

For biological experiments, the immunized mice sera replaced the standard anti-Dtxd antibody and the assay was performed as described. Antibody titers are the reciprocal serum dilution factor giving an absorbance value of 20% of the saturation value.

ELISA for Snake Venoms (Native and Modified): Snake venom samples were added to the ELISA plates and, after two hours at 37°C, they were blocked with 10% skimmed milk. After 30 minutes, the standard antibothropic-specific IgG or anticrotalic-specific IgG (both developed in horses) was added to the wells. The conjugate was added 30 minutes later, and, after a further 30 minutes, the substrate was added. After 15 minutes at RT, the reaction was stopped with H_2SO_4. The absorbance was automatically read at 450 nm in a Titertek Multiskan MCC/340.

For biological experiments, the immunized mice or immunized horse sera replaced the standard specific antivenom and the assay was performed as described. Antibody titers are the reciprocal serum dilution factor giving an absorbance value of 20% of the saturation value.

Toxin-Binding Inhibition Assay

The toxin-binding inhibition (ToBI) assay is composed by two parts: (1) serum neutralization followed by (2) ELISA (16,17). The liposomal formulation to be tested was injected in guinea pigs.

1. *Serum neutralization.* A 96-well plate (round-bottom wells) was blocked with 200 µL of 0.1% bovine serum albumin (BSA) in PBS for 60 minutes, at 37°C. The plate was washed and the sample sera (developed in guinea pig, injected with the formulation to be tested) was added in a serial dilution way (factor dilution of two), followed by the addition of antigen (0.1 Lf/mL, Lf = flocculation unit) to be tested (Dtxd, for example). The plate was, therefore, incubated for 60 minutes, at 37°C. After this time, the plate was incubated at 4°C, for 18 hours.

2. *Enzyme-linked immunosorbent assays.* The plate was sensitized with 0.1 UI/mL standard sera (developed in horses) antiantigen to be tested (e.g., Dtxd) in 50 mM carbonate buffer pH 9.6 and incubated for 18 hours, at 4°C [at the same time as in (a)]. The plate was washed and than blocked with 0.1% BSA in PBS. After 60 minutes of incubation at 37°C, it was added to each sera/antisera (100 µL) moisture (described in a) to the corresponding ELISA wells. After 90 minutes at 37°C, the plate was washed and the conjugate enzyme-antibody developed in horses (for example, anti-Dtxd-enzyme conjugate), diluted in PBS-Tween 80% to 0.1% BSA was added. The plate was incubated for two hours at room temperature. After washing, it was added with 100 µL of the substrate (0.416 mM tetramethyl benzidine in 22 mM acetate buffer, pH 5 containing 2.9 mM H_2O_2). After 15 minutes at room temperature, the reaction was interrupted with 1 N H_2SO_4 (100 µL). All of the plate washings were done with 0.05% Tween 80 in PBS.

Antibody titers (IU/mL) are the reciprocal serum dilution factor giving an absorbance value of 50% of the saturation value of the positive control (well containing the standard serum with known title) (16,17).

RESULTS AND DISCUSSION

The research on new delivery systems within vaccinology has started with the necessity to promote the development of vaccines simpler to deliver than existing ones with particular emphasis on reducing the number of doses needed to induce long-lasting protection. Particular attention has been paid to research on particulate systems with adjuvant and controlled release activities like liposomes. The majority of second-generation vaccines are poor immunogens and, consequently, need to be administered with carrier protein to increase their activities. To increase the immunogenicity of vaccines, we propose, initially, their co-encapsulation with a carrier protein as T-epitope source within an adequate delivery system. The 18 hsp protein from *M. leprae* has some sequences that are close to the heat shock proteins (18). Furthermore, the 18 hsp has sequences that bind to T cells, which may confer to the bound carried antigen a proper presentation to induce B/T cell immunity. The 18 hsp (as T-epitope) production in large scale and at a low cost was studied at our laboratory, with the aim of offering an alternative source of carrier protein (5,6). Our main idea is to co-encapsulate the 18 hsp with poor antigens within safe and pluripotent support, the liposomes. The protein is extremely resistant to large range of temperatures (60% of activity is retained at 80°C) (19). The N-acylation increases 4% of its ordered structure and decreased 2% of its β-T1 structure as observed by CD with the retention of its biological active conformation (19) and so it is possible to associate the protein externally to liposomes (8,19). It is known that, from literature, to be possible to produce differentially IgG or IgM (if properly manipulate the immune system). The 18 hsp, when encapsulated within liposomes, produces preferentially IgG and, when externally exposed, produces IgM (8). It was observed, through the replica analysis of liposomes containing 18 hsp prepared and lyophilized in the absence of trehalose, large aggregation process (Fig. 1A). In contrast, those liposomes prepared in the presence of trehalose are unilamellar, monodisperses, and without aggregation (Fig. 1B). This fact is corroborated by the observation in vivo, where, whenever mice are immunized with liposomes internally containing 18 hsp lyophilized in the absence of trehalose induces the formation of IgM, instead of IgG. It is interpreted as membrane disintegration with concomitant 18 hsp leak out and, consequently, changes on profile of antibody kinetics production. It means, that to act as adjuvant, the liposome must retain its membrane integrity, here preserved by the action of trehalose. The cryoprotectant action mediated by trehalose is an important improvement on liposome vaccine in the context of tropical countries. The capacity of 18 hsp

(A)   **(B)**

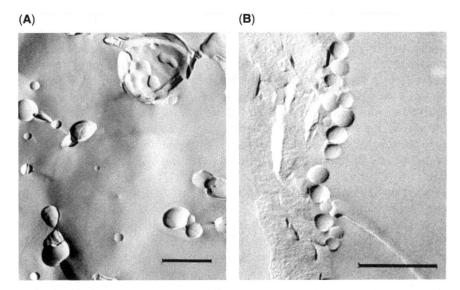

Figure 1 Effect of trehalose on liposome membrane. Liposomes prepared in the absence (**A**) or in the presence (**B**) of trehalose were lyophilized and freeze fractured. These liposomes (90–160 nm), composed by soy phosphatidylcholine:cholesterol: α-tocopherol (22:5:0.18 molar ratio, respectively), contain 18 hsp within their internal aqueous compartment.

to enhance the immune response to B-epitopes without chemical conjugation was investigated by co-encapsulating it with PSC within liposomes. Oligo deoxy nucleotides (ODNs) containing unmethylated CpGs motifs have been shown to act as adjuvants for protein and peptide vaccines (15,20). The immunopotentiation capacity of the 18 hsp was compared with that of the immunostimulatory oligonucleotide 1668 (ODN1668).

Glycoconjugate vaccines induce protective immunity in neonates and infants. However, polysaccharide-protein conjugation technology is antigen- and time-consuming and expensive, which are important drawbacks, especially for the development of vaccines. There is a definite need for alternative methods for enhancing the immune response to T-independent antigens. The ability of liposomes to act as carriers for the co-encapsulation of B- and T-epitopes eliminating the need of protein conjugation is real (21,22).

The liposome formulations containing the co-adjuvants (18 hsp or ODN1668) were, all of them, stable (in the presence of mannitol) and with an efficiency of encapsulation between 65% and 75%. The presence of PSC or ODN did not change appreciably the size of the liposomes. We decided to use mannitol, instead of trehalose, because while developing the process we must consider the final price of the formulation. If it takes into account the molarities of trehalose and mannitol required to protect the liposomal formulation, they are 400 and 295 mM, respectively. If we compare their prices, the trehalose costs 27 times more than mannitol. If we consider the molarities and prices/kg of mannitol and sucrose, the mannitol costs 1.9 times more than sucrose. It means that they are equivalent. The problem of using sucrose is that, frequently, it has ribonuclease as contaminant. So, taking into consideration all of these points, we decided to continue the use of mannitol as cryoprotectant. If we compare only the price of egg phosphatidylcholine and soybean phosphatidylcholine required to prepare 1 mL of liposomes, we conclude that the egg phosphatidylcholine (EPC) is 177 times more expensive than SPC. The SPC is approved to be used in humans, so we decided that we would continue to use it into our formulations.

The SPC:Chol:α-Toc 22:5:0.18 molar ratios liposome formulation remained stable in solution and in the presence of 295 mM mannitol, at 4°C during, at least 11 days after being prepared. When PSC was administered subcutaneously, both within liposomes containing 18 hsp or ODN1668, the antibody titers obtained after 13 days were of the same magnitude (15). The presence of both 18 hsp and ODN1668 within liposomes containing PSC did not significantly alter the antibody production against the polysaccharide, which indicates that their adjuvant effects are not additive. A booster was given after 90 days with free PSC. After 14 days of the booster, the antibody titers showed a secondary response. Confirming our expectations, we observed that the formulation containing 18 hsp conferred a memory response to the carried antigen—the *N. meningitides* serogroup C polysaccharide (15). These results were compared with those mice that were immunized with free PSC/PBS. These mice did not produce antibody against PSC. But, the most interesting fact was observed on those mice immunized with PSC encapsulated within liposomes: they produced antibody specific against the polysaccharide. So, the simple encapsulation of PSC within liposomes relieves the presence of both 18 hsp and ODN1668 (Fig. 2). The formulation was scaled up 10 times and its characteristics remained the same. Our laboratory is able to prepare up to 5 L/hr of liposome PSC formulation in GMP conditions.

The Dtxd is the first antigen encapsulated within liposomes (23). An enhancement on encapsulation efficiency for Dtxd of about 50% was observed, here, when the protein was encapsulated in buffer, pH 4.0. This was accompanied by changes on protein hydrophobicity and was observed by CD and fluorescence spectroscopies. Whenever the Dtxd exposes its hydrophobic residues at pH 4.0, it interacts better with the liposomal film than when its hydrophobic residues were buried (pH 9.0). The Dtxd partition coefficient in Triton-X100 and the acrylamide fluorescence quench were also pH dependent. Both were bigger at pH 4.0 than at pH 9.0 (Fig. 3). The relationship protein structure and lipid interaction is pH dependent and can

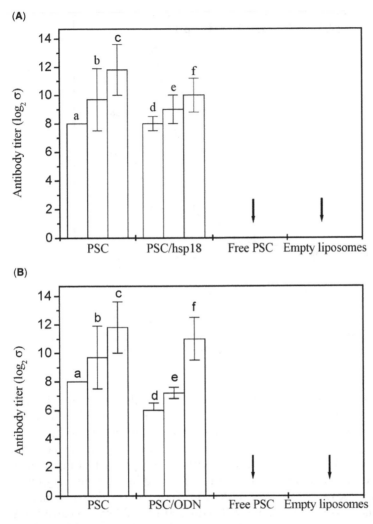

Figure 2 Specific anti-PSC production during the time. Genetically selected mice were immunized with liposome formulation of soy phosphatidylcholine:cholesterol:α-tocopherol (22:5:0.18 molar ratio, respectively), containing: (**A**) PSC or PSC + 18 hsp or (**B**) PSC or PSC + ODN1688. As controls (**A** and **B**) the mice were immunized with free PSC or empty liposomes. The bleedings were done before immunization and 9, 14, and 22 days after formulation injections. *Abbreviations*: PSC, The poly saccharide from Neisseria meningitides, sero-group C, ODN, oligodeoxyribonucleotides.

be easily maximized to enhance encapsulation of antigens in vaccine development. This was a biotechnological advance inexpensive, once this simple modification on the encapsulation protocol did not led to enhance of costs.

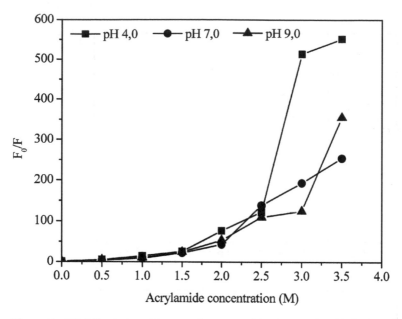

Figure 3 Diphtheria toxoid tryptophan quenching by acrylamide in function of pH. The fluorescence spectra were performed in the absence (F_0) or in the presence (F) of different acrylamide concentrations and in different pHs as indicated.

Horses have been used in immunization protocols for large-scale production of antisera against Dtxd and other toxins. Horses used to produce hyperimmune sera against Dtxd are injected with injections of 200 Lfs (flocculation units, 1Lf $= 2.5\,\mu g$ of Dtxd) of protein, followed by boosters of 200, 2500, 3000, 3600, 10,000 Lfs and finally with 20,000 Lfs of Dtxd (personal communication). The immunization with this toxoid induces tumor growth and local pain, because of the Arthus disease. To avoid toxoid effect on animals, here, the horses were immunized with 200 Lfs of Dtxd encapsulated within liposomes composed of soybean phosphatidylcholine:Chol (113:28 molar ratio) in a 295 mM mannitol, in 20 mM phosphate buffer, pH 7.2. After 100 days, the horses received a booster of the liposomal formulation containing 25,000 Lfs Dtxd or the Dtxd adsorbed in Al(OH)$_3$ as the control. The horses immunized with the liposomal formulation produced a high level of specific neutralizing antibodies (titer $= 2^{14}$) against Dtxd even 200 days after being immunized (Fig. 4). This liposomal formulation can substitute the traditional immunization protocols because it avoided toxoid toxicity and animal suffering, and showed excellent adjuvant properties. Economical and biotechnological benefits can be also envisaged because these formulations reduced the number of immunization doses and, consequently, the horse maintenance operational costs.

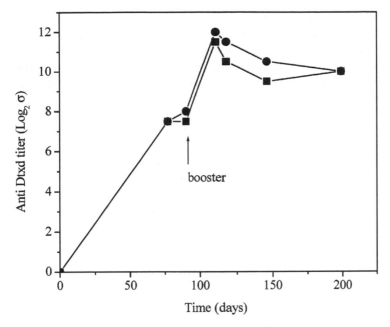

Figure 4 Production of specific anti-diphtheria toxoid (Dtxd) during the time. Horses were immunized with liposomes (●) composed by soy phosphatidyl-choline:cholesterol:α-tocopherol (22:5:0.18 molar ratio, respectively) containing Dtxd. Controls: (■) horses were immunized with Dtxd adsorbed in Al(OH)$_3$. See antibody titer definition in "Materials and Methods" section.

Horses have been immunized with snake venom to develop antisera which are used in the therapy of snake bites. The problems to be circumvented in the immunization protocol are low antibody production and venom toxicity that induces sufferings to the animals such as pain, local abscesses, hemorrhage, fistulae, and fibrosis. Immunization protocols for the antisera production, using large doses of 870 mg of venom/horse, were reported in the literature. The present immunization protocol, at the Instituto Butantan, involves two injections of 0.5 mg venom/horse, followed by two injections of 1.0 mg of venom/horse, one of 2.0 mg/horse, and one injection of 5 mg venom/horse (personal communication). In comparison with the data in the literature, the Instituto Butantan protocol is the one that uses the lowest amount of venom per horse.

Crotalic venom is a weak antigen; its antibody production is slow and unpredictable, with a wide variation in individual responses. The conspicuous individual variability related to antibody production in horses is observed even after immunization with purified venoms such as phospholipase A$_2$ from *Bothrops asper*. Liposomes would be an ideal way to enhance this antisera production because, in addition to their adjuvant property, they could decrease venom toxicity effects. It is known that membrane integrity is necessary to provide the correct adjuvant action of liposomes. The difficulty in using liposomes is their structural and biophysical instability in the presence of venoms. Our central idea in modifying the horse immunization protocol is the combination of three different tools: encapsulation of chemically modified venoms within stabilized liposomes.

We present here our results on encapsulating a pool of *Bothropic* venoms and *Crotalus durissus terrificus* venom within stabilized liposomes (LB and LC, respectively) (patent pending) (25).

Chemically modified venoms were solubilized in an inhibition buffer containing an excess of the inhibitor and a chelating agent. The structure of the venom was analyzed by UV or CD spectroscopies and ELISA. The liposomal formulation was composed of soybean phosphatidylcholine: Chol:D,L-α-Toc:*para*-bromo phenacyl bromide (113:28:1:10 molar ratio) in a 5 mM EDTA, 295 mM mannitol, in 20 mM phosphate buffer, pH 7.2 (patent pending) (25). To maintain the long-term stability in the dry state and to protect against membrane disintegration and fusion, mannitol was used as a cryoprotectant. Genetically selected mice were immunized with

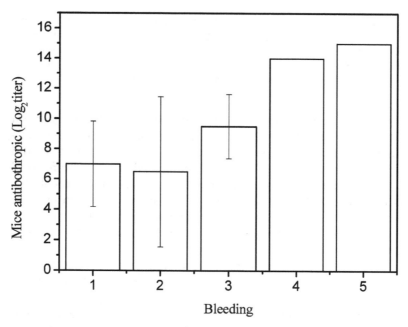

Figure 5 Production of specific anti-bothropic venom during the time. Genetically selected mice were immunized with 1 μg of modified bothropic venom within liposomes composed by soy phosphatidylcholine:cholesterol:α-tocopherol, pBB (22:5:0.18:1.0 molar ratios, respectively). The bleedings occurred one to five weeks after the first injection.

LB or LC (10 μg of venom). In addition, horses were immunized with the same liposomal formulations containing venoms.

Within a general context, the M_w of bothropic and crotalic venoms (pool of proteins) increased after chemical modification, which is in agreement with our expectation. The proteolytic and phospholipasic activities for these venoms were also measured. In the modified venoms, the most evident reduction was 94% of bothropic proteolytic and 81% of crotalic phospholipasic activities. In spite of differences in the α-helical content between natural and modified venoms, standard horse antisera recognized them. The encapsulation efficiencies were 59% and 90% for the crotalic and bothropic venoms, respectively, as followed by filtration on Sephacryl S1000. Light scattering measurements of the liposomal suspensions led us to conclude that both, LB (119 ± 47 nm) and LC (147 ± 56 nm), were stable for 22 days at 4°C. They remained stable even after dehydration and lyophilization.

The mice immunized with LB or LC did not show symptoms of venom toxicity. Both, LB and LC enhanced, by at least 30%, the antibody titers (on a log scale) 25 days after injection and total IgG titers remained high 91 days after immunization (Fig. 5).

The horses immunized with LB or LC did not show symptoms of venom toxicity (in kidney, liver, or heart). The horse antibody titles for those animals immunized with LB or LC was higher than the control and persisted longer as time passed (seven months) (Fig. 6).

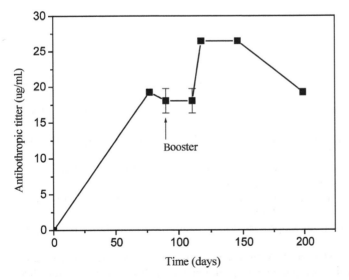

Figure 6 Antibothropic production during the time. Horses were injected with 0.5 mg of modified bothropic venom within liposomes composed by soy phosphatidylcholine:cholesterol:α-tocopherol, pBB (22:5:0.18:1.0 molar ratios, respectively). The booster was with 25 mg of modified bothropic venom within liposomes.

The formulation containing pBB inserted within the lipid membrane can be exploited for other animal venoms from, for example: bees, spiders, scorpions, or caterpillar (patent pending) (24). We intend to start studying the best encapsulation conditions of these venoms, mainly for that from bee, to be used in human desensitization. Among other proteins, the bee venom contains melitin and phospholipase. Both proteins interact with membranes, the phospholipase exerts an esterase action and melitin forms pores. To be encapsulated within liposomes, both proteins' actions must be inhibited. The liposome will be designed so as to preserve its membrane integrity and to avoid toxic effects on humans.

All the formulations described here can implement the vaccine industry because they are not expensive and only a few steps must be introduced on the production process.

ACKNOWLEDGMENTS

I would like to thank Marina Nanette for critical reading of the manuscript. Researchers from the Laboratório de Sistemas Biomiméticos (Instituto de Química- Universidade de São Paulo, SP) have been of kind help. These works were supported by: CNPq (477154/2003-4 and 474781/01-1), FAPESP (2000/14228-3 02/07293-9 and 94/5854-5), UNDP/World Bank/WHO Special Program for Research and Training in Tropical Disease (TDR), Paul Scherrer Institute (Villigen, Switzerland), and Fundação Butantan.

REFERENCES

1. Ramon G. Sur la toxine et sur anatoxine diphtheriques. Ann Inst Pasteur 1924; 38:1.
2. Rappuoli R. New Generation of Vaccines. 1st ed. New York: Marcel Dekker Inc, 1997:417.
3. Glenny AT, Hopkins BE. Diphtheria toxoid as an immunizing agent. Br J Exp Pathol 1923; 4:283.
4. Weiner AL. Liposomes for protein delivery: selecting manufacture and development process. Immunomethods 1994; 4:201.
5. Costa MHB, et al. Procedures for scaling up the recombinant 18hsp lepra protein production. Biotech Tech 1995; 9:527.
6. Sato RA, Costa MHB. Bioprocess design: study of a case. Biotech Lett 1996; 18:275.
7. Bueno da Costa MH, et al. Antibody response and T-epitope conformational analysis of a heat shock protein encapsulated within PLGA microspheres. Proc. 2nd World Meeting APGI/APV, Paris, France, 1998, 518 pp.
8. Bueno da Costa MH, et al. Conformational stability and antibody response to the 18kDa heat shock protein formulated into different vehicles. Appl Biochem Biotechnol 1998; 73:19.

9. Crowe JH. Interactions of sugars with membranes. Biochim Biophys Acta 1988; 947:367.
10. Strauss G, Schurtenberger P, Hauser H. The interaction of saccharides with lipid bilayer vesicles: stabilization during freeze-thawing and freeze-drying. Biochim Biophys Acta 1986; 858:169.
11. Tanizaki MM, et al. Purification of meningococcal group C polysaccharide by a procedure suitable for scale-up. J Microbiol Methods 1996; 27:19.
12. Campana RA, et al. Ionic interfaces and Diphtheria toxoid interactions. Prot Exp Purif 2004; 33:161.
13. Quintilio W, et al. Microquantification of proteins with low chromophore contents. Biotech Tech 1997; 11:697.
14. Quintilio W, et al. Large unilamellar vesicles as trehalose-stabilized vehicles for vaccines: Storage time and in vivo studies. J Control Release 2000; 67:409.
15. Bueno da Costa MH, et al. Heat shock protein micro-encapsulation as a double tool for the improvement of new generation vaccines. J Liposome Res 2002; 12:29.
16. Souza Matos DC, et al. Immunogenicity test of tetanus component in adsorbed vaccines by toxin binding inhibition test. Mem Inst Oswaldo Cruz 2002; 97:909.
17. Marcovistz R, et al. Potency control of diphtheria component in adsorbed vaccines by in vitro neutralization tests. Biologicals 2002; 30:105.
18. Booth RJ, et al. Antigenic proteins of *Mycobacterium leprae*—complete sequence of the gene for the 18 kDa-hsp protein. J Immunol 1988; 40:597.
19. Quintilio W, et al. The use of protein structure/activity relationships for the rational design of stable particulate delivery systems. Braz J Med Biol Res 2002; 35:727.
20. Quintilio W. Preservation of freeze-dried liposomes containing a heat shock protein by trehalose: a proposal for designing a stabilised vaccine vehicles and adjuvants. XI ièmes Journées Scientifiques du Groupe Thematique de Recherche sur les Vecteurs (G.T.R.V), Paris, December 7–8, 1998.
21. Carson DA, Raz E. Oligonucleotide adjuvants for T-helper 1 (Th1)-specific vaccination. J Exp Med 1997; 18:61.
22. Gregoriadis G, Wang Z, Barenholz Y, Francis MJ. Liposome-entrapped T-cell peptides help for a co-entrapped B-cell peptide to overcome genetic restriction in mice and induce immunological memory. Immunology 1993; 80:535.
23. Gregoriadis G. The immunological adjuvant and vaccine carrier properties of liposomes. J Drug Target 1994; 2:351.
24. Alison AC, Gregoriadis G. Liposomes as immunological adjuvants. Nature 1974; 252:252.
25. PCT PI0301513-0: Liposomes, liposomal formulations, its process of obtainment and uses thereof, process for the productions of hyperimmune serum and profilactic uses or curative method for the treatment of diseases.

21

Liposomal Therapeutics: From Animal to Man

Jill P. Adler-Moore

Department of Biological Sciences, California State Polytechnic University, Pomona, California, U.S.A.

Gerard M. Jensen

Gilead Sciences, Inc., San Dimas, California, U.S.A.

Richard T. Proffitt

Richpro Associates, Arcadia, California, U.S.A.

INTRODUCTION AND SCOPE

This chapter is by design limited in scope. To review the entirety of animal and human experience would fill many volumes, and require an immense multidisciplinary effort. First, we limit this discussion to parenteral liposomes used in conventional drug delivery applications. We, further, restrict ourselves to products that have been completely through the drug development process: from preclinical screening in animals, through the normal phases of clinical trials, culminating in large, randomized registrational trials, and subsequent approvals. Examined from this perspective, we explore the question—what information on ultimate clinical outcome could be adduced from earlier (or even later performed) animal experiments. Do the therapeutic index enhancements, realized either on the safety/toxicity or efficacy aspects, that were anticipated during early drug development, translate into the ultimate product profile? This is not always as straightforward as one might imagine. One does not design, and indeed cannot always ethically design, clinical trials as would be done from a simple scientific approach. Specifically, a liposomal drug is rarely compared directly and in the same manner to a free drug given in the same way. Nevertheless, some lessons can be learned from the products that have thus far made it to regulatory approval and commercialization. These lessons are important in the consideration for development of a new generation of liposomal therapeutics, and in building reasonable expectations on what can be learned in nonclinical programs.

REASONABLY WELL-ESTABLISHED ASPECTS

There are some aspects of human performance of liposomes that are reasonably well modeled in animals, at least to a basic degree. Many liposomal products have been examined in mice, rodents, dogs, and man and some rules for translation exist of basic pharmacokinetic properties such as volumes of distribution or elimination half lives (1). For example, a half-life in rats of between 10 and 20 hours will usually correspond to one in man of greater than 50 hours. Moving beyond this level of detail, however, proves difficult (2). The subtleties of liposomal drug accumulation in and clearance from tissues, and of drug release and cellular uptake, are well appreciated but not well understood. The so-called "enhanced permeability and retention" effect in this branch of science, wherein liposomes of the appropriate particle size, charge, and stability will preferentially accumulate in growing tumors is well known (3), but once again the details are not well understood. Lacking a microscopic understanding of the totality of drug distribution at the cellular and subcellular level (assuming such could be achieved with current methods), is not, however, an impediment in principle to drug development, or even commercialization, although aspects of this have become a focus of recent regulatory guidance development in the United States (4).

THERAPEUTIC INDEX ENHANCEMENT

Oncology

The key to success of any new oncology drug is, of course, useful efficacy with manageable side effects. For a drug-delivery system such as a liposome, what is added is improvement in principle in either or both of these aspects, so as to effect an improved product. In the cases of liposomal reformulation of generic drugs, i.e., those liposomal formulations which are attempts to improve an established anticancer drug, the fundamental question of efficacy is at least reasonably clearly defined from the beginning as being present. It is for these cases that one asks the question—how is this formulation different or improved from that already available, and does this difference or improvement lead to a product that provides sufficient novelty to justify expensive drug development programs?

Two such products have been developed through to marketing approval worldwide. Both of these are liposomal anthracyclines: DaunoXome (liposomal daunorubicin, Gilead Sciences, Inc.,) and Doxil (pegylated liposomal doxorubicin, known in Europe as Caelyx, Alza/Johnson & Johnson). Myocet, a different formulation of liposomal doxorubicin, has been approved in Europe but not as yet in the United States.

DaunoXome (5,6), as a product, was the daughter of VesCan (3), an [111]Indium tumor-imaging agent first developed more than 20 years ago. The imaging agent in the distearoylphosphatidylcholine and cholesterol liposome formulation was replaced by the anthracycline daunorubicin, loaded by a pH gradient. DaunoXome has been extensively studied in both preclinical and clinical models, examining both liquid and solid tumor types. DaunoXome has received marketing approvals for the treatment of advanced HIV-related Kaposi's sarcoma, which predominantly affects the skin. Kaposi's sarcoma did not directly figure into the nonclinical evaluation of DaunoXome, owing to a lack of relevant nonclinical models, and clinically, DaunoXome was compared to a cocktail of doxorubicin, bleomycin, and a vinca alkaloid (a so-called ABV regimen). Thus, for the purposes of the discussion here, although there is a wealth of information available, specific preclinical clinical correlations cannot be developed.

Some chemotherapeutic agents achieve wide usage not because of lowered toxicity relative to other agents, but rather because of different and thus potentially complementary toxicity profiles. An example can be found in the neurotoxic agents vincristine or cisplatin. Doxil (7–11) (pegylated liposomal doxorubicin) has achieved impressive usage in a variety of solid tumor settings. Doxil is marketed to treat patients with AIDS-related Kaposi's sarcoma; to treat patients with refractory (principally platinum) ovarian cancer, in this case displacing topotecan; and for patients with metastatic breast cancer, especially those at increased risk for cardiotoxicity. Doxil was shown in this setting to exhibit equivalent efficacy with an improved safety profile, particularly for cardiotoxicity. The dramatically decreased clearance of Doxil relative to free doxorubicin, the accumulation in tumor sites, and the slow release of drug therein, have all been studied extensively in nonclinical models and have at least at a fundamental level translated into what has been observed in man. The potential for improvement in cardiotoxicity profile was also noted. However, what has dominated dosing schedule and usage of Doxil clinically has been new dose limiting toxicities, different from that of the free drug: mucosal and cutaneous toxicities, including palmar/plantar erythrodysesthesia. These did not emerge clearly until they were noted in human studies. Nonclinical efficacy studies have been mixed, and no clear evidence of enhanced efficacy has emerged for Doxil. Nevertheless, the net effect of the Doxil formulation has been a highly useful and effective anticancer agent in major, solid tumor indications.

It has been posited (12) that efficacy enhancement by liposomal delivery may more likely be achieved in Man when cell cycle–specific anticancer drugs are formulated. This has led to the development of a liposomal vincristine formulation (Marquibo) (13). This drug remains in late stage clinical development for the treatment of non-Hodgkin's lymphoma and acute lymphoblastic leukemia. Other schedule dependent liposomal drugs in clinical development include liposomal lurtotecan (OSI-211) (14–16) and liposomal GW1843U89 (OSI-7904L) (17–19). In these cases, comparisons in performance to the free drug will not be possible, as the free drug formulations never reached commercialization.

Therefore, while there have been tremendous efforts and enormous amounts of data generated for liposome oncology drugs, simple rules have not emerged demonstrating clear nonclinical to clinical correlations, defined as we have, in the various studies. This is not in any way and indictment of the relevant development programs, but rather the result of the particular cases developed to date. Presently, the nonclinical evaluation of liposomal oncology drugs will not follow pathways wholly different from that utilized for new chemical entities. There is formulation development and evaluation of toxicity and efficacy. In the latter case, the field suffers from a general imperfection of correlation between efficacy studies, these days typically murine xenograft studies, and clinical outcome. Methods are being developed (20) wherein data are adduced from a large variety of different studies to provide for aggregate, and ostensibly improved, predictions of clinical outcome. It is unclear if these approaches will find success, or if such success as is obtained will apply to cases of pharmacokinetic modifying formulations such as liposomes, and for drugs with our without schedule dependency. Separately, characterization of passive tumor targeting, as noted above, has enjoyed many recent advances (21) as an adjunct with the revolution in studies of tumor angiogenesis and antiantiogenesis therapy.

Liposomal Amphotericin B—AmBisome—Toxicity Aspects

Amphotericin B remains the drug of choice for the treatment of life-threatening, systemic fungal infections. Nevertheless, the use of amphotericin B in the mixed micelle/surfactant deoxycholate formulation (d-AmB) has been limited by severe toxic side effects, including dose-limiting nephrotoxicity. Several relatively new formulations have received regulatory approval: amphotericin B colloidal dispersion (ABCD) (22), amphotericin B lipid complex (ABLC; Abelcet) (23), and liposomal amphotericin B (AmBisome) (24). The primary mechanism of action of both the toxic and antifungal activity of amphotericin B is thought to be a compromise of the barrier function of mammalian and fungal membranes, respectively. Amphotericin B forms pores or channels in sterol containing membranes, which cause leakage of cell constituents leading to cell death. A goal of the new formulations noted above was, inter alia, to improve the relative toxicity of an amphotericin B formulation and the success of this strategy depended on the propensity of the amphotericin B to partition into mammalian membranes in vivo, and this in turn was dependent on both the relative energetic stability of amphotericin B in the delivery vehicle, the pharmacokinetics, and the ease of transfer to a target mammalian membrane. (The d-AmB formulation provides relatively little energetic stability to the amphotericin B molecule outside of the conferral of water solubility; it thus can be viewed effectively as a "free" drug for the purposes of comparisons.) Most of the initial screening of these formulations came in the form of murine lethality studies (25). One can examine the relative toxicity of these formulations with enhanced precision in vitro by measuring the potassium release from red blood cells during incubation at 37°C (26). These data directly measure the availability of amphotericin B to mammalian membranes at physiological temperatures and were further shown (26) to correlate well with the high dose acute toxicity manifested in murine in vivo models and to be generally species-independent. As an example, Figure 1

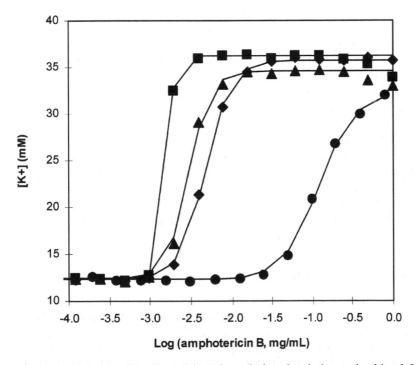

Figure 1 Titration of amphotericin B formulations in whole porcine blood. Incubation time and temperature were 12 hours and 37°C, respectively. (●), AmBisome; (■), deoxycholate amphotericin B; (♦), amphotericin B colloidal dispersion; (▲), amphotericin B lipid complex. *Source*: From Ref. 26.

shows results for incubation in porcine blood of serial dilutions of AmBisome, Abelcet (ABLC), Amphotec (ABCD), or Fungizone (d-AmB). After incubation and centrifugation, supernatants were evaluated for potassium and potassium release curves were generated. The formulations follow the order of apparent toxicity d-AmB > ABCD~ABLC≫AmBisome, with AmBisome requiring about 100-fold higher drug concentrations to achieve onset of potassium release relative to d-AMB.

Increase in the therapeutic index for a drug can be achieved by reduction of toxicity and improvement or retention of efficacy. For example, AmBisome was shown to be as effective as d-AmB for empirical antifungal therapy in patients with fever and neutropenia, and was shown to be associated with fewer breakthrough fungal infections, and exhibited less nephrotoxicity (27). To assess the ability to demonstrate this retention of efficacy in vitro in a manner analogous to the red cell toxicity assay studies of incubating the different amphotericin B formulations in *Candida albicans* cells were performed with measure of potassium release (26). The titration curves in these incubations were essentially superposable for all formulations and for 2- to 12-hour incubations. The results demonstrated physicochemically that the intrinsic availability of amphotericin B in AmBisome to partition into ergosterol containing yeast cell membranes was retained.

The potential for a difference in toxicity of AmBisome and Abelcet, noted above in vitro and in the acute murine lethality models, were possibly reflected in a double-blind, randomized study (28), wherein an improved safety profile of AmBisome over Abelcet was noted with respect to the frequency of chills/rigors and other infusion-related reactions, in terms of nephrotoxicity, and in other safety parameters, in a patient pool of febrile neutropenic patients at risk for fungal infections.

The in vitro and in vivo nonclinical toxicity results for AmBisome, other lipid-based formulations, and d-AmB, do appear to be useful in predicting clinical outcomes at least to some degree. This has become more important as it helps to lay the foundation for increased dosing levels in regimens designed to improve the efficacy of amphotericin B in the more difficult to treat fungal infections.

Liposomal Amphotericin B—AmBisome—Efficacy Aspects

The pharmacokinetics of amphotericin B have been markedly altered by its incorporation into the AmBisome formulation. These changes have been reported in both animal (29–31) and human (32,33) studies. There is a non-linear clearance of AmBisome from plasma associated with saturation of the reticuloendothelial system (RES) and a mean elimination half-life ranging from 5 to 24 hours depending on the dose and species. This nonlinear clearance results in the volume of distribution and the total clearance of AmBisome decreasing with increasing doses, and the area under the curve values increasing disproportionately to the relative increase in dose. There is an accompanying redistribution of the drug into non-RES organs including the kidneys, lungs, and brain (25,34). Since doses of AmBisome from 3 to 15 times higher than conventional amphotericin B are well tolerated by both animals (29–31) and humans (32,35), much higher concentrations of AmBisome than conventional amphotericin B can be delivered to sites of fungal infection in both RES and non-RES tissues (36).

The marked reduction in toxicity for AmBisome, as noted above, is not associated with a loss of the drug's broad-spectrum fungicidal activity when it is tested in vitro (37,38) or in vivo. Preclinical and clinical studies have shown that AmBisome is an effective therapy for extracellular as well as intracellular fungal infections in both immunocompetent and immunosuppressed hosts. In preclinical studies, AmBisome at doses ranging from 3 to 30 mg/kg has been reported to have therapeutic dose-dependent responses against many different fungal infections, including pulmonary blastomycosis (39), pulmonary paracoccidioidomycosis (40), systemic and pulmonary aspergillosis (41,42), systemic candidosis (43), meningeal cryptococcosis (44), histoplasmosis (45), and fusariosis (46). High doses of AmBisome given to animals have resulted in clearance of microbes from target tissues such as the kidneys (47), liver, spleen and lungs (48), and brains (49,50). Similarly, clinical studies have supported the use of high dose AmBisome for salvage therapy (35), candidiasis (51), cryptococcosis (52), aspergillosis (53), histoplasmosis (54), and leishmaniasis (55).

Pharmacokinetic data from animal studies have demonstrated that AmBisome administration results in the sustained presence of drug in the tissues for several days to weeks. This has led to investigations of loading doses (47) and intermittent dosing of AmBisome. These approaches have been successfully used in animal models for treating candidiasis (47), coccidioidomycosis (49), cryptococcosis (56), and histoplasmosis (45). In a clinical study (57) using intermittent AmBisome dosing of 5 mg/kg on alternate days, 83% (10/12) of pneumonia patients with suspected *Aspergillus* infection, were responders.

Prophylactic animal studies with AmBisome have also proven effective. In one study, a single dose of AmBisome at 5, 10, or 20 mg/kg, protected immunocompetent or immunosuppressed mice from lethal challenge with *C. albicans* or *Histoplasma capsulatum* (58). In another study, survival of mice infected with Aspergillus following hematopoietic stem cell transfer was significantly higher after prophylactic AmBisome treatment with 5 mg/kg compared with a dose of 1 mg/kg (59). In the clinic, prophylactic AmBisome at 1 mg/kg/day completely prevented invasive fungal infections in 40 liver transplant recipients, while 6 of 37 patients (16%) in the placebo control group developed fungal infections ($p < 0.01$) (60). In another clinical trial, neutropenic bone marrow transplant patients given AmBisome three times weekly at 2 mg/kg ($n = 74$) had no proven fungal infections compared to three in the placebo group ($n = 87$) and suspected fungal infections occurred in 42% and 46% of the AmBisome-treated and placebo patients, respectively. However, significantly fewer patients in the AmBisome-treated group became colonized with fungus compared to the placebo group (15 vs. 35, respectively, $p < 0.05$) (61).

In summary, the preclinical studies with AmBisome discussed above were able to predict to some extent how the drug would behave in humans. These studies also suggested new ways that the drug might be used to treat or prevent life-threatening fungal infections. Data from such models has been and will continue to be helpful in the design of clinical trials.

ACKNOWLEDGMENTS

Acknowledgment must go to those who have developed the liposome technology and data described here at, or in association with, Vestar/NeXstar/Gilead over the years—far too numerous to name.

REFERENCES

1. Fielding RM, Mukwaya G, Sandhaus RA. Clinical and preclinical studies with low-clearance liposomal amikacin (MiKasome®). In: Long Circulating Liposomes: Old Drugs, New Therapeutics. Berlin: Springer-Verlag and Landes Bioscience, 1998:213.
2. Fielding RM. Relationship of pharmacokinetically-calculated volumes of distribution to the physiologic distribution of liposomal drugs in tissues: implications for the characterization of liposomal formulations. Pharm Res 2001; 18:238.
3. Baldeschwieler JD, Schmidt PG. Liposomal drugs: from setbacks to success. Chemtech 1997; 27:34 (and references therein).
4. Center for Drug Evaluation and Research Draft Guidance for Industry. Liposomal Drug Products: Chemistry, Manufacturing, and Controls; Human Pharmacokinetics and Bioavailability; and Labeling Documentation. U.S. Department of Health and Human Services, Food and Drug Administration, August 2002.
5. Forssen EA, Ross ME. DaunoXome treatment of solid tumors: preclinical and clinical investigations. J Liposome Res 1994; 4:481.
6. Forssen EA, Coulter DM, Proffitt RT. Selective in vivo localization of daunorubicin small unilamellar vesicles in solid tumors. Cancer Res 1992; 52:3255.
7. Vaage J, Uster PS, Working PK. Therapy of human carcinoma xenografts with doxorubicin encapsulated in sterically stabilized liposomes (DOXIL): efficacy and safety studies. In: Long Circulating Liposomes: Old Drugs, New Therapeutics. Berlin: Springer-Verlag and Landes Bioscience, 1998:61.
8. Northfelt DW. Pegylated-liposomal doxorubicin (DOXIL) in the treatment of AIDS-related Kaposi's sarcoma. In: Long Circulating Liposomes: Old Drugs, New Therapeutics. Berlin: Springer-Verlag and Landes Bioscience, 1998:127.
9. Gabizon A. Pegylated liposomal doxorubicin: metamorphosis of an old drug into a new form of chemotherapy. Cancer Invest 2001; 19:434.

10. Uziely B, Jeffers S, Isacson R, et al. Liposomal doxorubicin: anti-tumor activity and unique toxicities during two complementary phase I studies. J Clin Oncol 1995; 13:1777.
11. Lyass O, Uziely B, Ben-Yosef R, et al. Correlation of toxicity with pharmacokinetics of pegylated liposomal doxorubicin (Doxil) in metastatic breast cancer. Cancer 2000; 89:1037.
12. Boman NL, Cullis PR, Mayer LD, Bally MB, Webb MS. Liposomal vincristine: the central role of drug retention in defining therapeutically optimized anticancer formulations. In: Long Circulating Liposomes: Old Drugs, New Therapeutics. Berlin: Springer-Verlag and Landes Bioscience, 1998:29.
13. Rodriguez MA, Sarris AH, East K, et al. A phase II Study of Liposomal Vincristine in CHOP with Retuximab for Elderly Patients with Untreated Aggressive B-Cell Non-Hodgkin's Lymphoma. American Society of Clinical Oncology, 2002 (Abstract 1132).
14. Moynihan KL, Emerson DL, Chiang SM, Hu N. Liposomal camptothecin formulations. United States Patent 6,740,33, 2004.
15. Emerson DL, Bendele R, Brown E, et al. Antitumor efficacy, pharmacokinetics, and biodistribution of NX211: a low-clearance liposomal formulation of lurtotecan. Clin Cancer Res 2000; 6:2903.
16. Desjardins JP, Abbott EA, Emerson DL, et al. Biodistribution of NX211, liposomal lurtotecan, in tumor-bearing mice. AntiCancer Drugs 2001; 12:235.
17. Ashvar CS, Chiang SM, Emerson DL, Hu N, Jensen GM. Liposomal benzoquinazoline thymidylate synthase inhibitor formulations, United States Patent 2004; 6,689,381.
18. Desjardins J, Emerson DL, Colagiovanni DB, Abbott E, Brown EN, Drolet DW. Pharmacokinetics, safety, and efficacy of a liposome encapsulated thymidylate synthase inhibitor, OSI-7904L [(S)-2-[5-[(1,2-Dihydro-3-methyl-1-oxobenzo[f]quinazolin-9-yl)methyl]amino-1-oxo-2-isoindolynl]-glutaric Acid] in mice. J Pharm Exp Ther 2004; 309:894.
19. Beutel G, Glen H, Schoffsski P, et al. Phase I study of liposomal thymidylate synthase inhibitor (TSI) OSI-7904L in patients with advanced solid tumors. Proc Am Soc Clin Oncol 2003; 22:140.
20. Teicher BA. Tumor Model Interpretation from Mice to Humans, 95th Annual Meeting of the American Society for Cancer Research, Program 332, Orlando, Florida, 2004.
21. Hobbs SK, Monsky WL, Yuan F, et al. Regulation of transport pathways in tumor vessels: role of tumor type and microenvironment. Proc Natl Acad Sci USA 1998; 95:4607.
22. Guo LSS, Fielding RM, Lasic DD, Hamilton RL, Mufson D. Novel antifungal drug delivery: stable amphotericin B-cholesteryl sulfate discs. Int J Pharm 1991; 75:45.
23. Janoff AS, Boni LT, Popescu MC, et al. Unusual lipid structures selectively reduce the toxicity of amphotericin B. Proc Natl Acad Sci USA 1988; 85:6122.
24. Adler-Moore JP, Proffitt RT. Development, characterization, efficacy and mode of action of ambisome, a unilamellar liposomal formulation of amphotericin B. J Liposome Res 1993; 3:429.

25. Proffitt RT, Satorius A, Chiang SM, Sullivan L, Adler-Moore JP. Pharmacology and toxicology of a liposomal formulation of amphotericin B (AmBisome) in rodents. J Antimicrobial Chemother 1991; 28B:49.

26. Jensen GM, Skenes CR, Bunch TH, et al. Determination of the relative toxicity of amphotericin B formulations: a red blood cell potassium release assay. Drug Deliv 1999; 6:81.

27. Walsh TJ, Finberg RW, Arndt C, et al. Liposomal amphotericin B for empirical therapy in patients with persistent fever and neutropenia. New Engl J Med 1999; 340:764.

28. Wingard JR, White MH, Anaissie E, et al. A randomized, double-blind comparative trial evaluating the safety of liposomal amphotericin B veruss amphotericin B lipid complex in the empirical treatment of febrile neutropenia. Clin Infect Dis 2000; 31:1155.

29. Lee JW, Amantea MA, Francis PA, ct al. Pharmacokinetics and safety of a unilamellar liposomal formulation of amphotericin B (AmBisome) in rabbits. Antimicrob Agents Chemother 1994; 38:713.

30. Boswell GW, Bekersky I, Buell D, Hiles R, Walsh TJ. Toxicological profile and pharmacokinetics of a unilamellar liposomal vesicle formulation of amphotericin B in rats. Antimicrob Agents Chemother 1998; 42:263.

31. Bekersky I, Boswell GW, Hiles R, Fielding RM, Buell D, Walsh TJ. Safety and toxicokinetics of intravenous liposomal amphotericin B (AmBisome) in beagle dogs. Pharm Res 1999; 16:1694.

32. Walsh TJ, Yeldandi V, McEvoy M, et al. Safety, tolerance, and pharma-cokinetics of a small unilamellar liposomal formulation of amphotericin B (AmBisome) in neutropenic patients. Antimicrob Agents Chemother 1998; 42:2391.

33. Heinemann V, Bosse D, Jehn U, et al. Pharmacokinetics of liposomal amphoter-icin B (AmBisome) in critically ill patients. Antimicrob Agents Chemother 1997; 41:1275.

34. Groll AH, Giri N, Petraitis V, et al. Comparative efficacy and distribution of lipid formulations of amphotericin B in experimental *Candida albicans* infection of the central nervous system. J Infect Dis 2000; 182:274.

35. Walsh TJ, Goodman JL, Pappas P, et al. Safety, tolerance, and pharmaco-kinetics of high-dose liposomal amphotericin B (AmBisome) in patients infected with aspergillus species and other filamentous fungi: maximum tolerated dose study. Antimicrob Agents Chemother 2001; 45:3487.

36. Adler-Moore JP, Proffitt RT. Effect of tissue penetration on AmBisome efficacy. Curr Opin Invest Drug 2003; 4:179.

37. Adler-Moore JP, Proffitt RT. AmBisome: long circulating formulation of amphotericin B. In: Long Circulating Liposomes: Old Drugs, New Therapeutics. Berlin: Springer-Verlag and Landes Bioscience, 1998:185.

38. Anaissie E, Paetznik V, Proffitt R, Adler-Moore J, Bodey GP. Comparison of the in vitro antifungal activity of free and liposome-encapsulated amphotericin B. Eur J Clin Microb Infect Dis 1991; 10:665.

39. Clemons KV, Stevens DA. Therapeutic efficacy of a liposomal formulation of amphotericin B (AmBisome) against murine blastomycosis. J Antimicrob Chemother 1993; 32:465.

40. Clemons KV, Stevens DA. Comparison of a liposomal amphotericin B formulation (AmBisome) and deoxycholate amphotericin B (Fungizone) for the treatment of murine paracoccidioidomycosis. J Med Vet Mycol 1993; 31:387.
41. Leenders ACAP, de Marie S, ten Kate MT, Bakker-Woudenberg IA, Verbrugh HA. Liposomal amphotericin B (AmBisome) reduces dissemination of infection as compared with amphotericin B deoxycholate (Fungizone) in a rat model of pulmonary aspergillosis. J Antimicrob Chemother 1996; 38:215.
42. Francis P, Lee JW, Hoffman A, et al. Efficacy of unilamellar liposomal ampho. B in treatment of pulmonary aspergillosis in persistently granulocytopenic rabbits: the potential role of bronchoalveolar D-mannitol and serum galactomannan as markers of infect. J Infect Dis 1994; 169:356.
43. van Etten EW, Otte-Lambillion M, van Vianen W, ten Kate MT, Bakker-Woudenberg IAJM. Biodistribution of liposomal amphotericin B (AmBisome) and amphotericin B-desoxycholate (Fungizone) in uninfected immunocompetent mice and leucopenic mice infected with *Candida albicans*. J Antimicrob Chemother 1995; 35:509.
44. Clemons KV, Stevens DA. Comparison of fungizone, amphotec, amBisome and abelcet for treatment of systemic murine cryptococcosis. Antimicrob Agents Chemother 1998; 42:899.
45. Graybill JR, Bocanegra R. Liposomal amphotericin B therapy of murine histoplasmosis. Antimicrob Agents Chemother 1995; 39:1885.
46. Ortoneda M, Capilla J, Pastor FJ, Pujol I, Guarro J. Efficacy of liposomal amphotericin B in treatment of systemic murine fusariosis. Antimicrob Agents Chemother 2002; 46:2273.
47. Adler-Moore JP, Olson JA, Proffitt RT. Alternative dosing regimens of liposomal amphotericin B (AmBisome) effective in treating murine systemic candidiasis. J Antimicrobial Chemother 2004; 54:1096.
48. Gangneux JP, Sulahian A, Garin YJ, Farinotti R, Derouin F. Therapy of visceral leishmaniasis due to *Leishmania infantum*: experimental assessment of efficacy of AmBisome. Antimicrob Agents Chemother 1996; 40:1214.
49. Clemons KV, Sobel RA, Williams PL, Pappagianis D, Stevens DA. Efficacy of intravenous liposomal amphotericin B (AmBisome) against coccidioidal meningitis in rabbits. Antimicrob Agents Chemother 2002; 46:2420.
50. Ibraham AS, Avanessian V, Spellberg B, Edwards JE Jr. Liposomal Amphotericin B, and not amphotericin B deoxycholate, improves survival of diabetic mice infected with Rhizopus oryzae. Antimicrob Agents Chemother 2003; 47:3343.
51. Scarcella A, Pasquariello MB, Giugliano B, Vendemmia M, de Lucia A. Liposomal amphotericin B treatment for neonatal fungal infections. Pediatr Infect Dis J 1998; 17:146.
52. Leenders ACAP, Reiss P, Portegies P, et al. Liposomal amphotericin B (AmBisome) compared with amphotericin B both followed by oral fluconazole in the treatment of AIDS-associated cryptococcal meningitis. AIDS 1997; 11:1463.
53. Leenders AC, Daenen S, Jansen RL, et al. Liposomal amphotericin B compared with amphotericin B deoxycholate in the treatment of documented and suspected neutropenia-associated invasive fungal infections. Br J Haematol 1998; 103:205.

54. Johnson PC, Wheat LJ, Cloud GA, et al. U.S. National Institute of Allergy and Infectious Diseases Mycoses Study Group. Safety and efficacy of liposomal amphotericin B compared with conventional amphotericin B for induction therapy of histoplasmosis in patients with AIDS. Ann Intern Med 2002; 137:105.
55. Sundar S, Jha TK, Thakur CP, Mishra M, Singh VP, Buffels R. Single-dose liposomal amphotericin B in the treatment of visceral leishmaniasis in India: a multicenter study. Clin Infect Dis 2003; 37:800.
56. Albert MM, Stahl-Carroll L, Luther MF, Graybill JR. Comparison of liposomal amphotericin B to amphotericin B for treatment of murine cryptococcal meningitis. J Mycol Med 1995; 5:1.
57. Bohme A, Hoelzer D. Liposomal amphotericin B as early empiric antimycotic therapy of pneumonia in granulocytopenic patients. Mycoses 1996; 3:419.
58. Garcia A, Adler-Moore JP, Proffitt RT. Single-dose AmBisome (Liposomal amphotericin B) as prophylaxis for murine systemic candidiasis and histoplasmosis. Antimicrob Agents Chemother 2000; 44:2327.
59. BitMansour A, Brown JM. Prophylactic administration of liposomal amphotericin B is superior to treatment in a murine model of invasive aspergillosis after hematopoietic cell transplantation. J Infect Dis 2002; 186:134.
60. Tollemar J, Hockerstedt K, Ericzon BG, Jalanko H, Ringden O. Prophylaxis with liposomal amphotericin B (AmBisome) prevents fungal infections in liver transplant recipients: long term results of a randomized, placebo-controlled trial. Transplat Proc 1995; 27:1195.
61. Kelsey SM, Goldman JM, McCann S, et al. Liposomal amphotericin (AmBisome) in the prophylaxis of fungal infections in neutropenic patients: a randomised, double-blind, placebo-controlled study. Bone Marrow Transplant 1999; 23:163.

Index